UAV Swarm Networks

UAV Swarm Networks: Models, Protocols, and Systems

Edited by

Dr. Fei Hu, Dr. Xin-Lin Huang, and Dr. DongXiu Ou

CRC Press is an imprint of the
Taylor & Francis Group, an **informa** business

First edition published 2021
by CRC Press
6000 Broken Sound Parkway NW, Suite 300, Boca Raton, FL 33487-2742
and by CRC Press
2 Park Square, Milton Park, Abingdon, Oxon, OX14 4RN

ISBN: 9780367457396 (hbk)
ISBN: 9781003039327 (ebk)

Typeset in Times LT Std
by KnowledgeWorks Global Ltd.

Contents

Preface

Unmanned aerial vehicles (UAVs) have been used in many important applications. For instance, they can be used to monitor city events by using equipped cameras, where each UAV monitors a certain area of the city. Amazon plans to use UAVs to deliver goods to customers' homes. UAVs can fly in harsh environments (such as the North/South Poles) to collect information about the weather conditions, to understand the Earth's evolution. Military departments can deploy UAVs to collect battlefield information and save soldiers.

Although a single UAV could achieve certain tasks, such as delivering goods, many applications require the cooperation of multiple UAVs, which form a UAV network. For example, each UAV only covers a small range in a large city, and multiple UAVs share the complementary image information to obtain the complete picture. Many UAVs can form a swarm to perform challenging tasks, such as forming a specific network shape to attack enemies. In disaster recovery scenes, multiple UAVs fly to different places to detect/save people and coordinate schedule/coverage allocations.

There are many interesting R&D issues in a UAV swarm network. First, we need to control the topology/formation of the entire network by asking questions like: How does each UAV move along a specific trajectory to avoid physical collisions with other UAVs while successfully reaching the desired location? How do the UAVs go around an obstacle or a radio jamming area (which has strong radio interference signals)?

Second, efficient communication protocols are needed to achieve seamless, high-throughput data sharing among UAVs. UAVs have special mobility modes that could be used to enhance the ad hoc routing performance. For instance, a UAV can predict a destination node's moving trend, and it always uses the relay nodes that are in the destination node's moving direction to relay the data. UAVs often form multiple swarm groups for different coverage areas or mission requirements. Those group-based communications could utilize many existing cluster-based routing protocols to achieve intra-cluster/inter-cluster communications. A large-scale UAV swarm network requires a low-overhead routing scheme to quickly deliver data many hops away. The corresponding transport layer congestion control scheme should be created to avoid traffic congestion anywhere in the large swarm network.

Today's UAV network requires intelligent management. Autonomous UAVs need accurate situation awareness via sensor data processing. Reinforcement learning algorithms could be used to generate proper actions based on state analysis and suitable reward/cost computation models. Although general machine learning schemes, such as support vector machines (SVMs), the hidden Markov model (HMM), Bayesian regression algorithms and so forth could solve general classification/regression issues, a large-scale UAV swarm network requires deep learning (DL)-based algorithms to handle hundreds of parameters/patterns. The DL models can also be used to predict mobility trends and UAV network congestion hotspots.

Because UAVs are often used in many critical applications such as disaster recovery, battlefield monitoring and so forth, reliability and security have to be addressed in communication protocol designs. Reliability means that any data in transmission have to be reliably sent to the final destination node. Some natural factors, such as radio interference, traffic congestion, UAV mobility and so forth could cause bit errors and packet loss. The system should be able to distinguish among different loss reasons and adopt different strategies to achieve transmission reliability. For example, if the loss is due to congestion (queue overflow), the source rate may be reduced. If the loss is due to UAV mobility, a backup routing path may be used to forward the data in case the main path has broken links.

Unlike reliability, which mainly deals with natural factors, security handles various intentional network attacks. A typical security issue is radio jamming. The jammer could use intelligent methods to adapt to the UAV traffic patterns. For example, a jammer may switch its radio frequency if the UAVs are using multi-channel transmission modes. The exact jamming region boundary needs to be detected. The UAVs may use directional antennas, frequency hopping, spread spectrum as well as other methods to overcome the impact of jamming attacks. Along with jamming, the UAVs may face message eavesdropping, a man-in-the-middle attack, data injection and so forth. Cryptography-based methods could be used to overcome those attacks.

In this book, we have covered the abovementioned topics in UAV swarm networks in five different parts:

Part I. UAV Network Architecture: We first provide the big picture of communication topology management, which summarizes the most typical UAV interconnection topologies and their communication performance requirements. Then we introduce intelligent UAV movement control by using the latest machine learning model, DL. The UAVs can move along different trajectories with the shortest movement delay and obstacle-avoidable path. DL can learn the minimum-cost path based on the history training data. Then we move to a critical topic, formation control, which aims to form different swarm shapes based on the mission requirements. The formation control has some similarities to birds' flocking, which controls the movements of a large amount UAVs based on birds' migration behaviors. Part I also explains 5G-oriented UAV network design. Overall, Part I of this book aims to provide a fundamental knowledge map for readers on UAV topology control.

Part II. UAV Mobility: There are dozens of UAV mobility models, e.g., random or mission-oriented, single-UAV-based or group-based, 2D or 3D, and so forth. Typically, UAV mobility behaviors show good predictability because most applications have clear mission requirements and require UAVs to move to specific places or stay in a region. Therefore, we explain the use of machine learning algorithms to predict the next-phase mobility pattern (velocity, position, etc.). In another chapter we discuss various mobility models that may be used to enhance routing performance. Part II will also explain how UAV communication protocols can be designed for computation purpose. This chapter serves as a transition into Part III, which will focus on routing protocols.

Part III. Communication Protocols: Data need to be shared among UAVs. However, it is challenging to deliver data in real time across many hops of UAV links. Therefore, we describe several efficient schemes to meet routing quality of service (QoS), such as skeleton routing, which aims to identify the relatively stable parts in a large swarming network. This is practical because in most swarming scenarios the core/middle parts are relatively more stable, and the marginal parts move more drastically. Those stable nodes could form a routing skeleton, and other nodes just try to deliver data to one of the skeleton nodes, which then forward data quickly to the destination node's nearby skeleton node. In Part III, we also introduce other routing schemes, such as bio-inspired data relay using the principle of virtual pheromones, which leaves "pheromones" in some special UAVs with good positions for faster transmissions.

Part IV. Reliability and Security: This part focuses on dependable UAV systems. On one hand, they are secure, which means that they can resist various external network attacks such as jamming; on the other hand, they are also resilient; that is, they could overcome the impact of some natural events such as communication noise. Many efficient countermeasure solutions are given in these chapters to achieve a trustworthy UAV network.

Part V. Hardware and Software Implementations: This section selects some interesting topics on practical UAV system design, such as its hardware and software integration for a testing platform, C-band communication system and its modulation/demodulation issues, a network operating system (called SwarmOS) and so forth.

This book is suitable to the following audiences: (1) industry UAV R&D engineers, administrators or technicians who would like to grasp the latest trends in UAV communications; (2) college graduate students or researchers who may want to pursue some advanced research on large-scale UAV swarming and networking technologies; (3) government agencies that determine future society development in this exciting field; and (4) other interested readers with a strong desire to understand the challenges of designing a QoS-oriented UAV network.

Because this book includes materials from different UAV experts, there might be some editing issues. Thank you for your interest in this critical topic – UAV swarming network design.

About the Authors

Dr. Fei Hu received his first Ph.D. in signal processing from Tongji University, Shanghai, China, in 1999, and a second Ph.D. in electrical and computer engineering from Clarkson University, New York, in 2002. He is currently a full Professor with the Department of Electrical and Computer Engineering, University of Alabama, Tuscaloosa, AL. He has published over 200 journal/conference papers and books/chapters in the field of wireless networks and machine learning. His research interests are wireless networks, machine learning, big data, network security and their applications. His research has been supported by the National Science Foundation, U.S. Department of Energy, U. S. Department of Defense, Cisco and Sprint.

Dr. Xin-Lin Huang is currently a full professor and the vice-head of the Department of Information and Communication Engineering, Tongji University, Shanghai, China. He received his M.E. and Ph.D. in information and communication engineering from Harbin Institute of Technology (HIT) in 2008 and 2011, respectively. His research focuses on cognitive radio networks, multimedia transmission and machine learning. He has published over 70 research papers and holds 8 patents in these fields. Dr. Huang was a recipient of the Scholarship Award for Excellent Doctoral Student granted by the Ministry of Education of China in 2010, Best Ph.D Dissertation Award from HIT in 2013, Shanghai High-Level Overseas Talent Program in 2013, and Shanghai Rising-Star Program for Distinguished Young Scientists in 2019. From August 2010 to September 2011, he was supported by the China Scholarship Council to do research in the Department of Electrical and Computer Engineering, University of Alabama, as a visiting scholar. He was invited to serve as the session chair for the IEEE International Conference on Communications 2014. He served as a guest editor for *IEEE Wireless Communications* and as chief guest editor for the *International Journal of MONET* and *WCMC*. He serves as associate editor for IEEE Access and is a Fellow of the East Asia Institute.

Dr. DongXiu Ou is a full professor at the Transportation Information Institute at Tongji University, China. She has accomplished many projects under the support of the China Natural Science Foundation. She holds multiple national patents and her work has been published in dozens of international and national publications in the field of advanced transportation information management via cyber-physical system control. Dr. Ou has also been given the Young Faculty Teaching Award in Shanghai.

Part I

UAV Network Architecture

1 Communication Topology Analysis upon a Swarm of UAVs: A Survey

Chen Wang[†], *Jin Zhao*
[†]Department of Computer Science,
The University of Alabama, Tuscaloosa, AL
Department of Electrical and Computer Engineering,
The University of Alabama, Tuscaloosa, AL

CONTENTS

1.1 INTRODUCTION

Unmanned aerial vehicles (UAVs), commonly known as drones, are aircraft without a human pilot aboard. Today, UAVs are of great interest in broad areas of applications, such as military reconnaissance, firefighter operation, police pursuit and so forth. To prevent the ill-utilization of UAVs, more and more nations have issued rules and regulations on the design, manufacture, distribution and setting up no-fly zones against UAVs for privacy purposes. The more and more advanced technologies enable the UAVs to perform tasks with longer distance, more accurate maneuvering, more efficient communication qualities and so forth. In this case, the U.S. Federal Aviation Administration (FAA) and the U.S. Army have issued well-built standards defining the UAV applications in detail [1, 2]. Research has been done based on those regulations and standards for a more reliable and robust aero-vehicle design.

One of the most technological difficulties lies in the topology strategies of network communications.

Currently, the most common flying strategy of UAVs is the single-UAV system [3]. The UAV is set off individually, controlled and communicated by a human through a single-phase, two-way channel. Because the single-UAV (agent [4]) is relatively short ranged [5], once the command has been issued by the controller, the drone executes and gives feedback via a wireless network. This end-to-end communication method has three possible design flaws: (1) the communication quality is dependent on the travel distance of the drone, where the increased distance of the UAV yields a poorer connection; (2) the short-range response time limits this drone from tasks that require a relatively long distance [3]; and (3) the drone is not able to respond to the change of aerial environment intelligently and give feedback in time, so that if something happened to the drone that terminated the commanding channel, the UAV-at-large might not be able to retract back to the user.

Under this circumstance, a new applicable tactic has been suggested in coping with the abovementioned disadvantages, namely multi-UAV systems. A brief comparison between single- and multi-UAV system is illustrated in Figures 1.1 and 1.2. The search range of single-UAV forms a 2D geographical map with relative coordinates x, y. Due to the limitation of its design, it is possible to leave a traceable blind spot on the map. In contrast, a multi-UAV utilizes a group of small UAVs working simultaneously on a mission such that the field of view (FOV) formed has a depth of zero or a very limited blind spot.

From [3, 4, 6], there are many advantages of multi-UAV systems: (1) economy, (2) flexibility, (3) continuity, (4) speediness, (5) higher accuracy, (6) sustainability, and (7) ease of problem solving. Nonetheless, despite the listed advantages, the most challenging design issue deals with communication. Traditionally, an individual drone is exclusively connected to ground stations, while no mutual communication

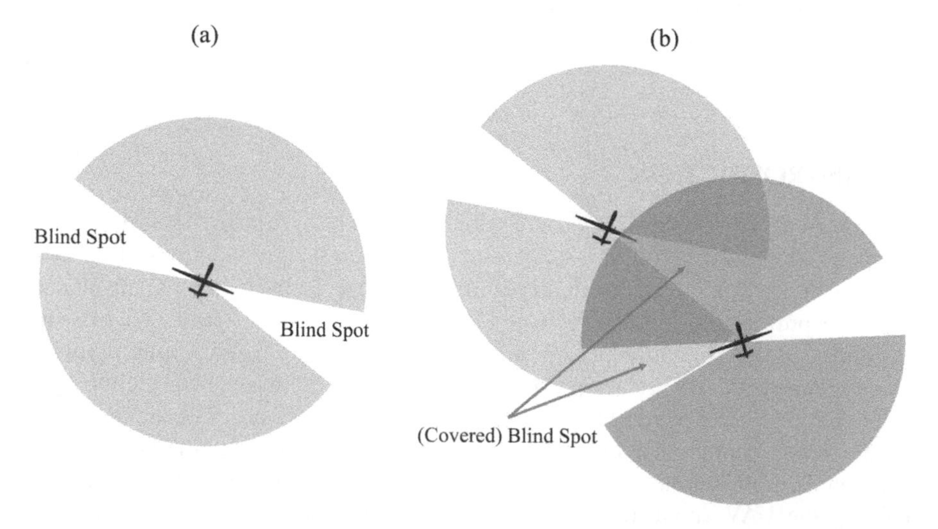

FIGURE 1.1 Blind spot illustration between (a) single-UAV and (b) multi-UAV systems.

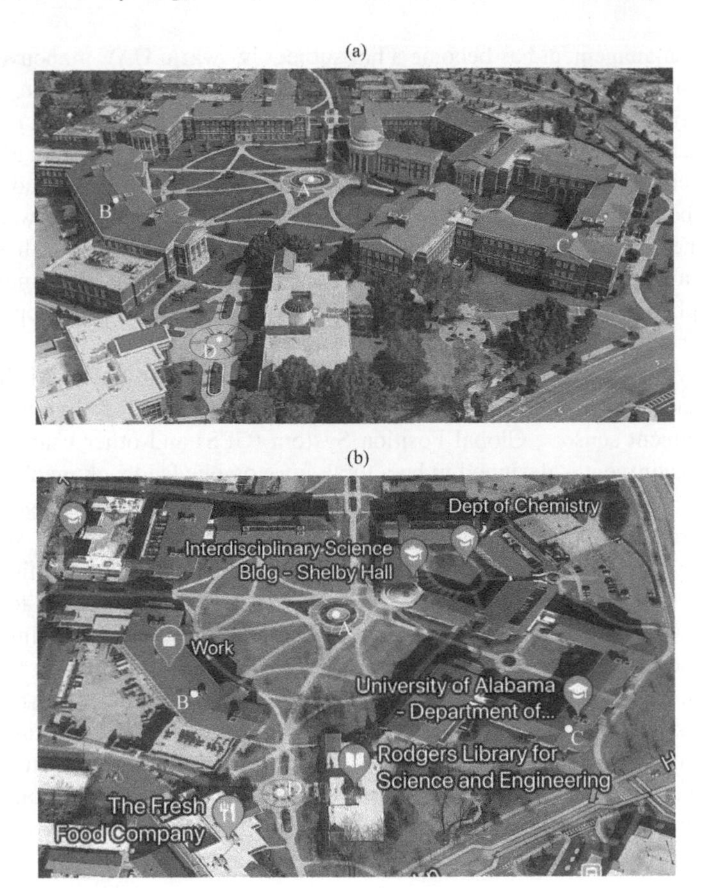

FIGURE 1.2 These are (a) 3D and (b) 2D geographical maps captured by single- and multi-UAV systems, respectively.

could be made between two independent drones. In this scenario, for the multi-UAV system topology, every time the communication has been made between two individual drones, the communication must go through the ground station, which will cause data traffic overflow on the ground and slow down communication efficiency. The exposed communication information requires more sophisticated data security methods to protect the data against intentional interceptions.

To solve this problem, Purta *et al.* suggested a multi-hop communication method, from which the movement of each agent is determined by its own assigned tasks and by the behaviors of others. The balanced icosystem rule ensures that no more than two agents will be working on the same task at once, while others in the same swarm will either be idling or working on something else [4]. In Lidowski *et al.* [7], the geographic greedy perimeter stateless routing (GPSR) is used for the UAV search mission protocol (USMP). Each agent is able to track its neighbor's location with the inner routing and conflict resolution rule designs. This method helps avoid existing conflicts and prevent route overlying.

Routing management has become a hot subject in swarm UAV maneuvering tactics, the major advantages of which are more movable and more flexible. Thus, the ground station–based control is no longer feasible due to the ranging and responding time limitations. Chen *et al.* [6], suggested two possible algorithms for an ideal drone package delivering system. The task assignment is agent based instead of group based for better time-management purposes. The provided algorithms were able to achieve linear growth in package delivery with respect to time: more drones were deployed, and more tasks were done in a limited period of time. This method also can be applied to the searching missions in cases of emergency, catastrophic events and similar civil activities.

Recently, Amazon proposed a future delivery system called Prime Air [8], which manages short-range delivery using drones with facial and voice recognition, optical and ambient sensors, Global Position System (GPS) and other usable features. The task assignment is designed to be a total autonomous feedback machine so that it can automatically communicate with customers, vendors and other agents via an inner secured network. The commanding dependency from a ground station has been largely weakened due to the use of multi-layer ad hoc network architecture.

This is a survey nature chapter providing the current existing network topology of the multi-UAV system. For the remainder of this chapter, Section 1.2 explains the existing major network strategies as well as the pros and cons of this survey; Section 1.3 provides an actual field test on different communication types, indicating the feasible applications on each type with multi-UAV systems; and Section 1.4 initiates a brief discussion on the current development on UAV communication techniques, including the achievements as well as caveats and bottlenecks. In the end, Section 1.5 draws a conclusion on the UAV networking strategies emphasizing multi-UAV application, then a future research suggestion is made on the acquired progress.

1.2 NETWORK AND ROUTING STRATEGIES OF EFFECTIVE UAV COMMUNICATIONS

A growing body of literature shows various communication strategies among UAVs and ground stations. The most commonly used networking method is the ad hoc network. An ad hoc network is a type of temporary computer-to-computer connection. In ad hoc mode, a wireless connection can be set up directly to another computer without connecting to a Wi- Fi access point or router. This connection makes it possible to use long-range routing if power permits. Because of this, the traveling distance will no longer be a big concern. This section names several popular yet different routing and connection methods as well as their pros and cons.

1.2.1 Optimal Routing Algorithm

With the improvement of UAV technology, UAV swarms have begun to be used as a package delivery network. Although the vehicle routing problem (VRP) has been fully studied at this stage, it is extremely difficult to study the routing on the UAV system. The routing of the system can transmit and receive new commands and location information almost instantaneously. In [6], the authors introduced the routing algorithm of the UAV parcel delivery system and proposed a theoretical model. The model suggested in the chapter uses a global clock with discrete time t for synchronization, then the parcels are mapped to the topological map with nodes $n,(i)$ and edges $e,(i)$ with weighting $d_{e},(i)$. The starting and destination nodes for a parcel are associated with estimated delivery time t. The job assignments are based on pseudorandom selection according to the relative geographical location of each agent. The closer agent figuratively gets a higher priority for success in "job hunting."

This algorithm can be deployed to multiple applications such as the vehicle routing problem, the goal of which is to minimize route costs. In a way this constitutes a classical traveling salesman problem (TSP). TSP is mainly aimed at the shortest possible route but with a single-drone scenario; with a swarm of UAVs, the task assignment is actually based on the weighting of a variable group n, e, d_e with respect to t. A basic rule of thumb in this case is that at the very beginning of each delivery period, this assignment is mostly arbitrary because the probability function Pr $\{n, e, d_e, t\}$ is at the same level. Therefore, the optimal routing algorithm performs as a supervisor to assign tasks to its agents as well as monitoring the work progress.

A greedy routing algorithm also is used on a variety of network graphs. It is an algorithmic paradigm that follows the problem-solving heuristic of making the locally optimal choice at each stage [9] with the intent of finding a global optimum. In many problems, a greedy strategy does not usually produce an optimal solution; nonetheless, a greedy heuristic may yield locally optimal solutions that approximate a globally optimal solution in a reasonable amount of time. As mentioned above, the TSP problem applying the greedy strategy does not intend to find a best solution; instead, it terminates in a reasonable number of steps. Finding an optimal solution to such a complex problem typically requires an unreasonable amount of steps. In mathematical optimization, greedy algorithms optimally solve combinatorial problems with the properties of matroids, and they give constant-factor approximations to optimization problems with submodular structure.

Despite the advantages of the proposed optimal routing algorithm, there are some non-negligible caveats with this method. First, the Open Graph Drawing Framework (OGDF) was considered a major code library, which has limited the system within a simulation phase. Second, with more nodes in the system, this algorithm was overwhelmed by the random generated packages. The performance was sabotaged by the increasing number of test nodes, which caused the entire system to be less controllable. Finally, the cost function was not taken into consideration to show if this application is economically or geographically viable.

1.2.2 Communication Architectures and Protocols

In the last decade, during the explosive development of drones, decentralized communication architectures have been adapted to operate a one-to-many mode of operating multiple drones by providing timely air-to-air and air-to-ground information exchange. This chapter presents four different architectures for communication and networking for network drones, in which UAVs can be automatically or remotely grounded by ground crews to save lives when replacing humans in hazardous

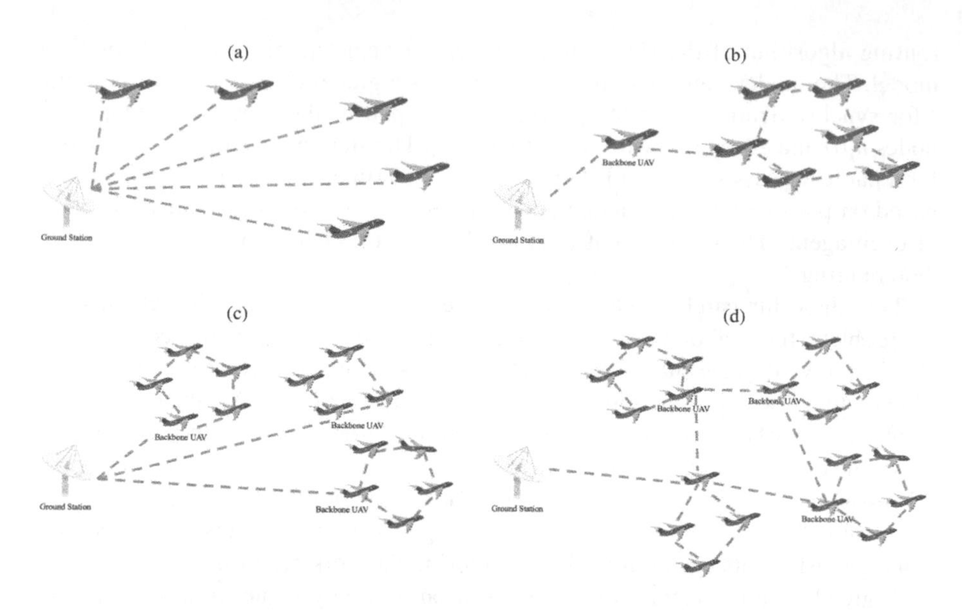

FIGURE 1.3 Communication architectures between UAVs and the ground station. (a) Centralized. (b) Ad hoc. (c) Multi-group. (d) Multi-layer ad hoc networks.

environments, divided by centralized communications and decentralized communications, including the (1) centralized UAV network, (2) UAV ad hoc network, (3) multi-group UAV network and (4) multi-layer UAV ad hoc network.

In this case, typical applications for drones include surveillance and reconnaissance, collaborative search and acquisition and tracking. However, at present, drones have their drawbacks. The drones in operation require one or more operators to control them, so the research focus is on drone autonomy. In the future, multiple drones with a higher degree of autonomy will be integrated into a single team to collaborate autonomously through more advanced network protocols to achieve common tasks.

Because most of the multi-UAV systems require instant communication, the communication architectures and protocols are under reformation from decade to decade. Li *et al.* introduced four major communication architectures for networking UAVs, and the topological differences are shown in Figure 1.3 and Table 1.1. The multi-layer ad hoc network works best among all four types with interchangeable network connections, and the supervising powers were distributed to the backbone UAVs such that the ground station only performs as information processing [10]. In this way, the computation and communication loads were significantly reduced, which also facilitates a more robust and reliable ad hoc networking system.

1.2.3 Multi-Hop Communications

Multi-hop routing is a type of communication in radio networks in which the network coverage area is larger than the radio range of the single nodes. Therefore, to reach some destinations, a node can use other nodes as relays [11]. Because the transceiver

TABLE 1.1
Communication Architectures between UAVs and Ground Stations

Communication Type	Connection Architecture	Pros	Cons
Centralized	Centralized connection	Direct communication with ground	Limited distance
		Overall supervising	No connection between UAVs
			Blind spots exist
Decentralized	UAV ad hoc	Interconnection between UAVs	Limited distance
		Coverage area extended	Single loop failure
Multi-group		Direct communication with ground	Semi-centralized
		Blind spots killed	Lack robustness
		Different flight and communication types	
Multi-layer ad hoc		Freely route between UAVs	Backbone UAV interchangeable
		Reduced computation load	No single point failure
		Reduced communication load	

is the major source of power consumption in a radio node and long distance transmission requires high power, in some cases multi-hop routing can be more energy efficient than single-hop routing [12].

In [4], the author applied the principles of the Dynamic Data-Driven Application System (DDDAS) as the communication strategy for a bunch of small, inexpensive drones to discover static targets in the search space. This technology could be used in rescue missions after catastrophic events, such as landslides, earthquakes and so forth. To extend the communication range of the system, a novel method was presented by Purta *et al.* [4] for an efficient communication path determination among multiple nodes. The model utilized the attenuation of wireless signals to transmit a signal between two distinct nodes, and the signal power receiving P_r is reversely proportional to the distance D:

$$P_r \propto \frac{P_t}{D_k} \tag{1.1}$$

where P_t is the transmitting signal power and $K \geq 2$ for two or more nodes.

The performance of this proposed model was simulated in a semi-realistic communication environment. The challenge here is to find the best multi-hop configuration for drones so that they can find the most targets, avoid conflicts as much as possible, and still be effective. This can be achieved by using the weighted shortest path problem of the *Dijkstra* algorithm, which is the transmission cost of the distance. Tests have shown that multi-hop communication performs well in target discovery, avoiding collisions with other drones and idealized communication environments.

```
function Dijkstra (Graph, Source):
create vertex set Q;
for each vertex v in Graph do
    dist[v] = INFINITY;
    prev[v] = UNDEFINED;
    add v to Q;
    dist[source] = 0;
end
while Q is not empty do
    u = vertex in Q with min dist[u];
    remove u from Q;
    for each neighbor v of u do
        alt = dist[u] + length(u, v);
        if alt < dist[v] then
            dist[v] = alt;
            prev[v] = u;
        else
            display ERROR;
        end
    end
end
return dist[], prev[];
```

Algorithm 1: *Dijkstra's* algorithm in pseudo code

Different from [6], the cost function was actually taken into account for evaluating the "best route" possible for mission-oriented UAV swarms. The *Dijkstra* algorithm can be expressed in pseudocode as in Algorithm 1. In the presented application, the purpose is to create an interconnection scenario among agents and to avoid collisions as much as possible. From the conducted experiment, the discovered targets are increasing exponentially with the increasing number of agents; in the meantime, the chance of collisions is significantly reduced.

From these results, the multi-hop communications help successfully prevent collisions between agents when the swarm is organized. However, it was observed that if the target verification is obscure or in an inefficient ordering, this would cause the leader to change direction sharply, thus giving its followers less time to avoid collision. In addition, this method is only able to detect the static target, which has largely limited the usage of this system. Finally, the multi-layer structure has relatively higher cost in computation and execution; hence, the communication range was limited such that the connection will not overflow. The target-found ratio (TFR) and collision-avoid ratio (CAR) are listed in Table 1.2.

1.2.4 Flight Ad Hoc Networks

UAVs are important in areas that are isolated from the ground and require complex tasks. Cooperation between multiple drones also is important. Communication is one of the most challenging problems for multi-UAV systems; therefore, higher

TABLE 1.2
Comparison of System Performance between Ordinary Established and Multi-Hop Communications

Communication Type	Targets	Target Found	Collision Avoided
Ordinary established	Small	Low	High
Multi-hop	Small	Median	Low
Ordinary established	Median	Low	High
Multi-hop	Median	Low	Median
Ordinary established	Large	Low	High
Multi-hop	Large	Low	High

coverage and accuracy are required to realize communication by two or more UAV nodes directly or via relay nodes. People need to deploy a network model to form a flight ad hoc network (FANET). The portable and flexible communication network between UAVs is often referred to as the FANET. Due to the high-speed mobility of UAVs, as well as environmental conditions and terrain structures, existing mobile ad hoc routing protocols are no longer applicable.

To overcome these obstacles, Gankhuyag *et al.* [13] proposed using a combination of position, trajectory information and geolocation route prediction position to achieve a longer transmission range. In addition, to increase path life and packet transmission success, the solution reduces path rebuild and service downtime. The simulation results in [13] verified that this scheme can significantly improve the performance of the FANET.

In [3], the existing FANET routing protocols are divided into six categories: (1) static routing protocols, (2) proactive routing protocols, (3) reactive routing protocols, (4) hybrid routing protocols, (5) position/geographic based routing protocols, and (6) hierarchical routing protocols, which are rigorously analyzed and compared according to various performance criteria to select an appropriate routing protocol.

Along with the routing protocols, researchers are following up with the antenna settings for a better connection that will be able to sustain long-range communication [13] using purposed adaptive and non-adaptive antennas, which showed an improved trend of average routing lifetime. Not only is the flying distance longer with FANETs, but there is also a higher data delivery ratio. Because the connecting time is proportional with distance, the feasibility and robustness of the flying ad hoc networks (FANET) and mobile ad hoc networks (MANET) are compared. Conventionally, FANET demands optimized link-state routing (OLSR) to track its variants in the flying range. It yields a serious problem of overwhelming managing requirements. Thus, a potential topology of FANET constitutes a nonlinear optimization problem [1] that formulated as:

$$\begin{aligned} \min \quad & f(X_{V_G}, X_{V_M}, X_{V_R}, \Re)\forall X_{V_R} \in \mathbf{S}^{|V_R|} \\ \text{s.t.} \quad & \max_{k=1,\dots,|\mathbf{p}_v|-1} d\left(\mathbf{p}_v(k), \mathbf{p}_v(k+1)\right) \le d_0 \forall v \in V_M \\ & \min_{u,v \in V} d(u,v) \ge d_{\text{safe}} \end{aligned} \tag{1.2}$$

where the system is considered with a single ground communication module (GCM), M is numbers of mission-critical UAVs (mUAVs) and N is numbers of communication relay UAVs (rUAVs), which are denoted by $V_G = \{g\}$, $V_M = \{m_k\}_{k=1}^{M}$ and $V_R = \{r_k\}_{k=1}^{R}$, respectively [14]. $\mathbf{X}_{V_i}$ is the location of nodes in V_i, $\mathfrak{R}$ is the number of routing paths for each mUAV, $\mathbf{p}_v$ is the routing path from the corresponding mUAV to the GCM and d_{safe} is the threshold distance for safety concerns.

The result of Eqn. (1.2) yields a satisfactory flying and collision-safe test. More UAVs form a hyperbolic-shape curve in the first quadrant, which indicates a robust and reliable routing scheme for FANETs. Further analysis should be done on testing the communication time and effective routing range.

1.3 FEASIBILITY AND ROBUSTNESS ANALYSIS UNDER DIFFERENT NETWORKING SCENARIOS

A good understanding of the propagation channel between UAVs and ground stations has been established in the previous section regarding the communication strategy feasibility analysis. According to the FAA Unmanned Aircraft Systems Regulations [15], there are three major conditions that need to be taken into consideration: (1) location, (2) height and speed, and (3) routing distance. In this section, those conditions will be discussed in detail for the feasibility and robustness of UAV swarms.

1.3.1 Networking Influence on Location

Similar to real estate agents' mantra of location, location, location, as for the rapid development of UAV for recreational and educational uses, the aircraft control becomes an important aspect for safety concerns. The FAA [15] describes the flying object's set-off regulation for different purposes such as recreational, commercial and/or government uses. The UAVs will affect the radar images near military facilities, airports and location-sensitive areas, and the communication system of the UAV swarms might cause radio-frequency (RF) interference for other communication channels.

An ad hoc network is considered a dynamic network that can be created anywhere with just two basic nodes that do not require any centralized infrastructure [16]. The main issue for the security connection is to set up a close-loop network so that the effects on "outsiders" can be shrunk to a minimum. Aircraft commonly utilize ad hoc networks. If there are no shielding techniques between UAVs and the aircraft–ground system, the UAVs are strictly banned near certain locations. Without losing generality, the ad hoc communication interference can be described as in Figure 1.4. Typically, the connections are set up among ground facilities (e.g., airport), ground stations and in-air aircraft. However, illegally flying UAVs also sends out a signal under the same networking protocols. This unexpected signal, if not well shielded, might cause normal communication failure or even aircraft accidents.

Under this circumstance, more than 150 nations all over the world issued detailed regulations specifying UAV no-fly zones (or no-drone zone according to the FAA [17]) for safety and privacy concerns. Thus, the location selection would bring potential communication issues including, but not limited to, networking efficacy, closed-loop connections, shielding and so forth.

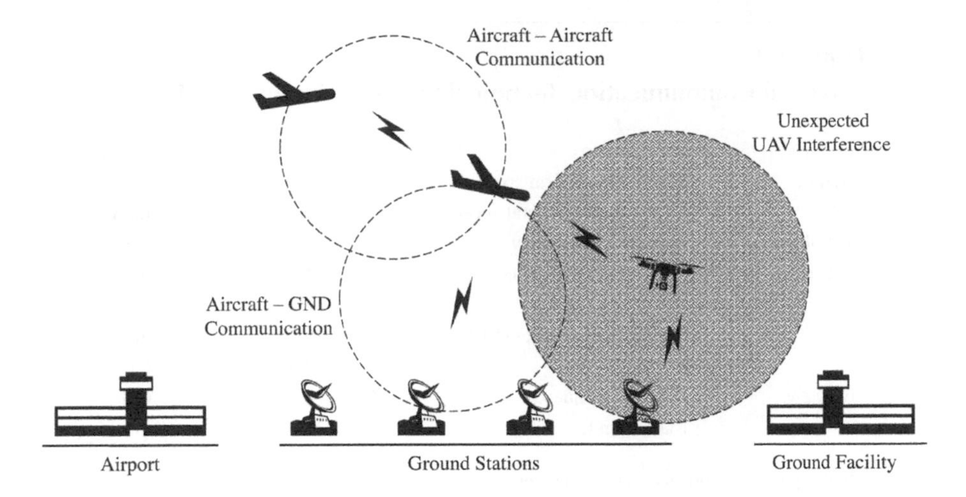

FIGURE 1.4 Typical aircraft–ground system connections under the ad hoc networking protocol with unexpected UAV signal invasion. GND, ground.

1.3.2 UAV Height and Speed with respect to Communication Efficiency

Most accessible UAV manufacturers in the market have specifications detailing the capability and sustainability of their products, among which these are the two most important features under the regulative constraints: (1) equivalent isotropic radiated power (EIRP [18]) and (2) flight attitude index (FAI). Although these two aspects are listed as different categories, more often FAI is closely associated with EIRP because of communication efficiency. Currently, the state-of-the-art communication method available is the integration of ad hoc and multi-hop networks. This integration sums up the advantages of both but also removes part of the limitations. First, the multi-hop network provides security for the connection such that the other outside signal emissions will not interfere with the UAV swarm. In contrast, flying agents also have the ability to communicate among their "companions" within the swarm as well as the controller; thus, the network further enhances the data protection and limited radiated emission. Second, the ad hoc network is well known for its multi-point direct connection mechanism such that the importance of the ground station has been weakened. This leads to a wider routing distance and longer flying range. In addition, multi-hop communication was shown to have difficulty finding large targets without collisions, but agents in ad hoc networks consider targets in a single mission as teamwork. Hence in this integration, the ad hoc networks play a better role in the "scavenger hunt" than the multi-hop run by itself.

Under the constraints of the U.S. Military Standard with radiated emission (MIL-STD-461F RE102, [2]), the frequency range of electromagnetic emissions is 10 kHz to 18 GHz, which is 10 times higher than the intentionally generated frequency from specs of most commercial UAV drones. Abridged communication specs of the DJI Phantom 4 Advanced UAV [19] are listed in Table 1.3. From the table, the normalized working communication frequency is rated at 2.4 GHz with an EIRP at 26 dBm,

TABLE 1.3
A Typical Communication Technical Specs of Commercial UAV

Category	Spec Name	Spec Value	Units
Aircraft.	Maximum vertical speed	4	m/s
	Maximum horizontal speed	45	mph
Controller	Operating frequency	2.400	GHz
Multi-hop	Transmission distance	4.3	mi
		2.483	GHz
	Transmission power (EIRP)	26	dBm
		2.483	GHz
Live view	Working frequency	2.4	GHz
	Maximum latency	220	ms

EIRP, equivalent isotropic radiated power.

which means a 26-dB decrease in level of signal equivalents to a 400-mW decrease in power. Using P for power in milliwatts and x for signal intensity in decibel-milliwatts, the following equivalent expressions may be used:

$$x = 26\log_{10}\left(\frac{P}{400\text{ mW}}\right) \tag{1.3}$$

The spectral analysis under different power density is shown in Figure 1.5. The further away the agents inside the swarm are, the sparser the power density would be.

1.3.3 Flying Distance Constraints under Networking Protocols

To obtain the live control ability of the UAV swarm, the flying distance must be taken into serious consideration. Normally for centralized or multi-group communication types as seen in Figure 1.3, the flying distance is largely guided by the inverse-square law between signal intensity, which was stated in Section 1.3.2, and is mathematically notated as

$$\text{intensity} \propto \frac{1}{\text{distance}^2} \tag{1.4}$$

$$\frac{\text{intensity}_1}{\text{intensity}_2} = \frac{\text{distance}_1^2}{\text{distance}_2^2} \tag{1.5}$$

In other words, distance is always square-reciprocally proportional to communication efficiency. Here, let the total radiated signal be denoted as P from a member of the flying UAV swarm; this signal is transmitted and distributed over an infinitively

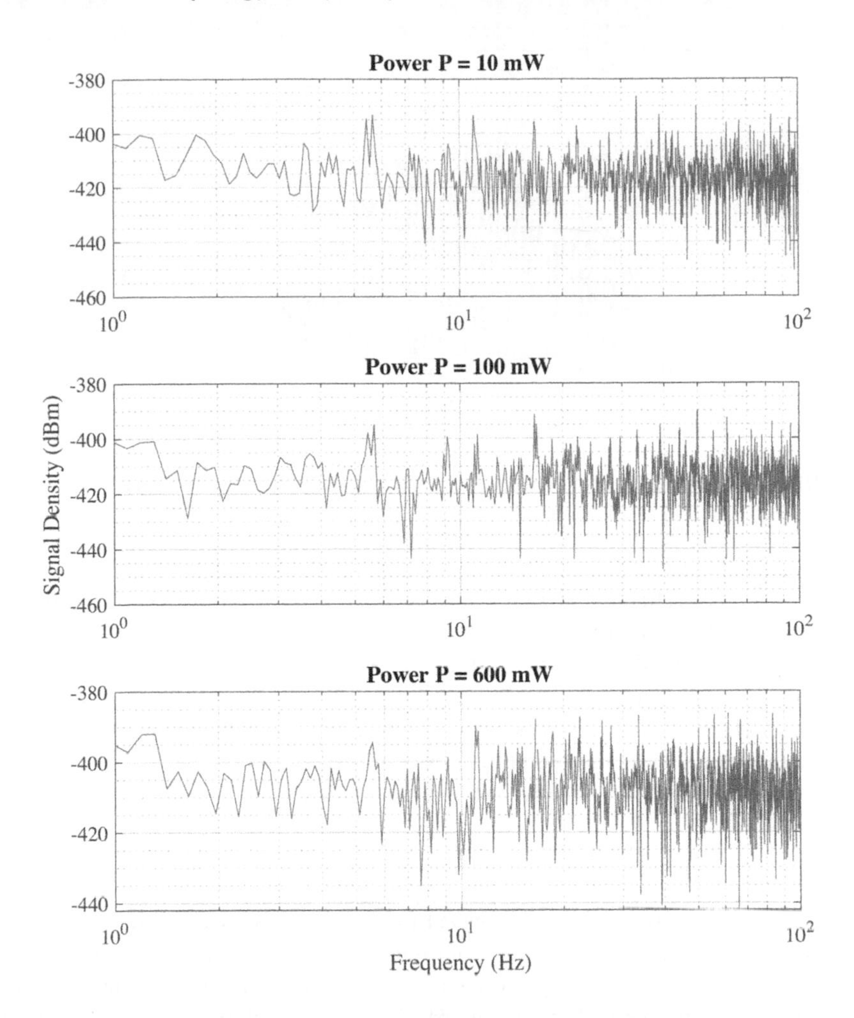

FIGURE 1.5 Spectral analysis of UAV signal intensity with respect to power following a decay rate of 26 dB per decade.

large spherical surface from the sender agent. The surface area A of a sphere of a radius r is defined as $A \triangleq 4\pi r^2$, thus the intensity I (power per unit area) of radiation at distance r is

$$I = \frac{P}{A} = \frac{P}{4\pi r^2} \tag{1.6}$$

From Eqn. (6), if the signal intensity is measured in decibels, the decay rate would be 6.02 dB per doubling of distance [20].

From the field test, the flying distance is represented by the number and density of nodes that spread in the sky via ad hoc networks. The distribution of nodes follows a normal distribution that is centralized around the target. Higher node density

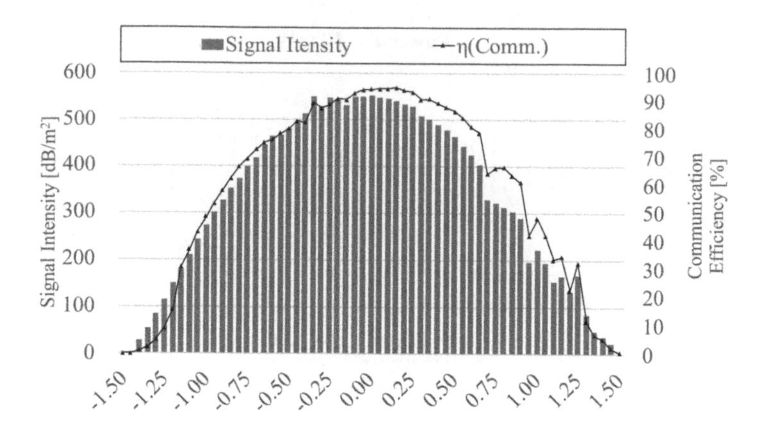

FIGURE 1.6 Signal intensity versus communication efficiency with distance change.

results in a more reliable and frequent network communication, but if the routing algorithm was not well developed, there is also a greater chance for UAV collisions. The visualized node–density relationship is shown in Figure 1.6. From the figure, the horizontal bar is normalized distance d, when $d = 0$, the distance between sender and receiver (S&R) is 0; similarly, when $d = \{+1.5, -1.5\}$, the S&R distances are both at maximum. In this way, when signal intensity η is at its maximum, the communication efficiency η is approximately 94.68%; meanwhile when $I \rightarrow 0$, $\eta \rightarrow 0$.

1.4 DISCUSSIONS

In [21], an optimal routing method based on *Dijkstra's* shortest path algorithm is proposed for UAV networks by incorporating predicted network topology. The core idea is to incorporate the anticipated locations of intermediate nodes during a transmission session into the path selection criterion. The simulation results confirm the superior delay performance of the proposed algorithm compared with the conventional routing algorithms. The enhancement is more significant when the network size becomes larger, the relative node velocities rise, and the average waiting times in transmit buffers extend. The potential future research direction is to evaluate the impact of imperfect prediction by characterizing the probability of selecting non-optimal links.

In [22], Khaledi *et al.* propose a distance-based drone network greedy routing algorithm. The algorithm is based entirely on the UAV's local observations of its surrounding subnets, so neither central decision makers nor time-consuming routing setup and maintenance mechanisms are required. To obtain a more accurate simulation result, the authors obtained the analysis limit of the expected hop count of the packet traversal and verified the accuracy of the developed analytic expression. Finally, the distance-based greedy routing algorithm shows considerable improvement over the centralized shortest path routing algorithm compared with the approach described in [21].

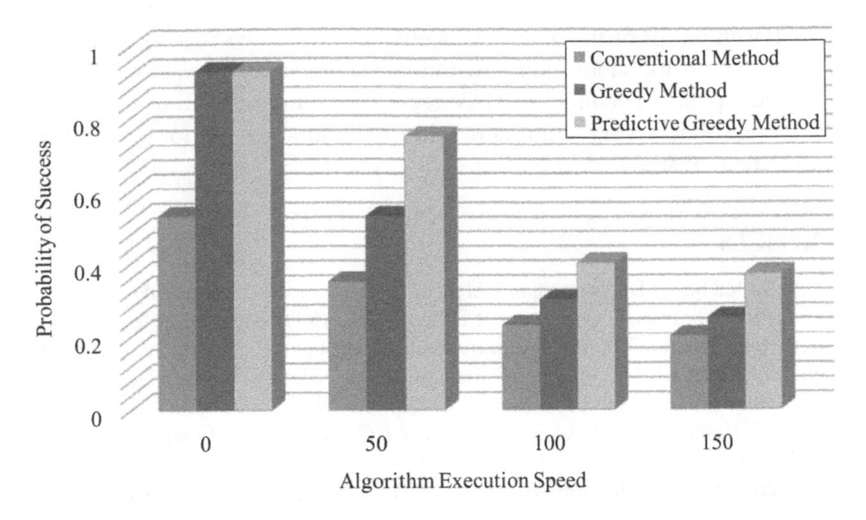

FIGURE 1.7 Probability of delivery success for predictive greedy algorithm, static greedy algorithm and conventional *Dijkstra's* algorithm.

Figure 1.7 compares the performance of the greedy algorithm proposed by the author with the traditional *Dijkstra* shortest path algorithm, and it evaluates the predicted position information of the proposed algorithm at different average node speeds. The results show that for all average node speeds, predictive greedy methods, static greedy algorithms and traditional shortest path algorithms, the success probability of transmission decreases in turn. Predicting location information reduces the probability of selecting a node with an empty progress area. When the network is more dynamic, more nodes will leave the communication range of their neighbors, so that the probability of success will decrease.

1.5 CONCLUSIONS

UAV networks have emerged as a promising technique to rapidly provide wireless coverage over a geographical area. The feasibility and reliability of drones depend to a large extent on the effectiveness of communication skills. Based on some typical research cases, this chapter first discusses feasible communication methods and applications, given the possible insight on the development of UAV network with higher reliability, higher security and lower dependency on ground stations. The pros and cons of different network strategies and actual field test results are thereafter described based on actual test data, the results of which showed a clear outcome of communication efficiencies with respect to (1) networking protocols, (2) UAV speed and height and (3) flying distance. Flying in swarm has now become a new trend due to its many advantages, despite some non-negligible caveats such as collision potential, task management burden, node distribution topology and so forth. Furthermore, the most recent research and analysis has pointed out the difficulties in a more complicated communication method. Based on all of the above, this chapter concluded that the construction of the drone network should keep to the route of feasibility and robustness by obeying the increasing number of UAV regulations and restrictions. A more reliable and controllable UAV communication technique awaits for a continuously expanding field of study.

REFERENCES

1. U.S. United States Department of Defense, "Environmental engineering considerations and laboratory tests," Department of Defense, MIL-STD-810F, 2000.
2. U.S. Department of Defense, "Requirements for the control of electromagnetic interference characteristics of subsystems and equipment," MIL-STD-461G, 2007.
3. M. Tareque, M. Hossain, and M. Atiquzzaman, "On the routing in flying ad hoc networks," *Proceedings of the 2015 Federated Conference on Computer Science and Information Systems (FedCSIS)*, vol. 5, pp. 1–9, 2015.
4. R. Purta, S. Nagrecha, and G. Madey, "Multi-hop communications in a swarm of UAVs," *Proceedings of the 2013 Agent-Directed Simulation Symposium, Society for Computer Simulation International*, vol. 1, pp. 5–13, 2013.
5. A. Agogino, C. Parker, and K. Tumer, "Evolving large scale UAV communication systems," *Proceedings of the 2012 14th Annual Conference on Genetic and Evolutionary Computation*, vol. 9, pp. 1023–1030, 2012.
6. M. Chen and J. Macdonald, "Optimal routing algorithm in swarm robotic systems," *Course for the Department of Computer Sciences, California Institute of Technology*, vol. 5, no. 1, pp. 1–8, 2014.
7. R. Lidowski, B. Mullins, and R. Baldwin, "A novel communications protocol using geographic routing for swarming uavs performing a search mission," *2009 IEEE International Conference on Pervasive Computing and Communications*, vol. 4, pp. 1–7, 2009.
8. Amazon, "Amazon Prime Air," https://www.amazon.com/Amazon-Prime-Air/b?node=8037720011, 2016.
9. P. Black, "Greedy algorithm," https://xlinux.nist.gov/dads//HTML/greedyalgo.html, in *Dictionary of Algorithms and Data Structures* [online], 2005. Accessed March 2019.
10. J. Li, Y. Zhou, and L. Lamont, "Communication architectures and protocols for networking unmanned aerial vehicles," *2013 IEEE Globecom Workshops (GC Wkshps), 2013.*
11. M. Kakitani, G. Brante, R. Souza, and A. Munaretto, "Comparing the energy efficiency of single-hop, multi-hop and incremental decode- and-forward in multi-relay wireless sensor networks," 2011 IEEE 22nd International Symposium on Personal, Indoor and Mobile Radio Communications, pp. 970–974, 2011.
12. S. Fedor and M. Collier, "On the problem of energy efficiency of multi-hop vs one-hop routing in wireless sensor networks," *21st International Conference on Advanced Information Networking and Applications Workshops (AINAW'07)*, vol. 2, pp. 970–974, 2007.
13. G. Gankhuyag, A. Shrestha, and S. Yoo, "Robust and reliable predictive routing strategy for flying ad-hoc networks," *IEEE Access*, vol. 5, no. 3, pp. 643–654, 2017.
14. D. Kim and J. Lee, "Topology construction for flying ad hoc networks (FANETS)," *2017 International Conference on Information and Communication Technology Convergence (ICTC)*, vol. 1, pp. 153–157, 2017.
15. Federal Aviation Administration, "Unmanned aircraft systems (UAS)," https://www.faa.gov/uas/. Accessed April 2019.

16. V. Kumar, A. Rana, and S. Kumar, “Aircraft ad-hoc network (AANET),” *2014 International Journal of Advanced Research in Computer and Communication Engineering*, vol. 3, pp. 6679–6684, 2014.
17. Federal Aviation Administration, “No drone zone,” https://www.faa.gov/uas/resources/community_engagement/no_drone_zone/. Accessed October 2018.
18. E. Williams, G. Jones, D. Layer, and T. Osenkowsky, *National Association of Broadcasters Engineering Handbook: NAB Engineering Handbook*, 10th ed. Burlington, MA: Taylor Francis, 2013.
19. Shenzhen DJI Technology Co., Ltd, “Phantom 4 Advanced specs,” https://www.dji.com/phantom-4-adv/info#specs/. Accessed August 2017.
20. O. Gal and R. Chen-Morris, “The archaeology of the inverse square law: (1) metaphysical images and mathematical practices,” *History of Science*, vol. 5, no. 3, pp. 391–414, 2005.
21. A. Rovira-Sugranes and A. Razi, “Predictive routing for dynamic uav networks,” *2017 IEEE International Conference on Wireless for Space and Extreme Environments (WiSEE)*, vol. 1, pp. 43–47, 2017.
22. M. Khaledi, A. Rovira-Sugranes, F. Afghah, and A. Razi, “On greedy routing in dynamic UAV networks,” *2018 IEEE International Conference on Sensing, Communication and Networking (SECON Workshops)*, vol. 2, pp. 1–5, 2018.

2 Unmanned Aerial Vehicle Navigation Using Deep Learning

Yongzhi Yang[†], *Kenneth G. Ricks, Fei Hu*
Department of Electrical and Computer Engineering, The University of Alabama, Tuscaloosa, AL
[†]Department of Electrical and Computer Engineering, The University of Alabama, Tuscaloosa, AL (email: yyang108@crimson.ua.edu).

CONTENTS

2.1 INTRODUCTION

Unmanned aerial vehicles (UAVs) have been developed for decades. The first UAV was created in 1849, when unmanned balloons were used to deliver explosives. Due to their features, UAVs also can be used in transportation, military and private industry activities. For example, UAVs can be used as a target, communication relay, transport vehicle, disaster relief platform and so forth. The advantages of UAVs include light weight, small size, low cost and high concealment. However, they are also sensitive to interference and human factors. The popularity of UAVs has seen exponential growth in the last few years, especially UAV control and navigation. Many different sensor packages and algorithms have been developed to address all aspects of navigation and control.

Deep learning has played an important role in many research areas in recent years. Deep learning techniques are used for big data processing and for solving complex tasks. Deep learning techniques have been utilized since 1971, and the convolutional neural networks (CNNs) were created in the 1980s and improved in the

1990s by more efficient training methods. With the development of graphic processing units (GPUs), more processing power can be applied to large datasets contributing to the advancement of deep learning approaches.

This chapter will discuss the use of deep learning in UAV navigation. Section 2.2 discusses object detection and tracking using one single camera. The method presented integrates background subtraction, deep learning and optical flow. First, low-resolution video data are transformed into a sequence of frames by using background subtraction. Then, a deep neural network (DNN) is trained to classify the object candidates. Finally, optical flow is used to track the object, and a Kalman filter is used to improve target detection. Section 2.3 discusses UAV autonomous landing via reinforcement learning (RL). Two topics are discussed separately in Sections 2.3.1 and 2.3.2. Section 2.3.1 introduces deep reinforcement learning (DRL) to detect the landing target and complete a safe landing. First, the landmark is detected by using a low-resolution, down-looking camera that controls the UAV to position itself over the target. Next, the UAV descends quickly in vertical dimension maintaining its position over the target. Finally, it uses a closed-loop controlling model to continuously and precisely land on the target. Section 2.3.2 discusses a simulation of fast reinforcement learning (fast RL) to achieve autonomous landing. Least-squares policy iteration (LSPI) is the model used in solving this problem. Section 2.4 briefly discusses obstacle avoidance in a limited environment. Compared with simultaneous localization and mapping (SLAM) and structure from motion (SfM), using DRL with a monocular camera is much better. The partially observable Markov decision processes (POMDP) model is introduced in this section. Q-value is the solution of the POMDP model. Then, an optimal Q-value is calculated to successfully avoid obstacles. Deep recurrent Q-network (DRQN) with long short-term memory (LSTM) approximates the optimal Q-value from the recent observation.

2.2 OBJECT DETECTION AND TRACKING

Object detection and tracking is a problem in UAV navigation. During UAV operations, the UAV scans its surrounding environment to check if any moving objects exist. If there is any moving object, then the UAV will target and follow the object; this is called object detection and tracking. In this section, we will discuss how to use a single camera to detect and track moving objects. Figure 2.1 shows three steps of how to achieve this function. As shown in Figure 2.1, the first step is resolving the video data into a sequence of frames and estimating the background motion between two subsequent images. The method used for estimating background motion is called the perspective transform model [1]. The second step uses a deep learning classifier to detect the moving object candidates. Using deep learning can greatly improve the accuracy of classifying the moving objects. By scanning all moving object candidates, the Lucas-Kanade optical flow algorithm [2] can be used to find the local motion, and then spatio-temporal characteristics are used to locate actual objects [3]. Finally, to decrease the noise, Kalman filter tracking [4] can be used. The description of each step will be discussed in more detail below.

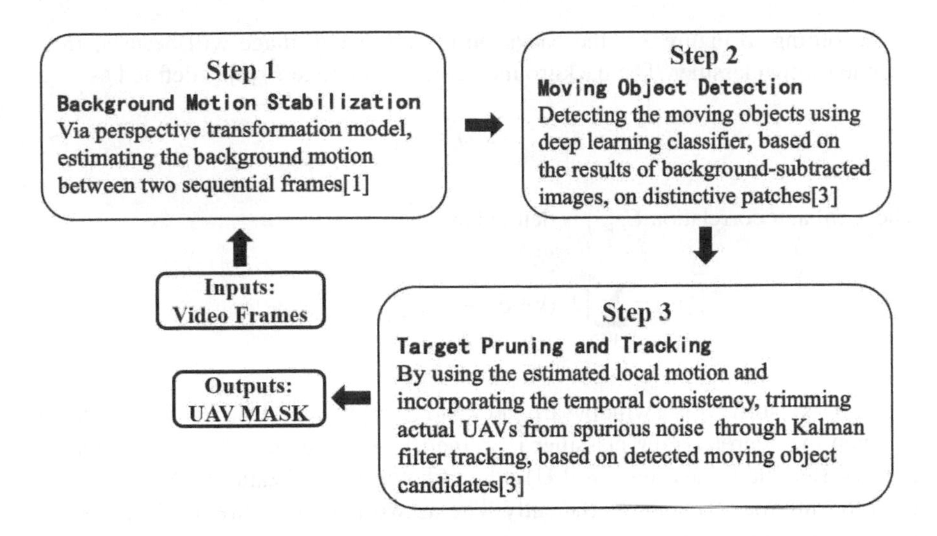

FIGURE 2.1 Overview of proposed method for object detection and tracking.

As shown previously, the perspective transform model [1] is used to estimate background motion. Unlike other global transformation models, such as rigid or affine transformation models, the perspective transformation model can take into account projection based on the distance from the camera, which is beneficial to compensate for the background motion at a far distance from a camera [3]. This method resolves the video into a sequence of frames. For each point $p_{t-1} \in \mathbb{R}^2$, corresponding points $p_t \in \mathbb{R}^2$ in the current frame X_t must be calculated using block matching [5]. The next step is estimating the perspective transformation $H_{t-1} \in \mathbb{R}^{3\times3}$ from X_{t-1} to X_t [3].

The formula of H_{t-1} [3] is shown in Eqn. 2.1:

$$H_{t-1} = \arg\min_{H} \sum_{P_t \in \mathbf{P_t},\, P_{t-1} \in \mathbf{P_{t-1}}} \left\| P_t - H \circ P_{t-1} \right\|_2^2 \tag{2.1}$$

where $\mathbf{P_t}$ and $\mathbf{P_{t-1}}$ are a set of corresponding points in X_t and X_{t-1} and $\circ$ is the warping operation [3]. Matrix H [3] is shown in Eqn. 2.2:

$$H = \begin{bmatrix} h_{11} & h_{12} & h_{13} \\ h_{21} & h_{22} & h_{23} \\ h_{31} & h_{32} & 1 \end{bmatrix} \tag{2.2}$$

where h_{11}, h_{12}, h_{13}, h_{21}, h_{22}, h_{23} represent the affine transformation matrix (such as sheer and translation, scale, rotation), and h_{31}, h_{32} are the additional parameters that allow a perspective projection to the vanishing point [3].

According to Figure 2.1, the background subtracted image will be used in the deep learning classifier. The background subtracted image E_t [3] is defined as

$$E_t = \left| X_t - H_{t-1} \circ X_{t-1} \right| \tag{2.3}$$

The local autocorrelation C_t [3] is defined as

$$C_t(s) = \sum_W \left[E_t(s + \delta s) - E_t(s) \right]^2 \approx \delta s^T \Lambda_t(s) \delta s \tag{2.4}$$

where δs is a shift, W is a window around s and Λ_t represents the precision matrix [3].

Then the deep learning classifier is trained using three CNNs to detect moving objects. Rectified linear unit (ReLU) [6] and batch normalization (BN) [7] also are used to train the DNN more efficiently. The network architecture for deep learning is shown in Figure 2.2.

Then to detect a moving object candidate, define the motion difference $d_t^{(n)}$ [3] as

$$d_t^{(n)} = \frac{1}{M} \sum_{m=1} M \left(h_t^{(n,m)} - u_t^{(n,m)} \right) \tag{2.5}$$

where $h_t^{(n,m)}$ means interpolated motion vector from the perspective transform H_t between X_t and X_{t+1} at the point $r_t^{(n,m)}$[3].

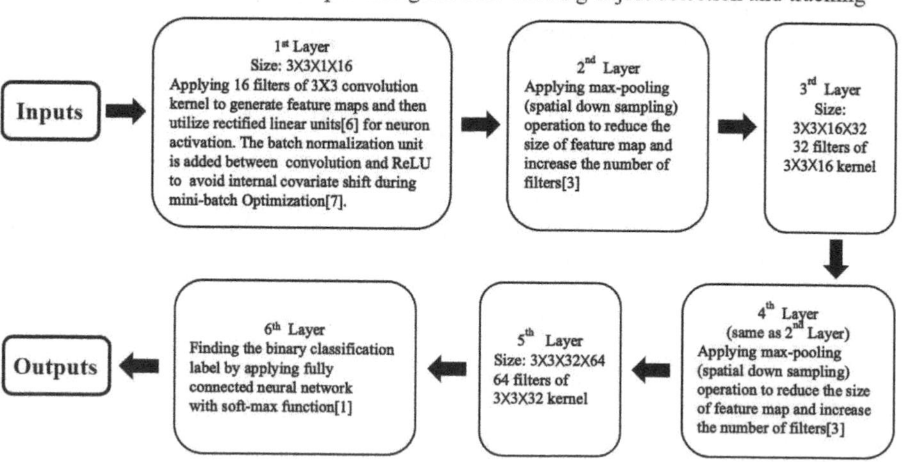

FIGURE 2.2 Network architecture for deep learning for object detection and tracking. ReLU, rectified linear unit.

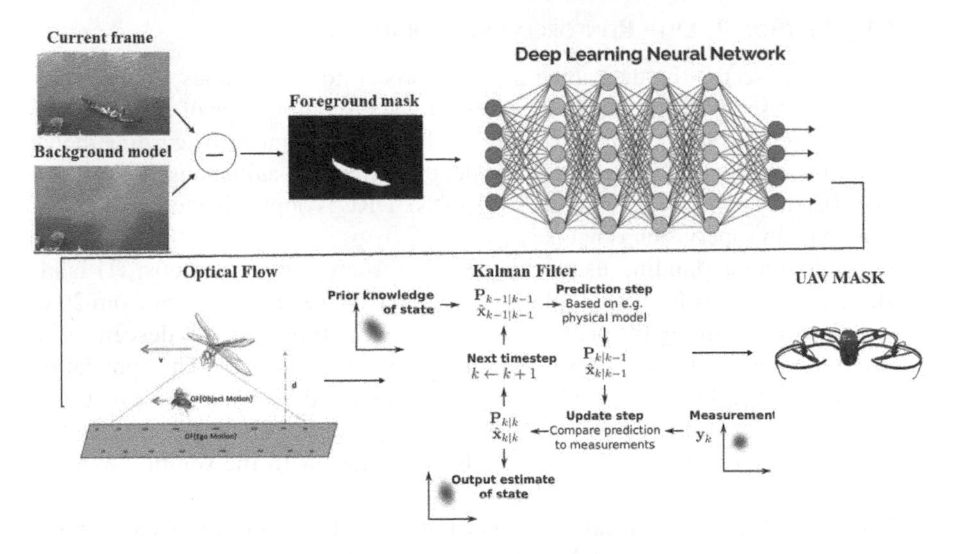

FIGURE 2.3 General diagram of object detection and tracking.

Define $y_t^{(n)}$ as a binary label where the positive value indicates that the moving object is the target [3] as

$$y_t^{(n)} = \begin{cases} 1, & \text{if } T_L < \left\| d_t^{(n)} \right\|_2 < T_H \\ 0, & \text{otherwise} \end{cases} \tag{2.6}$$

where T_L and T_H are the empirical low and high threshold for pruning [3].

Using a Kalman filter [4] improves target detection. To initialize the Kalman filter, the authors find the corresponding objects from optical flow matching in L previous frames and start tracking if the classification labels $y_{t-1}^{(n)},\ldots,y_{t-L}^{(n)}$ are positive [3].

As discussed previously, deep learning has been applied to UAV navigation and control, and this approach shows potential for improving overall performance. An overview of moving object detection and tracking is shown in Figure 2.3.

2.3 AUTONOMOUS LANDING

Autonomous landing is another popular topic in UAV navigation, due to the increasing demands on transportation and package delivery. Autonomous landing requires the UAV to detect and land on the target landmark within limited time and space. To solve this problem, there are several approaches. In this section we will discuss two different ways to achieve UAV autonomous landing using RL.

2.3.1 Method 1: Deep Reinforcement Learning

We begin this section by describing a new approach to autonomous UAV landing using DRL presented by the authors in [8]. DRL uses a combination of deep learning and RL approaches to combine a training model and a dynamically learning model to solve this problem [9]. Specifically, for deep learning the authors use CNNs, and for the RL, they apply deep Q-networks (DQNs). DRL is applicable to this problem because human supervision is not necessary.

For autonomous landing using this solution, there are three steps: (1) landmark detection in the horizontal plane (x,y-plane), (2) vertical descent from 20 to 1.5 m while maintaining the position over the landing target and (3) descent from 1.5 m to the landing target. These steps are shown in Figure 2.4. The application of RL to the general landing problem involves steps 1 and 2 only, as a conventional closed-loop controller can easily be used to complete the landing in step 3. The remaining paragraphs describe the details associated with the solution to steps 1 and 2.

For step 1, the UAV translates at a constant altitude (20 m in this case) in the X,Y-plane until it is positioned directly over the landing target. A low-resolution, down-looking camera mounted on the UAV is the only sensor required for this approach. The authors apply DQNs, Markov decision processes (MDPs) and two CNNs to solve this step. The input of one CNN is a stack of four 84 × 84 gray scale images acquired by the down-looking camera [8]. The key method is using the ReLU as an activation function [10]. The layers of the CNN are shown as Figure 2.5.

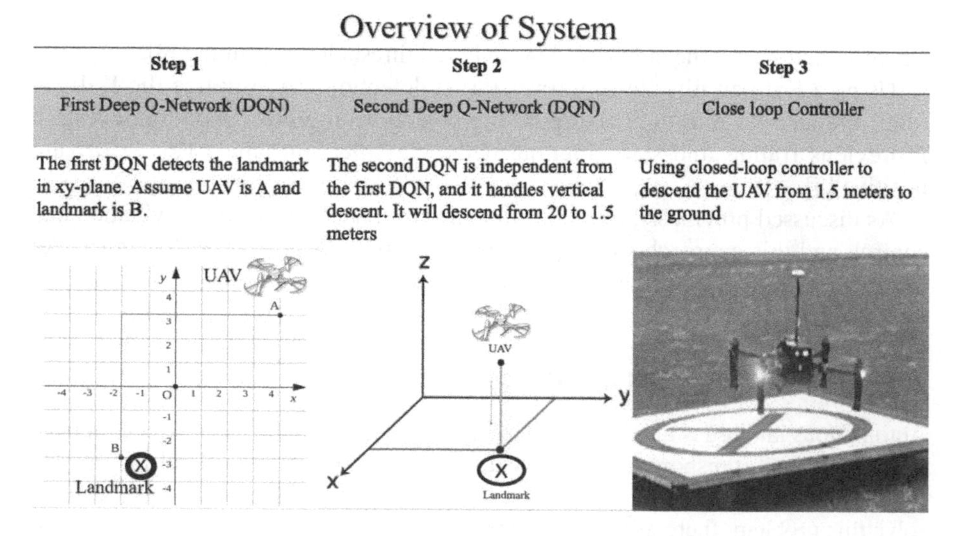

Overview of System
Step 1
Step 2
Step 3
First Deep Q-Network (DQN)
Second Deep Q-Network (DQN)
Close loop Controller
The first DQN detects the landmark in xy-plane. Assume UAV is A and landmark is B.
The second DQN is independent from the first DQN, and it handles vertical descent. It will descend from 20 to 1.5 meters
Using closed-loop controller to descend the UAV from 1.5 meters to the ground
UAV
A
B
Landmark
Z
y
X
UAV
Landmark

FIGURE 2.4 Overview of landing system using deep reinforcement learning.

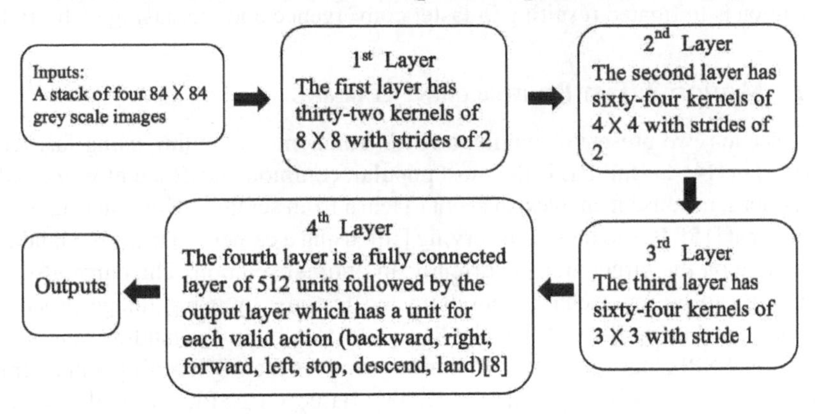

FIGURE 2.5 Different layers of CNNs for UAV landing.

The loss function $L_i(\theta_i)$ [8] is used to adjust the parameters of the DQN, and the function is shown in Eqn. 2.7:

$$L_i(\theta_i) = E_{(s,a,r,s')\sim U(D)}\left[\left(Y_i - Q(s,a;\theta_i)\right)^2\right] \tag{2.7}$$

where $D = (e_1, \ldots, e_t)$ is a dataset of experiences e, and $e_t = (s_t, a_t, r_t, s_{t+1})$ is used to uniformly sample a batch at each iteration i [8]. In the loss function, network Q is used to estimate actions at runtime, and the target Y_i is shown in Eqn. 2.8 [8]:

$$Y_i = r + \gamma \max_{a'} Q\left(s', a'; \theta_i^-\right) \tag{2.8}$$

where $Q\left(s', a'; \theta_i^-\right)$ is used to generate the target.

After positioning the UAV over the target in the horizontal plane, the UAV must vertically descend from 20 to 1.5 m, while maintaining its position over the target. During this decent phase, which is a form of Blind Cliffwalk [11], the agent often encounters many negtive rewards and few positive rewards due to the small size of the target and the resulting state space. A form of buffer replay called partitioned buffer replay can be used to solve this problem. It discriminates between rewards and guarantees a fair sampling between positive, negative and neutral experiences [8]. Overestimation [12] is a common problem related to reward sparsity. To solve this issue, a double DQN [13] is used. The double DQN is defined as

$$Y_i^d = r + \gamma Q\left(s', \arg\max_{a} Q(s', a'; \theta_i); \theta_i^-\right) \tag{2.9}$$

Using this target instead of the one in Eqn. 2.8, the divergence of the DQN action distribution is mitigated resulting in faster convergence and increasing stability [8].

2.3.2 METHOD 2: FAST REINFORCEMENT LEARNING

In this section, we present a solution to UAV autonomous landing using fast RL as described in [14]. So far, RL is the most popular, common and efficient way to adapt UAV actions, because it enables an agent to learn from scratch by interacting with its environment [15]. Based on visual serving [16], using a camera to track the landmark and landing on the target area is currently an optimal solution. The current state of the system can be represented from extracted features by using image processing techniques, and using LSPI [17] as the RL model to achieve the landing automation. In addition, LSPI converges faster with fewer samples than Q-learning and no initial tuning of parameters is required [14, 17, 18]. Based on LSPI, the goal is to work on a continuous state-action space by extending the algorithm. LSPI does not need any initial parameters and it can work on mobile applications. It does not face any overshooting, oscillation or divergence problems. This section discusses how LSPI is extended to work on a continuous action space [14].

The continuous action space is set as a simulation environment. The Unified System for Automation and Robot Simulation system (USARSim system) [19] is used to simulate a suitable environment. Using an extended version of the USARSim software, which is known as altURI [20], provides images through the widely used OpenCV vision processing library [21]. OpenCV is one of the graphics tools used to capture vision components. There are three steps to detect the landmark target by using OpenCV, as shown as Figure 2.6. The state space is limited by parameters (r, θ, ϕ), as shown in Figure 2.7.

Eqn. 2.10 demonstrates how to use the LSPI algorithm to estimate approximate Q-value:

$$Q(s,a) \approx \hat{Q}^{\pi}(s,a,\omega) = \sum_{i=1}^{k} \phi_i(s,a)\omega_i = \Phi(s,a)^T W \tag{2.10}$$

where ϕ_i is the ith basis function, ω_i is its weight in the linear equation and k is the number of basis functions [14]. Using Eqn. 2.10, the temporal difference (TD) update equation [15] can be rewritten to solve $Q(s,a)$ shown in Eqn. 2.11:

$$\Phi W \approx R + \gamma P^{\pi} \Phi W \tag{2.11}$$

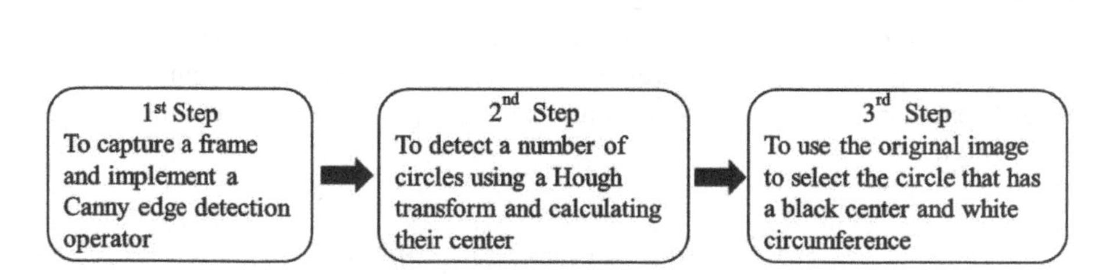
1st Step
To capture a frame and implement a Canny edge detection operator
2nd Step
To detect a number of circles using a Hough transform and calculating their center
3rd Step
To use the original image to select the circle that has a black center and white circumference

FIGURE 2.6 Three steps for detecting target using OpenCV.

State Space Parameters

Parameters	r	θ	φ
Definition	radial distance of that agent in the space from the center of the target	the azimuthal angle in the xy-plane from the x-axis of the target	the polar angle from the z-axis
Range	$0 \leq r \leq 10$ meter	$0 \leq \theta < 360$ degree	$-90 \leq \varphi \leq 90$ degree
Goal State	$0 \leq r \leq 0.1$ meter	$0 \leq \theta < 10$ degree	$-5 \leq \varphi \leq 5$ degree
	Others		
Reward	(1) 120 if it gets to the goal, (2) −1500 if it finishes outside state space, (3) equal to a value, this value decreases as φ or θ increases		
Control	Lateral Velocity: [−5, 5] Linear Velocity: [−5, 5]		

FIGURE 2.7 State space parameters.

where Φ represents the basis function for all state action, and it is an $|S||A| \times k$ matrix. Reformulated Eqn. 2.11 becomes

$$\Phi^T(\Phi - \gamma P^\pi \Phi)\omega^\pi = \Phi^T R \tag{2.12}$$

where P is a stochastic matrix that contains the transition model of the process, and R is a vector that contains the reward values [14]. The weights, W, can be represented by

$$W = [\Phi^T(\Phi - \gamma P^\pi \Phi)]^{-1}\Phi^T R = A^{-1}b \tag{2.13}$$

where A and b are offered by the environment samples. If a set of samples is given, based on Eqns. 2.10–2.13, the optimal solution with the best weight can be calculated easily.

Thus, by combining the policy-search efficiency of the approximate policy iteration with the data efficiency of approximate estimation of the Q-value function, the authors obtain the LSPI algorithm [22]. The π is the approximated policy that can maximize the corresponding Q-function. When the optimal W is calculated, the policy improvement will start as

$$\pi(s \mid \omega) = \arg\max_a \phi(s,a)^T \omega \tag{2.14}$$

This is how LSPI extends the approach to work with a continuous action space.

2.3.3 Summary of Autonomous Landing Methods

In summary, there are many ways to achieve autonomous landing by using deep learning. The two methods introduced in this section are the most popular approaches. Although both of them utilize deep learning models, the routes to achieve the goal

TABLE 2.1
Compare Two Approaches for Autonomous Landing

Compare	Method 1	Method 2
Method, model or algorithm	Deep reinforcement learning (DRL), Deep Q-networks (DQNs), double deep Q-networks, Markov decision processes (MDPs)	Fast reinforcement learning (fast RL), least-squares policy iteration (LSPI), OpenCV, USARSim system, altURI
Advantages	This method only needs low-resolution images with one down-looking camera. It uses three steps to detect and land on the target and has high accuracy. It is also simple to operate because the three individual steps are independent.	Method 2 needs fewer samples to train the model, which means the speed of training is improved. Instead of separate horizontal and vertical movements to land on the target, this method moves directly down toward the target resulting in faster landing.
Disadvantages	Method 1 achieves the goal with slower speed because all the steps are serialized. Also, it needs more samples, and it relies on an onboard GPU increasing the cost of the UAV.	This method relies on a good environment. In addition, the accuracy depends on the optimal value tuned by agent. The agent has to learn the correct direction (right and left for the lateral velocity, or forward and backward for the linear velocity) and optimal value for turning [14]. All the steps are integrated together making it much more difficult to tune.

are totally different. Both methods have their pros and cons, as shown in Table 2.1. Researchers can make their choice based on their needs. If speed of the UAV to land on the target is preferred, then the second approach is the better choice. However, this approach is more complicated and requires an in-depth understanding of the model to tune the directions of travel and turning values.

2.4 OBSTACLE AVOIDANCE IN POOR ENVIRONMENT CONDITIONS

Prior to RL approaches, UAV obstacle detection and avoidance were dominated by SLAM [23] and SfM [24]. Both of these approaches use various sensor packages to detect obstacles, map the environment, localize the UAV within the environment and help to plan a path through the environment. Sensor packages typically used include Kinect [25], light detection and ranging (LiDAR), sound navigation and ranging (SONAR), monocular cameras, stereo cameras and optical flow sensors [24, 26–29].

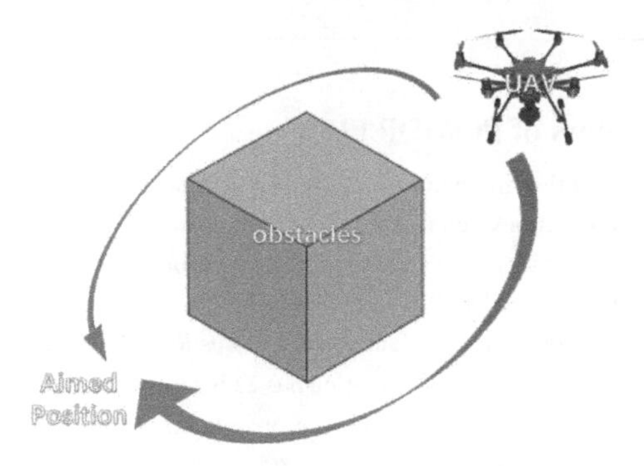

FIGURE 2.8 One simple model of UAV obstacle avoidance.

Because of the high computational load associated with sensing the environment, creating a depth map and dynamically determining the best path through the environment, all in real-time, SLAM and SfM proved to be limited in their effectiveness for UAV navigation in three dimensions, although they are still very popular for 2D ground vehicles. To improve 3D UAV obstacle avoidance, additional work has been done including a behavior arbitration scheme to obtain the yaw and pitch angles for the UAV to avoid an obstacle and for navigation in general [30]; trajectory planning using obstacle bounding boxes and depth estimation [31]; two different CNN architectures, one for depth and surface normal estimation and the other for trajectory prediction [32] and an unconventional approach using a collection of UAV crash datasets [33]. These last four approaches [30–33] are attempts by the research community to get away from the larger, more expensive sensor packages required by SLAM and SfM; instead they apply DL to the problem using only a monocular camera. The DL approaches have limitations because they typically do not handle the dynamic nature of obstacle avoidance very well. Table 2.2 summarizes the various sensors that have been used in the research and lists their pros and cons.

TABLE 2.2
Disadvantages of Using Some Sensors to Achieve Obstacle Avoidance

Kinect, LiDAR, SONAR	They are all complicated and expensive. Because of the nature of sophisticated sensors, some of them require an independent power supply, and the others are heavy. In either situation, they are not good choices for UAVs.
Stereo cameras	They have the same weight issue as Kinect, LiDAR and SONAR sensors, and another disadvantage, range limitation. For long-range obstacle avoidance, their performance will drop dramatically.
Monocular camera	Even though a monocular camera is a low-cost and lightweight sensor suitable for UAVs, measuring the range through RGB images is a challenge. Monocular cameras offer 2D images of the surrounding environment. Although a great deal of research is ongoing to improve environmental mapping using deep learning, it is not the best option to achieve obstacle avoidance.

TABLE 2.3
Meaning of Parameters of POMDP [26]

S	Set of states of the environment, referred to as "state space."
A	Set of feasible actions, referred to as the "action space."
P	Transition probability function that models the evolution of states based on actions chosen and is defined as P: $S \times A \times S \rightarrow [0, 1]$.
R	Reinforcement or the reward function defined as R: $S \times A \rightarrow \mathrm{R}$.
Ω	Set of observations and an observation $o \in \Omega$ is an estimate of the true state S.
O	O: $S \times A \times \Omega \rightarrow [0, 1]$ is a conditional probability distribution over Ω, while $\gamma \in (0,1)$.
γ	Discount factor; mostly, the range is $\gamma \in (0,1)$.

Considering the limitations of DL approaches, the research community has begun to apply RL to this problem. We discuss the performance of a DRQN with temporal attention, which is utilized by a deep RL robotic controller for effective obstacle avoidance for UAVs in cluttered and unseen environments. This approach enables the UAV controller to collect and store relevant observations gathered over time [26].

A monocular camera is used in this algorithm, even though it has limited field of view. In this case, the monocular RGB image is the input to the model without depth or other sensory inputs, which means the observation is the predicted depth map obtained from the monocular image.

The POMDP model can be used to represent the UAV obstacle avoidance and navigation problem. The parameters of the POMDP model are (S, A, P, R, Ω, O, γ), and the details are shown in Table 2.3.

The reward function serves as a feedback signal to the UAV for the chosen action [26]; π^*: $\Omega \rightarrow A$ is the optimal solution for obstacle avoidance. By determining an optimal policy, the UAV controller is able to select an action at each timestep t that maximizes the expected sum of discounted rewards, which is denoted as [26]

$$\mathrm{E}\left[\sum_{t=1}^{\infty} \gamma^t R(s_t, a_t)\right] \tag{2.15}$$

To solve this problem, one classic approach is to learn an optimal policy using Q-values. A Q-value can be represented by $Q^\pi(s, a)$, where π determines the expected sum of discounted rewards obtained by taking the action a on state s and following the policy π thereafter [26]. The optimal solution for POMDP is the optimal solution of Q-value, which is represented as

$$Q^*(s,a) = \max_\pi Q^\pi(s,a) \tag{2.16}$$

However, this algorithm has a limitation of dimensions. This is because iterative learning the Q-values for a huge state space requires maintaining and updating

Q-values for all unique state-action pairs, which turns out to be computationally infeasible [26]. Therefore, weights, w, are needed to solve this issue while calculating Q-values, and this Q-value is defined as $Q(s, a|w)$. The loss function is defined as $L_i(\omega_i)$:

$$L_i(\omega_i) = \mathrm{E}_{(s,a,r,s')\sim D}\left[\left(r + \gamma \max_{a'} Q(s',a';\omega_i^-) - Q(s,a;\omega_i)\right)^2\right] \quad (2.17)$$

where ω_i^- represents weights of the target network, which is an older copy of network weights lagging behind a few iterations [26], and s, a, r, γ are the parameters of POMDP, which are shown in Table 2.3. Presenting a memory-augmented CNN architecture estimates the Q-values from the observations.

DRQN is the model used for estimating Q-values. The recurrent network possesses the ability to learn temporal dependencies by using information from an arbitrarily long sequence of observations, while the temporal attention weighs each of the recent observations based on their importance in decision making [26]. DRQN with LSTM [34] approximates the Q-value as

$$Q(o_t, h_{t-1}, a_t) \quad \mathbf{(2.18)}$$

where o_t is the latest observation from the sequence of recent L observations, $o_{(t-1-L)}, \ldots, o_t$, h_{t-1} is the hidden state of the recurrent network and is determined as $h_{t-1} = \mathrm{LST}\,M(h_{t-2}, o_{t-1})$ [26]. Using temporal attention [35] in this model evaluates the informativeness of each observation in the sequence because it optimizes a weight vector with values depicting the importance of observations at the previous instants [26]. The model is shown as Figure 2.9. Therefore, the Q-value can be predicted based on the model discussed previously.

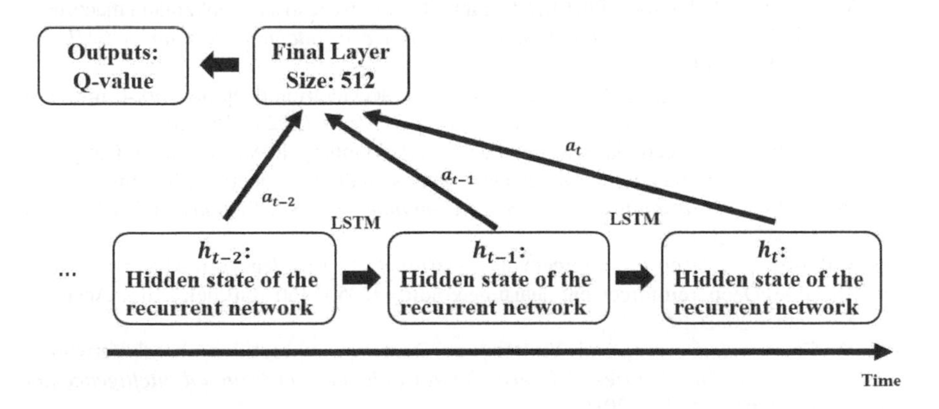

FIGURE 2.9 Model of DRQN with LSTM.

2.5 CONCLUSIONS

This chapter focuses on UAV navigation using deep learning and RL. Precise object detection and tracking, autonomous landing and obstacle avoidance are three topics discussed here. Object detection and tracking is defined as a UAV scanning its surrounding environment to check if any moving objects exist. If an object exists, then the UAV targets the object and follows the object. To achieve this, three steps are discussed in Section 2.2, including background motion stabilization, moving object detection and target pruning and tracking. The autonomous landing problem is defined as when a UAV can land on the target landmark itself using a simple camera. Solving this problem requires the UAV to target the landmark first, then control its landing onto the target. Two different methods, both using RL, are discussed in Section 2.3. Also, obstacle avoidance is discussed in Section 2.4. The obstacle avoidance problem usually happens in limited environments or complex environments in which there is not much space for UAV operation. The disadvantages of some methods, such as SLAM and SfM, are presented. SLAM and SfM are two great methods to achieve obstacle avoidance for ground vehicles; however, they are not suitable for UAVs. Then, we presented the POMDP model, and showed how to use DRQN with LSTM to approximate optimal Q-values to optimize the POMDP model.

REFERENCES

1. I. Carlbom and J. Paciorek, "Planar geometric projections and viewing transformations," *ACM Computing Surveys (CSUR)*, vol. 10, no. 4, pp. 465–502, 1978.
2. B. D. Lucas, T. Kanade *et al.*, "An iterative image registration technique with an application to stereo vision," *Proceedings of the 7th International Joint Conference on Artificial Intelligence*, 1981.
3. D. H. Ye, J. Li, Q. Chen, J. Wachs, and C. Bouman, "Deep learning for moving object detection and tracking from a single camera in unmanned aerial vehicles (UAVs)," *Electronic Imaging*, vol. 2018, no. 10, pp. 466–1, 2018.
4. G. Welch, G. Bishop *et al.*, "An introduction to the Kalman filter," *SIGGRAPH*, 1995.
5. K. Hariharakrishnan and D. Schonfeld, "Fast object tracking using adaptive block matching," *IEEE Transactions on Multimedia*, vol. 7, no. 5, pp. 853–859, 2005.
6. V. Nair and G. E. Hinton, "Rectified linear units improve restricted boltzmann machines," *Proceedings of the 27th international conference on machine learning (ICML-10)*, pp. 807–814, 2010.
7. S. Ioffe and C. Szegedy, "Batch normalization: accelerating deep network training by reducing internal covariate shift," *arXiv preprint arXiv: 1502.03167*, 2015.
8. R. Polvara, M. Patacchiola, S. Sharma, J. Wan, A. Manning, R. Sutton, and A. Cangelosi, "Toward end-to-end control for uav autonomous landing via deep reinforcement learning," *2018 International Conference on Unmanned Aircraft Systems (ICUAS)*, IEEE, pp. 115–123, 2018.
9. Wikipedia, "Deep reinforcement learning," https://en.wikipedia.org/w/index.php?title=Deep_reinforcement_learning&oldid=920502460 [online], 2019. Accessed November 20, 2019.
10. X. Glorot, A. Bordes, and Y. Bengio, "Deep sparse rectifier neural networks," *Proceedings of the Fourteenth International Conference on Artificial Intelligence and Statistics*, pp. 315–323, 2011.

11. T. Schaul, J. Quan, I. Antonoglou, and D. Silver, "Prioritized experience replay," *arXiv preprint arXiv: 1511.05952*, 2015.
12. S. Thrun and A. Schwartz, "Issues in using function approximation for reinforcement learning," in *Proceedings of the 1993 Connectionist Models Summer School Hillsdale, NJ. Lawrence Erlbaum*, 1993.
13. H. Van Hasselt, A. Guez, and D. Silver, "Deep reinforcement learning with double q-learning," *Thirtieth AAAI Conference on Artificial Intelligence*, 2016.
14. M. Shaker, M. N. Smith, S. Yue, and T. Duckett, "Vision-based landing of a simulated unmanned aerial vehicle with fast reinforcement learning," *2010 International Conference on Emerging Security Technologies*, IEEE, pp. 183–188, 2010.
15. R. S. Sutton, A. G. Barto *et al.*, *Introduction to reinforcement learning*, Cambridge, MA: MIT Press, vol. 135, 1998.
16. T. Martínez-Marín and T. Duckett, "Fast reinforcement learning for vision-guided mobile robots," *Proceedings of the 2005 IEEE International Conference on Robotics and Automation*, IEEE, pp. 4170–4175, 2005.
17. M. G. Lagoudakis and R. Parr, "Model-free least-squares policy iteration," in *Advances in Neural Information Processing Systems*, pp. 1547–1554, 2002.
18. P. Wang and T. Wang, "Adaptive routing for sensor networks using reinforcement learning," *The Sixth IEEE International Conference on Computer and Information Technology (CIT'06)*, pp. 219–219, 2006.
19. S. Hughes and M. Lewis, "Robotic camera control for remote exploration," *Proceedings of the SIGCHI conference on Human Factors in Computing Systems*. ACM, pp. 511–517, 2004.
20. M. Smith, M. Shaker, S. Yue, T. Duckett *et al.*, "Alturi: a thin middleware for simulated robot vision applications," 2011.
21. G. Bradski and A. Kaehler, "Opencv," *Learning OpenCV*, p. 1, 2008.
22. M. G. Lagoudakis and R. Parr, "Least-squares policy iteration," *Journal of Machine Learning Research*, vol. 4, Dec, pp. 1107–1149, 2003.
23. C. Leung, S. Huang, and G. Dissanayake, "Active slam using model predictive control and attractor based exploration," *2006 IEEE/RSJ International Conference on Intelligent Robots and Systems*, pp. 5026–5031, 2006.
24. D.-J. Lee, P. Merrell, Z. Wei, and B. E. Nelson, "Two-frame structure from motion using optical flow probability distributions for unmanned air vehicle obstacle avoidance," *Machine Vision and Applications*, vol. 21, no. 3, pp. 229–240, 2010.
25. Z. Zhang, "Microsoft kinect sensor and its effect," *IEEE Multimedia*, vol. 19, no. 2, pp. 4–10, 2012.
26. A. Singla, S. Padakandla, and S. Bhatnagar, "Memory-based deep reinforcement learning for obstacle avoidance in uav with limited environment knowledge," *arXiv preprint arXiv: 1811.03307*, 2018.
27. W. G. Aguilar, G. A. Rodríguez, L. Álvarez, S. Sandoval, F. Quisaguano, and A. Limaico, "Visual slam with a RGB-D camera on a quadrotor UAV using on-board processing." In: I. Rojas, G. Joya, and A. Catala, Eds. *Advances in Computation Intelligence, IWANN 2017. Lecture Notes in Computer Science*, Cham, Switzerland: Springer, vol 10306, pp. 596–606, 2017.
28. T. Gee, J. James, W. Van Der Mark, P. Delmas, and G. Gimel'farb, "Lidar guided stereo simultaneous localization and mapping (slam) for uav outdoor 3-d scene reconstruction," *2016 International Conference on Image and Vision Computing New Zealand (IVCNZ)*, IEEE, pp. 1–6, 2016.
29. H. Alvarez, L. M. Paz, J. Sturm, and D. Cremers, "Collision avoidance for quadrotors with a monocular camera." In: M. Hsieh, O. Khatib, and V. Kumar, Eds. *Experimental Robotics*. Cham, Switzerland: Springer, vol 109, pp. 195–209, 2016.

30. P. Chakravarty, K. Kelchtermans, T. Roussel, S. Wellens, T. Tuytelaars, and L. Van Eycken, "CNN-based single image obstacle avoidance on a quadrotor," *2017 IEEE International Conference on Robotics and Automation (ICRA)*, IEEE, pp. 6369–6374, 2017.
31. M. Mancini, G. Costante, P. Valigi, and T. A. Ciarfuglia, "J-mod 2: joint monocular obstacle detection and depth estimation," *IEEE Robotics and Automation Letters*, vol. 3, no. 3, pp. 1490–1497, 2018.
32. S. Yang, S. Konam, C. Ma, S. Rosenthal, M. Veloso, and S. Scherer, "Obstacle avoidance through deep networks based intermediate perception," *arXiv preprint arXiv: 1704.08759*, 2017.
33. D. Gandhi, L. Pinto, and A. Gupta, "Learning to fly by crashing," *2017 IEEE/RSJ International Conference on Intelligent Robots and Systems (IROS)*, IEEE, pp. 3948–3955, 2017.
34. S. Hochreiter and J. Schmidhuber, "Long short-term memory," *Neural Computation*, vol. 9, no. 8, pp. 1735–1780, 1997.
35. W. Pei, T. Baltrusaitis, D. M. Tax, and L.-P. Morency, "Temporal attention-gated model for robust sequence classification," *Proceedings of the IEEE Conference on Computer Vision and Pattern Recognition*, pp. 6730–6739, 2017.

3 Formation Control of Networked UAVs

Guiqin Yang[1], *Fei Hu*[2]
[1]School of Electronic and Information Engineering, Lanzhou Jiaotong University, Lanzhou, 730070, China
E-mail: 1261397198@qq.com
[2]Electrical and Computer Engineering, The University of Alabama, Tuscaloosa, AL

CONTENTS

3.1 INTRODUCTION

With the developments of electronics and software, there is an increasing deployment of unmanned aerial vehicles (UAVs) that are capable of performing autonomous coordinated actions. UAVs have attracted a great deal of research interest over the last decades. They have been widely exploited in military and civilian applications to carry out a number of high-risk tasks, such as transportation of heavy loads, search and rescue missions, surveillance and reconnaissance, forest fire detection, disaster monitoring, drag reduction [1], radiation detection and contour mapping [2], precision agriculture, weather forecasting and telecommunication relay [3]. Because UAVs are typically small, light, and have a low CPU capacity, a single UAV is only capable of conducting relatively simple missions. Compared with an individual UAV, a fleet of UAVs has extreme superiority in performing missions, such as high efficiency, low cost of fuel and time, wide coverage, improved robustness, fault-tolerant capabilities and high flexibility. In addition, the network of multiple UAVs may generate a new functionality via various formations of UAVs. For example, when UAVs are in a special formation, they could be used for surveillance or exploration purposes, and they synthesize an antenna with a dimension far larger than the case of a single UAV [4]. Moreover, some localization tasks may require that multiple UAVs share information (such as relative positions) with each other via wireless communications. The main task for a swarm of UAVs is to adjust their relative positions to perform autonomous coordinated actions.

In fact, the concept of formation is inspired by natural animal behaviors such as bird flocking or fish schooling. With certain formations in a group, birds or fish can enhance the survival of the individuals. By mimicking birds or fish's formation behaviors, the fleet of unmanned UAVs can fly in the formation to accomplish complex tasks, and the level of system autonomy can be improved.

When UAVs fly in a formation, how to design the formation control of multi-UAV systems is very important. Generally speaking, several tasks are required to control a formation: (1) moving the whole formation or the center of mass of the formation from one position to another one; (2) maintaining the relative positions of the UAVs during formation motions, so that the shape is preserved; (3) avoiding obstacles and collisions; (4) splitting the formation in some situations; (5) ensuring that a fleet of UAVs remains in a state in which the communication network can meet cooperative planning (or coordination) and data collection requirements and so forth.

There are two main basic control strategies for the formation control of multi-UAV systems: (1) *centralized* architecture wherein the control signals for all the UAVs are generated only in the leader UAV or in a central station, and (2) *distributed* or decentralized architecture in which each UAV computes the control signals based on its own position and neighbors' information [5, 6].

New developments in the fields of control and communications have produced many algorithms to control a fleet of UAVs. Algorithms of centralized and distributed (or decentralized) control have been developed for many specific application scenarios in the last few years. Today there are a few typical formation control approaches for multi-UAV systems, for example, leader–follower (LF)-based methods [7], virtual structure (VS)-based ones [8] and behavior-based ones [9].

Although these approaches can be applied to deal with the formation control problems of multi-UAV systems, each of them has certain weaknesses. For example, the LF-based approaches lack robustness due to the existence of the explicit leader and paucity of the feedback from the followers. VS-based approaches are not suitable for distributed implementation and not flexible for shape deformation. They also cannot achieve collision avoidance and require significant communications and computations. Behavior-based approaches cannot be easily modeled and have difficulties achieving the stability of the system.

In recent years, with the development of consensus theory, many researchers have proved that consensus solutions based on graph theory can be used to solve formation control problems. Moreover, some study results showed that the traditional methods, including LF-based, VS-based and behavior-based approaches, can be regarded as special cases of consensus-based approaches [10].

How to define practical architectures for the formations depends on the schemes of communications, sensing and control. With the development of information technology, using digital communication networks in the implementation of control structure brings tremendous advantages, for example, low maintenance cost, scalability, effectiveness and robustness. The communication network is one of the key components of an UAV system. It should ensure that command and control messages are exchanged properly and that remotely sensed data are transmitted back to the desired stations.

However, due to the rapid mobility of a multi-UAV team, the wireless communication channel performance is dynamic and difficult to be predicted accurately. Moreover, link failure or long delay in data transmissions is unavoidable in many cases. How to control and implement multi-UAV systems that are robust to uncertainties and interference is challenging. Control methods including motion planning should produce an effective network topology to ensure good connectivity. The models for the cooperative control and networking need to be integrated together to jointly optimize the control and routing strategies [11].

Moreover, the problem of collision and obstacle avoidance is of great importance in formation control of mobile UAVs for the system safety.

In this chapter, we provide a detailed survey on some typical methods for the formation control of multi-UAV systems, including three traditional approaches, such as LF-based, VS-based and behavior-based methods. We specifically discuss and review consensus problems for dynamic UAV networks as well as the collision avoidance problem.

The chapter is organized as follows. In Section 3.2, we briefly present two main types of control architectures to maintain the shape of a formation, while the formation flies as a cohesive whole. We elaborate the characteristics of autonomous formations controlled by centralized or distributed control structure. In Section 3.3, we first introduce several UAV operation modes, including man-in-the-loop and autonomous modes. We also discuss the communication paradigm that UAVs should rely on and several methods to control the network for a fleet of UAVs. Section 3.4 describes the notion differences between motion planning and path planning. In Section 3.5, we summarize some recent results on three traditional formation control methods, i.e., LF-based, VS-based and behavior-based approaches. Section 3.6 is dedicated to consensus-based algorithms for formation control and provides a detailed review of consensus problems for networks of dynamic UAVs. The chapter is concluded in Section 3.7.

3.2 UAV FORMATION ARCHITECTURES

As previously mentioned, it is often more effective to deploy small UAVs as a fleet to carry out many critical tasks. With the development of formation control models for multi-UAV systems, the question of what sort of sensing and control architectures are needed to maintain the shape of a formation, while the formation moves as a cohesive whole, has drawn more and more attention over the last two decades.

Using the digital communication networks, command and control messages (including UAV states, estimated target locations, waypoints or task allocation, etc.)

can be exchanged among the central station and UAVs, and remotely sensed mission data (including video, still images, atmospheric samples, etc.) can be transmitted to the designated centers for data processing, exploitation and dissemination purposes [11, 12]

Formation applications require precise, robust, flexible and scalable control algorithms/models. This requires reliable communication protocols for data transmissions. The design and control of the communication network depends on the system architecture and data transmission requirements.

There are two main strategies for the formation control of multi-UAV systems to ensure that the shape of a formation is preserved. One strategy is *centralized* architecture wherein the control signals for all the UAVs are generated in the leader UAV or in a central station. The other UAVs remain within the transmission range of the command and control station. In this case a hub-and-spoke network model can be adopted, similar to a cellular system [Figure 1(a)]. Centralized control strategies utilize a single controller to generate collision-free trajectories for the formation.

The other strategy is known as distributed control in which each UAV computes its own control signals based on its position and the neighbors' information [5]. A great deal of information needed by each UAV for cooperation assignments is generated from a local source; thus, much faster reactions might be obtained by keeping all of computations in local places. The UAV shares the results of local computations with other UAVs through communication links [Figure 1(b)].

In the case of *centralized* architecture, all UAVs are connected directly to a control center, and all communications among UAVs are routed through the control center. This may cause blockage of links, higher latency and requirement of higher bandwidth. In addition, as the UAVs are mobile, steerable antennas may be required to keep the antenna orienting toward the control station. Because the

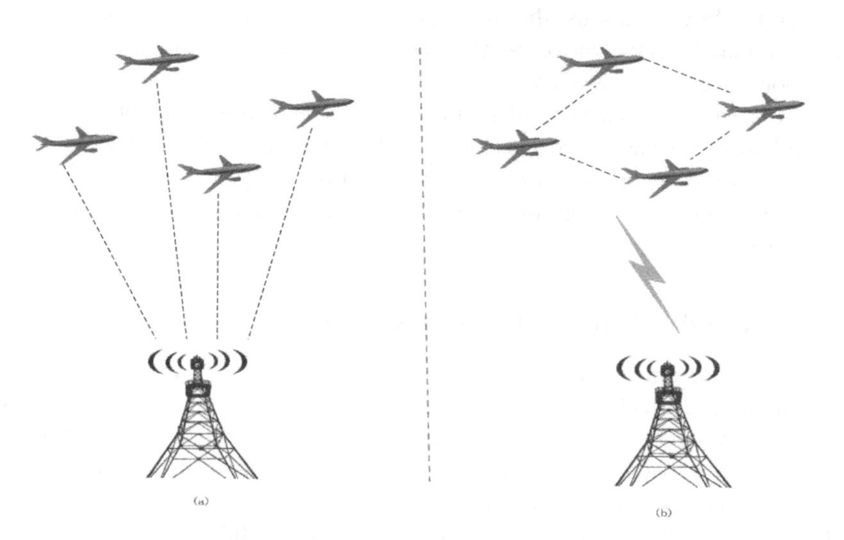

FIGURE 3.1 (a) Centralized architecture. (b) Distributed architecture.

downlink length is longer than the inter-UAV distance, centralized configurations suffer from high latency and all communications must pass through the control center. Also, if the control center fails, there will be no inter-UAV communications. For centralized solution strategies, all motion planning computations occur in a single location.

Centralized schemes also require high computational power and are not robust due to the heavy dependence on a single controller (thus not scalable to a large team size). In fact, the following cases may not be ideal for centralized approaches: (1) if large amounts of data need to be transmitted to the center's location, a centralized architecture may not be ideal and (2) the processing speed of the centralized mechanisms is throttled by the rate at which the important information reaches the computation center. If communications are slow and unreliable, it is critical not to allow a large amount of possibly irrelevant data to pass through the network. Moreover, in most civilian applications, normal operations do not require communications among UAVs to be routed through the control center.

In a distributed architecture, UAVs are interconnected and can usually communicate directly with each other on more than one link, and a small number of UAVs may connect to the control center in one link. Because we do not require all UAVs to reach the center, the communication latency could be reduced.

In distributed networks, the UAVs may move away, thus formations may break; therefore, the links may be intermittent. To tackle these issues, the network needs to be self-healing with continuous radio-frequency (RF) connection and flexible reconfiguration around a broken path. Compared with the centralized system, distributed networks are more flexible, reliable and offer better Quality of Service (QoS) performance. However, a critical requirement for distributed systems is that they need to rely on stable communications and reliable information sharing among all entities. If communication links are not sufficiently reliable, algorithm performance may degrade significantly.

Any formation architecture should be scalable. When the number of UAVs in the formation increases, the number of communication links required for a single UAV also grow linearly [4]. Collective animal behaviors, such as flocking of birds, schooling of fish, swarming of insects and herding of quadrupeds, give us an inspiration that an individual animal in a group tends to maintain close affiliations with its nearby neighbors. Distributed control laws will stably restore the shape of a formation, when the distances between UAVs become unequal to their predefined values.

The major differences between centralized and distributed architectures are given in Table 3.1.

As we can see, there are advantages and disadvantages for each system architecture. However, only in distributed systems are the resources allocated across the whole system. By doing so, they are remarkably robust to system faults and more transparent to user applications. With the adoption of distributed systems, their deployment and maintenance costs are low, and these systems will likely become a promising choice in the near future. At the same time, many of the applications of multi-UAV systems require greater topology flexibility, with inter-UAV information exchanged and relayed. This is the reason that the distributed architecture has drawn significant interests in many applications.

TABLE 3.1
Comparison of Centralized and Distributed Architecture Properties

Centralized System	Distributed System
Point-to-point	Multi-point to multi-point
Central control station is present.	Infrastructure-based system may have a control center; ad hoc system has no control center
Not self-configurable.	Self-configurable.
Single hop from UAV to central station.	Multi-hop communications.
UAVs cannot move freely.	In ad hoc systems, UAVs are autonomous and free to move. In infrastructure-based system, movement is restricted around the control center.
Low fault tolerance.	High fault tolerance.
Low scalability.	High scalability.
Links between UAVs and central stations are configured.	Inter-UAV links are intermittent.
UAVs communicate through central center.	UAVs relay traffic for other UAVs.
All of planning-relevant computations occur in a central center.	keeping all computations local.

One thing to note is that here we only mentioned two basic system architectures. In reality, there are also several other general architectures to deploy multiple UAVs as a team, such as hierarchical architecture or hybrid architecture.

3.3 COMMUNICATION NETWORKS FOR FLEETS OF UAVs

In principle, the operations of UAV can be categorized into three modes: (1) man-in-loop, which is a typical mode of operation; (2) man-on-loop, which is also called "semi-autonomous" and (3) autonomous (in this mode, the UAV operates without direct real-time human control and responds automatically to the changes in its operating environment [13]).

Certainly, a single UAV might be capable of all three modes of operations. The operation platforms of man-in-loop or man-on-loop have low levels of autonomy or are only semi-autonomous. Due to the increasing trend toward the development of autonomy and personnel risk minimization, the involvement of human operator might be reduced. It is often more effective to let these small UAVs operate autonomously and deploy them as a team in the formation, while maintaining certain positions, altitudes and angular velocity, with respect to each other. Over the last decades, by using UAVs as a formation, numerous autonomous cooperative task allocation and planning methods have been developed.

Because of the absence of pilots onboard, to ensure an adequate level of safety for UAVs to operate in shared airspace, the communication network protocols must be properly designed to manage the behaviors of the multi-UAV system. Proper control of the network is necessary for the system to function properly. Communication links must be highly reliable and exhibit high levels of integrity.

The levels of required communication performance can be measured by a paradigm called required communications performance (RCP) in the aviation community. RCP consists of four performance metrics: availability, continuity, integrity and latency. Without doubt, when UAVs fly in a formation, the command and control messages of the system that carries certain safety-critical data must be robust and reliable.

To control the network for a fleet of UAVs to ensure that (1) command and control messages are exchanged properly and (2) remotely sensed data can be transmitted back to desired locations, there are some control methods as follows [11]:

1. Motion planning to create an effective network topology.
2. Topology control algorithms for adapting the transmission power of UAVs. Depending on the applications with different densities of UAVs and area coverages, the transmission power of each UAV should be adjusted.
3. Deploying and controlling the relay nodes.
4. Task allocation algorithms to ensure communication connectivity.
5. Regarding topology control, there are several challenges to be addressed:
   a. Robustness to uncertainty. UAVs are deployed to execute a large amount of missions in both military and civilian environments. Because of the diversity of communication links and the rapid mobility of UAVs, it is difficult to predict wireless channel performance in dynamic environments. It is necessary for multi-UAV systems to be robust to uncertainties and interference.
   b. Limitation of bandwidth usage. The radio spectrum is becoming more crowded. The amount of information to be transmitted for all sensing and control functions must be minimized. It is necessary that the consensus in cooperative control and decision making must converge fast across a team of UAVs with as few message exchanges as possible. The command and control messages have low bandwidth requirements but must be exchanged with the minimal delay and errors for effective team coordination.
   c. To enhance resource utilization, the cooperative control algorithms and communication network protocols need to jointly optimize the control and routing strategies. For the large amount of promising applications, network protocols need to rapidly adapt to the dynamic environments of multi-UAV networks.

Communication networks have limitations that may significantly degrade the performance of the multi-UAV system. For instance, delay in data transmissions, especially in command delivery, may cause a formation of UAVs to be unstable and perform inefficiently, which can have dangerous consequences such as collisions among UAVs or other obstacles. On the other hand, delayed or dropped messages sent to UAVs, whether via centralized or distributed systems, can cause inconsistencies of situational awareness, resulting in inaccurate mission planning. For instance, message delays in distributed planning can prevent UAVs from reaching consensus on a plan and some UAVs may remain idle without accomplishing the missions.

3.4 FORMATION CONTROL, MOTION PLANNING AND PATH PLANNING

Recently the mechanisms for controlling the formation of mobile multi-UAV system have been extensively investigated. Generally, formation control and cooperative motion planning are two different research topics [14].

The goal of formation control is to generate correct control commands to drive multiple UAVs to achieve their movement tasks. The main concern of formation control includes control stability/robustness and UAV dynamics constraints (such as inter-UAV collision/deadlock avoidance, etc.)

With the high-level decision-making capability, the aim of cooperative motion planning is to provide accurate guidance information such as the optimal trajectories for the formation to ensure the good coordination of multiple UAVs. Cooperative motion planning is another active research area. The main factors of cooperative motion planning are obstacle-free safety distance, shortest distance, computational time and trajectory smoothness level.

There are some task similarities between formation control and cooperative motion planning in terms of formation shape forming, maintenance, variation, re-generation, collision avoidance, and UAV coordination. As a result, both tasks should be interactive with each other when being implemented in multi-UAV formation systems. For instance, when performing cooperative motion planning, to generate the trajectory for each UAV, the desired formation shape should be achieved. Meanwhile, the formation control strategy should be capable of evaluating the features of the generated trajectories and making decisions on whether to rigorously follow each individual path or change them together to avoid obstacles/collisions.

Sometimes motion planning also is referred to as path planning. Path planning focuses on a safe or collision-free path, disregarding dynamic characteristics such as velocity and acceleration; whereas motion planning takes UAV dynamics into account. Because the difference is relatively subtle, in many review papers, both motion and path planning have been used interchangeably.

3.5 FORMATION CONTROL MODELS

Different kinds of control strategies and motion planning approaches to the multi-UAV/agent formation problem have been widely studied over the last decades. There are three typical formation control approaches [15], that is, the LF-based, VS-based and behavior-based methods. They have received considerable interest. Most research works investigating formation control use one or more of these approaches in either a centralized, distributed or hybrid architecture.

3.5.1 Leader–Follower Formation Control

Wang [6] and Desai [7] have proposed the standard LF approach for UAV formation platforms. Based on the LF control approach, one UAV is chosen as the team leader with full access to the overall navigation information and acts as the reference point in the formation. Other UAVs in the formation are viewed as followers. To maintain the formation shape, the leader guides the operation of followers to maintain the desired distance and pose angle.

The LF approach is a widely adopted strategy because it is relatively simple to design and implement in practice. Because of the existence of the explicit leader UAV, the LF-based formation control approach is robust to random factors.

The LF approach usually adopts a centralized communication architecture that requires the follower UAVs in the formation to establish connections with the leader only. Mesbahi and Hadaegh proposed a graph theoretic formulation of LF approach [16], which revealed the intrinsic properties of the LF assignment.

For the LF approach, a communication link between the leader and the follower is ensured such that the leader can monitor the status of followers at all times, and they can exchange distance and pose information timely. Sometimes UAVs in an LF scheme always work independently, and the leader gets no response from followers. Certainly, information delivery from the followers to the leaders can improve the group robustness.

The overall amount of exchanged information in the LF scheme is typically less than the amount in a distributed or decentralized architecture, thus the communication efficiency is much higher. However, the primary shortcoming of the LF scheme is its high dependence on the leader UAV's performance. The formation is hard to control and maintain, especially when the leader UAV malfunctions or the communication link between the leader and the follower is disrupted.

3.5.2 Virtual Structure–Based Formation Control

The VS approach was proposed by Lewis and Tan [8]. The main concept of VS is to treat the formation shape as a VS or a rigid body, UAVs maintain a rigid geometric relationship with each other to minimize the position errors between the VS and actual formation positions, and the positions of the VS are determined by the status of the formation. To achieve this goal, the UAVs are controlled by the virtual force based on a bidirectional control scheme. Fault-tolerant capability is one of the most appealing advantages of the VS approach.

However, the VS method is not beneficial for formation modification tasks. The variation of the formation requires a redesign of the VS, and the computational burden of the formation is high. Because of the inflexibility of the formation control, the VS approach has limited capability in dealing with collision avoidance.

As in the LF approach, a robust communication channel is important for the VS approach because each vehicle is highly dependent on the data received in real time. Moreover, when a pair of vehicles are in the predefined communication range, the control actions of a vehicle may be calculated, not only based on its own position and velocity, but on the prescribed communication range to ensure collision avoidance between the pair of vehicles [18].

3.5.3 Behavior-based Formation Control

Balch and Arkin [9] first proposed the behavior-based formation control. The formation control problem was solved by utilizing a hybrid vector-weighted control function, which generates the control command based on various kinds of formation missions, such as moving to the goal, avoiding static obstacles, avoiding robots and maintaining formation. According to the general mission requirements, four different mission-control schemes were developed, and each scheme was assigned with a gain value in terms of the specific mission or traffic environment. The higher gain for controllers represents the higher importance for the corresponding behavior, and the final control is determined as the weighted combination of these gains. By implementing behavior-based formation control, the formation can be easily maintained and collision avoidance can be achieved.

However, the controller design for this approach is not based on kinematic/dynamic characteristics of the vehicles. Thus it is complex to analyze the system stability using mathematics, which makes it difficult to theoretically justify the performance of this control method. The paucity of system stability analysis makes it unsuitable for large-scale UAV networks. Despite this drawback, due to its capability of accomplishing different mission requirements through one control command, the behavior-based formation control is still popularly adopted by mobile robots and unmanned ground vehicle (UGV) platforms. The behavior-based method has a more flexible formation shape.

There is no single solution that is appropriate for all scenarios. In the open space, where the system stability is the priority, the LF or VS method could be used. While in a complex environment, the behavior-based method takes over the control. In the future, hybrid control schemes might be the trend. Yang et al. [19] have proposed a hybrid formation control method based on the LF scheme and behavior-based control strategy. It relies on the relative position bias between virtual formation targets and follower vehicles. Such a hybrid approach has both the features of an active behavior-based control scheme and a virtual target structure. The LF model generates and maintains the formation, whereas the behavior-based method and VS approach mainly focus on the motion planning of individual vehicles. Through the LF scheme, a supervision mechanism has been built between the leader and followers so the formation integrity can be provided when the number of controlled vehicles varies, and an interconnection between the leader and each follower is achieved so that the status of followers can be monitored full time by the leader.

As discussed above, although LF-, VS- and behavior-based approaches can be used for the formation control problems of multi-UAV systems, they have respective shortcomings. For instance, the LF-based approach lacks robust communications, and the VS-based approaches are not suitable for distributed implementation as they require many communications and computations. Behavior-based approaches are difficult to model and cannot guarantee the stability of the system.

In recent years, with the development of consensus theory, consensus control based on graph theory has been proved to be able to solve formation control problems. Moreover, some research results showed that the LT-based, VS-based and behavior-based approaches can be unified in the general framework of consensus control [10]. The consensus-based approach can overcome the weakness of the above three traditional formation control methods. In the rest of this chapter, we will emphasize consensus-based formation control strategies.

3.6 CONSENSUS-BASED FORMATION CONTROL FOR UAVS

3.6.1 Consensus Problem and Network Topology

For a team of UAVs, a consensus algorithm uses a set of protocols/rules to determine how the information is exchanged among UAVs to ensure that the entire system will converge to an equilibrium in terms of the quantities of interest such as plans, parameters and situational awareness.

By analyzing the network topology (or information flow), we can understand how the team of agents reaches consensus on the quantities of interest and how information propagates throughout the network [12].

Assume that links are considered available if UAVs are within a threshold distance of each other [20]. To intuitively analyze the networked multi-UAV, a directed graph (also called a digraph) or an undirected graph (balanced graph) can represent the network of multi-UAV systems. Let $G=(V,E,A)$ be a weighted directed graph (or digraph) of order (cardinality) k with the set of nodes $V=v_1,v_2,\ldots,v_k$, and the node indexes belong to a finite index set $\mathrm{I}=1,2,\ldots,k$. Here G consists of a set of edges E, and an edge of G is denoted by $e_{ij}=(v_i,v_j)$, where the edge has a direction associated with it, and the edge e_{ij} is said to be incident away from v_i and incident toward v_j. The in-degree of a node in a digraph is the number of edges that are incident toward it. Similarly, the out-degree of a node is the number of edges that are incident away from it. A node is called isolated if it has zero in-degree and zero out-degree. A directed graph is said to be strongly connected or disconnected if every node is reachable from every other node.

If and only if the in-degree of the node v_i of adigraph G is equal to its out-degree, i.e., $\deg_{\text{out}}(v_i)=\deg_{\text{in}}(v_i)$, we say that the node v_i is balanced. If and only if all nodes of a digraph G are balanced, we say the graph G is balanced.

Denote a weighted adjacency matrix $A=[a_{ij}]$ with non-negative adjacency elements a_{ij}. The adjacency elements associated with the edges of the graph are positive, i.e., $e_{ij}\in E\leftrightarrow a_{ij}>0$. We assume that $a_{ii}=0$ for all $i\in I$. The set of neighbors of node v_i is denoted by $N_i=\{v_j\in V:(v_i,v_j)\in E$.

Let $x_i\in\mathcal{R}$ denote the value of node v_i. We refer to $G_x=(G,x)$ with $x=(x_1,x_2,\ldots,x_k)^T$ as a network (or algebraic graph) with value $x\in\mathcal{R}^k$ and topology (or information flow) G, and switching topology means the set of neighbors $N_i=N_i(G)$.

If and only if the in-degree of the node v_i of a digraph G is equal to its out-degree, i.e., $\deg_{\text{out}}(v_i)=\deg_{\text{in}}(v_i)$, we say the node v_i is balanced. If and only if all of nodes of a digraph G are balanced, we say the graph G is balanced. Any undirected graph is balanced.

The value of a node may represent physical quantities including position, velocity, attitude, temperature, voltage and so on. If and only if $x_i=x_j$, we say v_i and v_j agree in a network. If and only if $x_i=x_j$ for all $i,j\in I, i\neq j$, we say the nodes of a network have reached a consensus. Whenever the nodes of a network are all in agreement, the common value of all nodes is called the group decision [21].

Suppose each node of a graph is a dynamic agent with dynamics

$$\dot{x}_i=f(x_i,u_i), i\in I \tag{3.1}$$

A dynamic graph is a dynamical system with a state (G,x) in which the value x evolves according to the network dynamics $\dot{x}_i = F(x,u)$, which is the column-wise concatenation of the elements $f(x_i,u_i)$, for $i = 1,2,\ldots,k$. In a dynamic network with switching topology, the information flow is a discrete state of the system that changes in time.

Assume $\chi : \mathcal{R}^k \rightarrow \mathcal{R}$ is a function of variables $x_1, x_2, \ldots, x_k$ and $a = x(0)$ denotes the initial state of the system. For a distributed system, the χ-consensus problem in a dynamic graph is a distributed method to calculate $\chi(\mathrm{x}(0))$ by applying inputs u_i, which only depend on the states of the node v_i and its neighbors.

In addition, the graph Laplacian L can be induced by the information flow G, which is defined by

$$l_{ij} = \begin{cases} \sum_{d=1,d\neq i}^{k} a_{id},\ j=i & \sum_{d=1,d\neq i}^{k} a_{id},\ j=i \\ -a_{ij},\ j\neq i & -a_{ij},\ j\neq i \end{cases} \tag{3.2}$$

It has been proved that graph Laplacians and their spectral properties are graph-related matrices that play an important role in convergence analysis of consensus and alignment algorithms [22, 23]. It has been proved that the second smallest eigenvalue of graph Laplacians, which is called algebraic connectivity, quantifies the speed of convergence of consensus algorithms. The stability analysis of multi-UAV systems must consider the connectivity property of the graph G, i.e., the way in which the individual systems communicate with each other.

Furthermore, because multi-UAV systems are networked systems, with their connections described by a graph, just like potential energy stored in a spring or in a potential field, we can assume that there exists "virtual potential energy" for a multi-agent system stored in a graph, and it is called the graph Laplacian potential. The concept of the graph Laplacian potential was first used for unweighted undirected graphs. Later, it was extended to weighted undirected graphs, general directed graphs and balanced graphs, and further to generalized, strongly connected digraphs.

For directed graphs, the Laplacian potential can be simply defined as

$$P_L = \sum_{i,j=1}^{k} a_{ij}(x_i - x_j)^2 \tag{3.3}$$

The graph Laplacian potential depends on the communication graph topology, and it can be used to construct Lyapunov functions, which are suitable for the analysis of cooperative control systems on graphs. Control protocols coming from such Lyapunov functions are distributed, relying only on the information from the UAV and its neighbors.

In recent years, based on communication graph properties, by using the generalized Laplacian potential, many researchers have studied various Lyapunov functions for analyzing consensus/synchronization problems, such as [24, 25] and so forth.

In many practical applications, topology control algorithms for multi-UAV networks are only used to control and adapt the transmission power of nodes. Next, we will discuss the relationship between the coverage and connectivity in UAV networks.

3.6.2 Coverage versus Connectivity

In UAV network systems, using fully autonomous algorithms to control the movement of UAVs should meet two objectives: (1) increasing spatial coverage area of interest to rapidly identify targets, and (2) keeping communication connectivity to enable real-time communications among UAVs. In practical applications, these two objectives might be contrary to each other.

Because of the major resource constraints of multi-UAV systems such as battery power, communication bandwidth and computation processing capabilities, the lifetime of UAVs and onboard transceiver transmission distances are both limited. A longer distance will cause higher propagation loss and more serious signal degradation. The distance configuration in the UAV network must not exceed the receiver's sensitivity and needs to be restricted to the boundaries of minimum signal-to-noise-ratio (SNR) or received signal strength indicators (RSSIs), respectively. A disruption-free connectivity is indispensable to control UAVs' behaviors. On the other hand, high spatial coverage is required for gaining required information from disjunct network regions that cover a large area of interest. There is a trade-off between coverage and connectivity.

3.6.3 Communication Constraints

When multi-UAV systems cooperatively accomplish a set of objectives, the communication network must always exchange a large amount of information such as command and control messages and remotely sensed data. Moreover, When UAVs fly in a formation, the formation must safely reconfigure itself in response to changing missions, UAV density and environment. Network performance depends on the quality of wireless links, the number of interconnected neighbors, and the ability to route information based on network topology. Moreover, the quality of wireless links in a UAV network may alter over time due to many reasons, such as Doppler effects, changes in communication distance, and so forth.

The dynamics and uncertainties of the communication network are impacted on system performance by the following things, all of which need to be carefully considered:

1. *Energy conservation.* Onboard units are typically equipped with limited energy supplies. One of the main goals of the design is to use energy conservation techniques at different levels of the wireless architecture, so that the limited energy is used as efficiently as possible to extend the functional lifetime of both individual UAVs and the network considerably.
2. *Limited bandwidth.* Network bandwidth is a limited resource. Bandwidth limitations can restrict the content and frequency of planning messages. For instance, although the theoretical bandwidth in industrial standards such as IEEE 802.11 can be as high as 54 Mb/sec, the practical rate is far worse, due to the radio interference caused by simultaneous communications.

Thus, a major problem in the design of multi-UAV networks is to maintain the network traffic capacity at a reasonable level. Hence, in distributed systems, cooperative decision-making methods need to efficiently select the pieces of information that really need to be shared with other UAVs, and the entire network reaches consensus with as few messages as possible to mitigate delays and conflicts.

3. *Unstructured and time-varying network topology.* In principle, UAVs may be arbitrarily deployed in the desired region, so the graph representing the communication links among the UAVs is always unstructured. Furthermore, because of the mobility of UAVs, the network topology may vary with time. In general, the quality of communication on wireless channels is strongly influenced by practical environmental factors, which can also be time varying. It is difficult to determine the appropriate values of fundamental network parameters such as the transmitting range for wireless connectivity.
4. *Delay in data transmission.* Delay in data transmission is unavoidable and may cause many problems. For example, communication delays can cause negative effects on multi-UAV path planning for collision avoidance applications. In distributed systems, even small network delays can cause UAVs to plan asynchronously, which makes consensus difficult to achieve.
5. *Network dropouts.* In distributed systems, multiple UAVs in the team become a cohesive whole to cooperate in a mission, and the team must plan and make decisions collectively. Therefore, for the networked team, cooperative decision-making means to require the team to reach an agreement on the tasks, plans and actions to accomplish the mission. Network dropouts may prevent UAVs from participating in cooperative decision-making, which can impact on plan executions.

In a nutshell, the imperfections in the communication network may degrade the multi-UAV system performance. The inability to transmit sensed data to the designed processing centers, or delayed/dropped messages sent to planning UAVs, whether in centralized or decentralized systems, can cause inconsistencies in situational awareness, and it may lead to flawed planning. Inadequate team control or failures to properly exchange command and control messages, even short delays, can lead to formation instability and cause a formation of UAVs to perform inefficiently or churn, which further results in disastrous failures of systems such as UAV collisions. In distributed systems, message delays in distributed planning also may prevent UAVs from reaching consensus on a plan and/or make them take actions on an undesirable plan.

Therefore, in practical applications, these communication limitations have implications on the system architecture, choices of algorithms, and performance of multi-UAV systems.

3.6.4 Consensus-based Formation Control for UAVs

According to all the aspects mentioned previously, we provide an overview of consensus-based formation control for UAVs.

From the control point of view, using the centralized approach to control a networked multi-UAV system is a natural way, provided that the state information of all agents can be obtained. A central station gathers the state information of all UAVs, makes the control decision, and sends the corresponding control commands to each UAV. However, in most applications of multi-UAV systems, some environments may not be ideal, thus they cannot use centralized approaches. For example, the complete state information cannot be obtained by a central station, because of communication issues, such as the limited range of transmission power. Another weakness of the centralized approach is that the complexity of the central station increases with the number of UAVs. Moreover, any variations in the network topology may require the redesign of the central controller.

In terms of the distributed control (or cooperative control), there is no central controller involved, and each UAV makes its own decisions by using the state information of itself and its neighbors. The distributed control approach has many advantages, such as scalability, flexibility and robustness.

For cooperative control of multi-UAV systems, there are two widely studied consensus problems:

1. In centralized architecture, our target is the LF consensus problem (also known as cooperative tracking problem, pinning control or synchronization tracking control). There is a leader node, which acts as a command generator and generates the desired reference trajectory, and whose behavior is not affected by the follower nodes. All followers are controlled to track the leader node.
2. In distributed architecture, we need to solve the leaderless consensus problem (also known as consensus problem, synchronization, rendezvous or cooperative regulator problem). Distributed controllers are designed for each UAV. All UAVs play an equal role and tend to reach a consensus.

Most early works focused on consensus of first-order or second-order integrators linear time-invariant (LTI) systems. Today the consensus of high-order linear systems, even high-order dynamic nonlinear systems, have been developed.

Although there are several advantages of using communication networks for data exchange among autonomous UAVs, such as inexpensive installation and maintenance, the presence of communication network also brings challenges in the controller design. Although we could take appropriate measures to control the communication network, packet latency and dropout in the data transmission are virtually unavoidable and may occur unexpectedly. Time delay or jitter can degrade the performance of the control systems and even destabilize them. Therefore, taking time delays into account in the design of formation controllers for networked multi-UAV systems also is a recent active research field. There exists a fundamental trade-off between the performance of reaching a consensus and the robustness to delays.

Moreover, collision/obstacle avoidance is a practical problem in multi-UAV motion coordination that should be considered in the design of a formation controller. To avoid collisions with obstacles and other UAVs, the notions of potential fields and repulsive forces are commonly used to develop relevant techniques.

In the last years, based on the consensus method, various formation control strategies for distributed multi-UAV coordination have been studied extensively. However, the control of formation depends on the individual UAV's dynamics and the structure of the whole network, which may be modeled by directed or undirected graphs. The interaction topologies among UAVs can be assumed to be fixed or dynamic. Because the communication channels may fail and new channels may be created, formations of multi-UAV systems are mostly time varying.

Here, we provide a short list of some typical studies about consensus-based formation control for UAVs, as shown in Table 3.2.

TABLE 3.2
Consensus-Based Formation Control for UAVs

Article Title	Topology	Graph	Delays	Collision Avoidance
Formation control for UAVs with directed and switching topologies [25]	Switching	Directed	Time varying	None
Formation control of VTOL UAVs with communication delays [26]	Fixed	Undirected	Under delay-dependent or delay-independent conditions	None
Consensus-based reconfigurable controller design for UAV formation flight [27]	Switching	Directed	Time varying	None
A decentralized receding horizon optimal approach to formation control of networked mobile robots [28]	Fixed	Full connected graph	Packed delaying network	Yes
Formation flight control of multi-UAVs System with communication constraints [29]	Jointly connected topologies	Undirected	Non-uniform time delays	None
Decentralized formation control of a swarm of quadrotor helicopters [30]	Fixed	fully connected	Without consideration	Yes
Cooperative control of UAV cluster formation based on distributed consensus [31]	Fixed or switching	Directed	Without time delay and with time delay	None
Consensus-based formation of second-order multi-agent systems via linear-transformation-based partial stability approach [32]	Fixed	Directed	Time invariant	None
Lyapunov, adaptive, and optimal design techniques for cooperative systems on directed communication graphs [24]	Robust to the graph topology	Directed	Without consideration	None

3.7 CONCLUSIONS

In this chapter, we have provided a comprehensive survey of formation control for networked UAVs and presented two main types of control architectures to maintain the shape of a formation, while the team flies as a cohesive whole. Several operation modes for UAV, such as man-in-the-loop, man-on-the-loop and autonomous are introduced. Effective and robust control of the communication network will be an almost universal requirement for networked UAVs. We introduced the communication performance paradigm that UAVs should rely on. The differences between formation control and motion planning, and path planning have been explained. We also summarized some recent results on three formation control schemes, i.e., LF-based, VS-based and behavior-based approaches. Finally, we described consensus-based algorithms for formation control and provided a review of consensus problems for networked UAVs. There are still many issues to be solved on the formation control of multi-UAV systems in the real world.

REFERENCES

1. W. R., Williamson, et al., An instrumentation system applied to formation flight, *IEEE Transactions on Control Systems Technology*, vol. 15, no. 1, pp. 75–85, 2007.
2. J. Han, et al., Low-cost multi-UAV technologies for contour mapping of nuclear radiation field, *Journal of Intelligent & Robotic Systems*, vol. 70, nos. 1–4, pp. 401–410, 2012.
3. A. S. Tan and K.-Y. Colin, UAV swarm coordination using cooperative control for establishing a wireless communications backbone, *9th International Conference on Autonomous Agents and Multiagent Systems*, pp. 1157–1164, 2010.
4. B. D.O. Anderson, B. Fidan, C. Yu, and D. van der Walle, UAV formation control theory and application. In: V. D. Blondel, S. P. Boyd, and H. Kimura, Eds. *Recent Advances in Learning and Control. Lecture Notes in Control and Information Sciences*. London: Springer, vol 371, pp. 1–20, 2008.
5. M. Hosseinzadeh Yamchi and R. Mahboobi Esfanjani, Distributed predictive formation control of networked mobile robots subject to communication delay, *Robotics and Autonomous Systems*, vol. 91, pp. 194–207, 2017.
6. X. Wang, V. Yadav, and S. N. Balakrishnan, Cooperative UAV formation flying with obstacle/collision avoidance, *IEEE Transactions on Control Systems Technology*, vol. 15, no. 4, pp. 672–679, 2007.
7. J. P. Desai, J. Ostrowski, and V. Kuma, Controlling formations of multiple mobile robots, *Proceedings of the 1998 IEEE International Conference on Robotics & Automation*, pp. 2864–2869, 1998.
8. M. A. Lewis and K.-H. Tan, Virtual structures for high-precision cooperative mobile robotic control, *Proceedings of the IEEE/RJS International Conference on Intelligent Robots & Systems*, vol 1, pp. 132–139, 1996.
9. T. Balch and R. C. Arkin, Behavior-based formation control for multirobot teams, *IEEE Transactions on Robotics and Automation*, vol. 14, pp. 926–939, 1998.
10. W. Ren, Consensus strategies for cooperative control of vehicle formations, *IET Control Theory & Applications*, vol 1, no. 2, pp. 505–512, 2007.
11. K. P. Valavanis and G. J. Vachtsevanos, *Handbook of Unmanned Aerial Vehicles*, Dordrecht: Springer Netherlands, 2015.
12. Z. Zhao and T. Braun, Topology control and mobility strategy for UAV ad-hoc networks: a survey, *Joint ERCIM eMobility and MobiSense Workshop*, pp. 27–32, 2012.

13. S. B. Heppe, Problem of UAV communications. In: K. P. Valavanis and G. J. Vachtsevanos, Eds. *Handbook of Unmanned Aerial Vehicles*, Dordrecht: Springer Netherlands, pp. 715–748, 2015.
14. Y. Liu and R. Bucknall. A survey of formation control and motion planning of multiple unmanned vehicles, *Robotica*, vol 36, no.7, pp. 1019–1047, 2018.
15. X. Dong, et al., Time-varying formation control for unmanned aerial vehicles: theories and applications, *IEEE Transactions on Control Systems Technology*, vol 23, no. 1, pp. 340–348, 2015.
16. M. Mesbahi and F. Y. Hadaegh, Formation flying control of multiple spacecraft via graphs, matrix inequalities, and switching, *Journal of Guidance, Control, and Dynamics*, vol 24, no. 2, pp. 369–377, 2001.
17. K. D. Do, Bounded controllers for formation stabilization of mobile agents with limited sensing ranges, *IEEE Transactions on Automatic Control*, vol 52, no. 3, pp. 569–576, 2007.
18. P.K.C. Wang, Navigation strategies for multiple autonomous mobile robots moving in formation, *Robot and Systems*, pp. 177–195, 1991.
19. F. Yang, F. Liu, S. Liu, and C. Zhong. Hybrid formation control of multiple mobile robots with obstacle avoidance, *Proceedings of the 8th World Congress on Intelligent Control and Automation*, pp.1039–1044, 2010.
20. P. Santi, Topology control in wireless ad hoc and sensor networks, *ACM Computing Surveys*, vol 37, pp. 164–194, 2005.
21. R. Olfati-Saber and R. M. Murray, Consensus problems in networks of agents with switching topology and time-delays, *IEEE Transactions on Automatic Control*, vol 49, no. 9, pp. 1520–1533, 2004.
22. M. Fiedler, Algebraic connectivity of graphs, *Czechoslovak Mathematical Journal*, vol 23, pp. 298–305, 1973.
23. R. Merris, Laplacian matrices of a graph: a survey, *Linear Algebra and Its Applications*, vol 197, pp. 143–176, 1994.
24. H. Zhang, F. L. Lewis, and Z. Qu, Lyapunov, adaptive, and optimal design techniques for cooperative systems on directed communication graphs, *IEEE Transactions on Industrial Electronics*, vol 59, no. 7, pp. 3026–3041, 2012.
25. Y. Qi, et al., Formation control for unmanned aerial vehicles with directed and switching topologies, *International Journal of Aerospace Engineering*, pp. 1–8, 2016.
26. A. Abdessameud and A. Tayebi, Formation control of VTOL unmanned aerial vehicles with communication delays, *Automatica*, vol 47, no. 11, pp. 2383–2394, 2011.
27. J. Seo, Y. Kim, S. Kim, and A. Tsourdos, Consensus-based reconfigurable controller design for unmanned aerial vehicle formation flight. *Proceedings of the Institution of Mechanical Engineers, Part G: Journal of Aerospace Engineering*, vol 226, no. 7, pp. 817–829, 2012.
28. M. Hosseinzadeh Yamchi and R. M. Esfanjani, A decentralized receding horizon optimal approach to formation control of networked mobile robots, *Optimal Control Applications and Methods*, vol 39, no. 1, pp. 51–64, 2018.
29. R. Xue and G. Cai, Formation flight control of multi-UAV system with communication constraints, *Journal of Aerospace Technology and Management*, vol 8, no. 2, pp. 203–210, 2016.
30. Sinan Oğuz, M. Altan Toksöz, and M. Önder Efe, Decentralized formation control of a swarm of quadrotor helicopters, *IEEE 15th International Conference on Control and Automation (ICCA)*, Edinburgh, Scotland, July 16–19, 2019.
31. J. Zhang, W. Wang, Z. Zhang, K. Luo, and J. Liu, Cooperative control of UAV cluster formation based on distributed consensus, *IEEE 15th International Conference on Control and Automation (ICCA)*, Edinburgh, Scotland, July 16–19, 2019.
32. X. Qu, et al., Consensus-based formation of second-order multi-agent systems via linear-transformation-based partial stability approach, *IEEE Access*, vol 7, pp. 165420–165427, 2019.

4 5G-Enabled UAV Communications

Iftikhar Rasheed[1], Fei Hu[1], Niloofar Toorchi[1]
[1]Electrical and Computer Engineering,
The University of Alabama, Tuscaloosa, AL, USA
Email: AL irasheed@crimson.ua.edu,
fei@eng.ua.edu, ntoorchi@crimson.ua.edu

CONTENTS

4.1 INTRODUCTION

By 2020 a new era of wireless communications had started because of the gradual maturity of 5G systems [1–3]. Various intelligent transportation system (ITS) applications desire advanced wireless access schemes that can support unmanned aerial vehicle (UAV) to UAV and UAV- to-infrastructure communications. We can define V2X communications between a UAV and other things/entities such as other UAVs (V2V), pedestrians (V2P), infrastructures (V2I) or network elements (V2N) [4].

ITS incorporates intelligent algorithms/modules to assist drivers and, consequently, transportation systems can be safer, more fuel-efficient and environment friendly. Traffic control applications require that the infrastructures have the communication capability to send periodic updates to remote UAVs and traffic control centers. Those applications exhibit some unique features, in terms of message generation patterns, data delivery performance, communication primitives and spatial/temporal dynamics, all of which challenge the existing wireless networking solutions [5–8].

Today, wireless access for vehicular environment (WAVE) that uses a dedicated short-range communication (DSRC) protocol, LTE, and many other wireless communication technologies, have played key roles in the development of future ITSs. However, they still cannot handle the large-scale, heterogeneous transportation management with an ultrahigh data rate (>10 Gbps) and ultralow routing delay (<5 ms) requirements [9, 10].

5G is a wireless mobile technology that can handle very high data rates for heterogeneous, ultradense networks. It can achieve much lower routing latency and higher end-to-end transmission reliability than LTE and 4G systems. Therefore, 5G is expected to provide the inherent support for V2V/V2I communication scenarios [11–13]. Encompassing its usage to support the UAV network's quality-of-service (QoS) requirements will open new markets for various service providers.

5G predominantly fits the high bandwidth and delicate QoS requirements of V2X applications. It will greatly enhance the Internet-based ITS applications such as 3D terrain map downloading, online weather querying and so forth. Still, how to implement 5G in practical V2X infrastructure devices to provide extremely high traffic safety services is still a challenging issue and has attracted many researchers recently. This chapter will provide a comprehensive review and assessment of the 5G capabilities in terms of supporting the features of UAV applications. A brief layout of a 5G-enabled UAV communication system using two communication modes (V2V/V2I) can be seen in Figure 4.1.

4.2 KEY CAPABILITIES REQUIRED FOR 5G-UAVS

The capabilities required for UAV communication are listed as follows:

- *Scalability*: UAV networks should have good scalability, i.e., having the ability to add or remove nodes on demand without involving much processing overhead.
- *Computation and monitoring capability*: UAV networks should support high computation capability, because a large amount of data may need to be collected and processed for intelligent transportation control. Many sensors embedded in the vehicle may be used to monitor the surrounding environment and driving styles to detect any potentially dangerous scenarios [14]. Internet-connected vehicles should cooperate with each other to improve the performance of some applications such as city traffic management. In addition to in-vehicle processing units, there should be network infrastructures with high computation capabilities to process a large amount of data gathered from surrounding objects, similar to the fog computing structure [15].

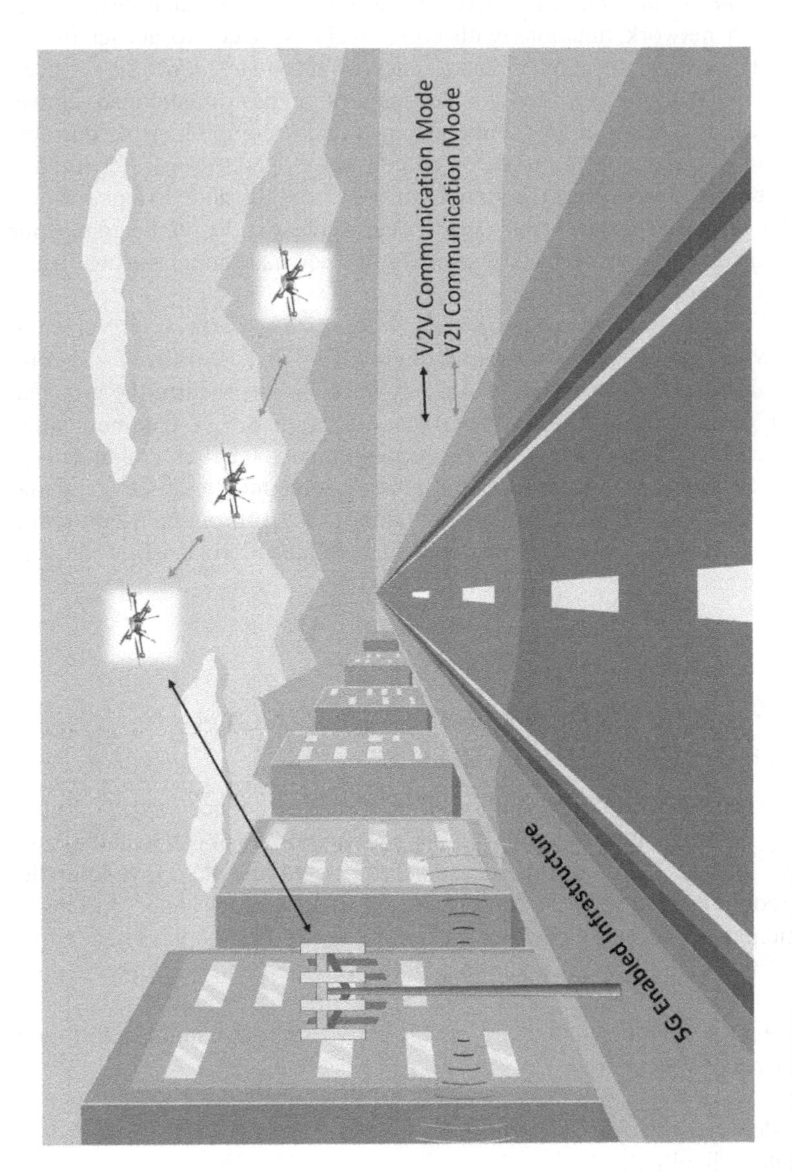

FIGURE 4.1 5G-enabled UAV communication.

- *Real-time assessment*: UAVs can experience frequently changing traffic conditions; hence, the environment should be assessed in a real-time fashion [16]. Effective online learning algorithms may be used to provide the real-time responses for complex road conditions.
- *Commercial benefits*: UAV networks should have the capability of integrating network functions with commercial services to attract investments for transportation enhancement. In fact, many vehicular services, especially those provided for passengers' entertainment and convenience, can bring large commercial profits. A comprehensive business model and cost analysis model should be developed to accurately predict the required resources, the number of users and the added value for possible vehicular services. Such models can represent the predicted revenue for involved investors and, consequently, persuade more parties to use the ITS [16].
- *Reliable, fast and high-throughput communication structure*: Vehicular networks should support both distributed V2V and centralized V2I communications. For example, emergency applications require the propagation of much information to nearby vehicles via a V2V communication system, while traffic assessment/management is better conducted by the centralized V2I devices. Successful implementation of vehicular networks should be based on low-latency, high-throughput communication technologies in conjunction with reliable/collision-free network protocols.

4.3 5G: EMERGING PROMISING TECHNOLOGY FOR UAV COMMUNICATIONS

It is predicted that a total of 50 billion devices will be connected by 2020 [1, 21], indicating the necessity of a wireless mobile technology that can handle very high data rates for heterogeneous, ultradense networks. Although LTE can handle the latency required for most of the current services (on the order of about 15 ms [22]), it is anticipated that many new services will require very small latency (on the order of 1 ms) such as two-way gaming, tactile Internet, and virtual reality and so forth.

5G, the fifth generation of wireless mobile networks, is able to meet the following requirements [9, 23–25]:

- Wide variety of data rates up to 100 Gbps (it guarantees 10 Gbps with very high availability).
- Very low latency, high reliability and improved coverage to support critical applications (e.g., telemedicine and UAV communications). In 5G network, the end-to-end delay will be on the order of 1–5 ms [22, 26].
- Enhancing and supporting the existing technologies, including GSM, LTE, LTE-Advanced and so forth [26].

- High network scalability and flexibility to support a huge number of heterogeneous devices (100 times larger than the current wireless networks) with very low complexity and power consumption.
- Device-to-device (D2D) communication: LTE Release 12 and beyond support D2D communications controlled by a macrocell base station (BS). 5G will extend this feature to allow nodes to control D2D communications [26].

4.3.1 Cellular UAVs (C-UAVs)

The C-UAV is a new UAV communication standard that was launched in June 2017 [8]. This UAV system incorporates the cellular system that utilizes the LTE V–based cellular network. A C-UAV is an important step toward a 5G-based UAV communication scenario [1, 17, 18]. It attempts to deliver consistent communications and provisions for short- and long-distance communications between vehicles as well as between a vehicle and an infrastructure. In fact, a C-UAV can deliver direct transmissions between vehicles and infrastructures, and with other roadside users. In the C-UAV scenario, vehicles also can operate autonomously. There are two major modes of operations for C-UAVs: direct communications and network-based relay communications. In this cellular mode a vehicle is able to receive information like traffic, road conditions, accidents and weather and so forth. A C-UAV empowers improved transportation system. In [8] and [17] the authors have shown that C-UAVs can lead to platooning of vehicles to form convoys; thus, they can be very energy efficient. Moreover, autonomous driving, collision avoidance and queue warnings would help improve the transportation system. C-UAVs would allow vehicles to see around corners to avoid pedestrians and obstacles [20]. More importantly, a C-UAV is the first step toward 5G-based UAVs. Further detailed comparisons can be seen in Table 4.1.

4.3.2 From Cellular-UAV to 5G-UAV

A 5G UAV is built on the concept of a C-UAV [27, 28]. The 5G system will bring new capabilities for autonomous vehicles as well as connected vehicles [29]. It can achieve very high throughput, i.e., up to multi-Gpbs. Its wider bandwidths with advanced antenna systems would provide an inherent support for applications like autonomous driving. A 1-ms end-to-end latency can be achieved by using a malleable frame structure. It has added functionalities like new uplink resource spread multiple access (RSMA). Non-orthogonal access means that 5G meets the lower latency criteria more easily and efficiently than any of the previous or available technologies [17, 28].

Moreover, some features are available in 5G, such as edgeless connectivity and new methods of establishing a connection, e.g., multi-hop communication, to lengthen the coverage, and inherently support effective D2D connections. High availability/reliability is important in 5G, which would mean multidimensional connectivity to deliver multiple links for failure tolerance and mobility situations (like in UAV scenarios), thus achieving ultrareliable communications that can be time multiplexed with nominal traffic through puncturing. These features make

TABLE 4.1
Comparison between 802.11p and C-UAV

Metrics	802.11p	C-UAV
Purpose [1, 8, 17, 18]	Real-time communication between vehicles to roadside users and V2I	Real-time communication between vehicles to roadside users and V2I
Market deployment [1, 8, 17–19]	From 2017	Late 2020
Cellular network [8, 17–19]	Can be integrated with a cellular network	Hybrid model
Modulation [1, 8, 17, 18, 20]	OFDM with CSMA	SC-FDM with semi-persistent sensing
Origin [1, 17, 18]	Wi-Fi	LTE
Latency [1, 8, 17, 18]	Depends on packet length. 10–20 ms	More per bit energy. 5–15 ms
Concurrent transmissions [8, 17, 18]	No	Yes
Symbol time [8, 17, 18]	8 ms	71 ms
Coding [17, 18]	Convolution coding	Turbo code
Time synchronization [8, 17, 18]	Very light asynchronous	Very confine synchronous
Retransmissions [8, 17, 18]	No	Yes

5G very suitable for UAV communications [1, 13, 27, 28]. For example, for the case of autonomous UAVs, many researchers have pointed out that autonomous UAVs will become a reality, but this is not possible without 5G. The existing 4G-based solutions are not able to provide high-speed support for autonomous driving. For this reason, Rel-16 5G NR C-UAV has incorporated the real-time local updates, trajectory sharing, high throughput with sensor sharing and synchronized driving [28].

4.4 5G AS A KEY-ENABLING TECHNOLOGY TO SUPPORT UAV COMMUNICATIONS

5G becomes an enabling technology in the UAV communication environments due to the features:

4.4.1 Mobility and Coverage

5G has large spectrum allocations at millimeter wave (mmWave) frequency bands. It provides ultrahigh bit rates (>10 Gbps) and large coverage areas by using highly directional beamforming antennas. It also has high capacity due to the support of many users' simultaneous communications in both licensed and unlicensed spectra [30]. Moreover, positioning of gNBs and multi-radio access technologies (RATs) would solve the issues of poor, recurrent and short-lived connectivity in the previous standards such as 802.11p [31]. Thus, 5G enables V2X communications, especially V2I, at very high speed [12, 30]. The 5G infrastructure supports D2D, which is a viable method for network fragmentation and brings extended connectivity to UAV networks.

4.4.2 Market Infiltration

5G is available commercially by 2020, and a great deal of research is being done in terms of using its various features for practical applications [11]. As mentioned previously, 5G achieves a higher obstacle penetration rate compared with LTE and 802.11p [6, 32]. Vehicles using UAV infrastructure can use 5G to provide the clients with high data rates and more functionalities.

Here we list a few commercial applications that could use 5G to significantly enhance their performance. One application is called a 3D terrain map. It provides drivers with high- resolution, 3D and colorful terrains, as well as the real-time traffic distributions in different roads/city regions. However, such a terrain map needs a big-data transmission network to reach each vehicle in real time. 5G provides more than 10 Gbps of data speed, and it has the capability of using different wireless networks (such as cellular, D2D, WiMax, etc.) to relay the data with high throughput.

Another example is traffic jam prediction in a large city. Using New York city as an example, there are over 10 million people in the city. The spatio-temporal traffic dynamics in the entire city needs an ultrahigh-speed network to collect the vehicle moving states. The 5G's spectrum aggregation schemes can be used to fuse the spectrum fragments and form a wider spectrum band for high-rate transmissions. 5G can also use the cognitive radio networks to achieve dynamic spectrum access. The 5G servers have ultrahigh-speed CPUs, which help with fast machine learning executions to generate traffic jam prediction results in real time. Thus, each driver can be timely notified and prepared for possible traffic jams in different roads and know which path to go to avoid those jamming areas.

4.4.3 Network Capacity

5G provides the maximum downlink and uplink data rates at more than 20 and 10 GB/sec, respectively, which means that it can support the data transmission tasks from many vehicles in each macrocell/microcell. Such data rates are very high compared with the LTE or 802.11p capacities, whose data rates are only up to 300 and 27 Mb/sec, respectively [7]. Moreover, LTE has some problems with its network support for UAV communication, such as the lack of protocols for vehicles/infrastructure link privacy protection, whereas the 5G standard has the compatibility and built the security/privacy support for vehicle ad hoc networks (VANETs), and so forth.

4.4.4 Heterogeneous Architecture

5G is able to support different network sizes, transmission power levels, various RATs, an unprecedented number of smart devices of various types and, more importantly, backhaul connections [33]. The main goal of 5G is to provide seamless and pervasive communication between anybody [i.e., human-to-human (H2H)], anything [such as human-to-machine (H2M), D2D, V2V], anywhere, anytime and anyhow, i.e., by whichever service or network topology [34]. Cooperative awareness messages (CAMs), proxy CAMs, decentralized environmental notification messages (DENMs) and basic safety messages (BSMs) all have the corresponding topologies defined for message transfer. Those topologies will be discussed in detail in the next section.

4.4.5 Channels and Transport Modes

The wo basic modes of transmission are uplink and downlink. For uplink, the dedicated or common channels can be established, whereas for downlink, unicast or broadcast methods can be used. The D2D communications can be utilized as 5G components for short-distance data relay.

5G allows the seamless integration of different wireless access techniques. This means that the dynamic spectrum access (DSA) can be used in some particular places without strict spectrum rules. For example, because Las Vegas is located in a large desert, as long as some spectrum bands are not used there, the local government may request the use of those free bands to relieve the heavy burden of the wireless access users in the city. DSA is a good option here.

4.4.6 Status Mode of the Device

The status of the any network terminal also influences the link latency. Therefore, if we want to utilize the limited resources in a better way, we can configure 5G cellular networks such that the devices will go to idle status if no communication task is assigned. Before sending any data, the device needs to perform the connection setup so that it can switch to the connected mode for data transfer. This may cause the communications to take a longer time than the simple transmission mode.

4.5 D2D over 5G Cellular Architecture to Support 5G-UAV Communication

5G cellular network is able to support large amounts of data traffic, and many novel communication models and platforms are being developed to enable the 5G network to function efficiently for complex UAV applications. The 5G network with D2D support can enable highly reliable end-to-end UAV data transmissions [35]. 5G provides low-latency UAV communication via its cooperative communication modes. By sending messages that hold the critical information, such as road traffic density, accident location or harsh road conditions, D2D can help vehicles to quickly exchange information within a short distance through direct node-to-node data transmission without the involvement of cellular towers [36]. D2D communication

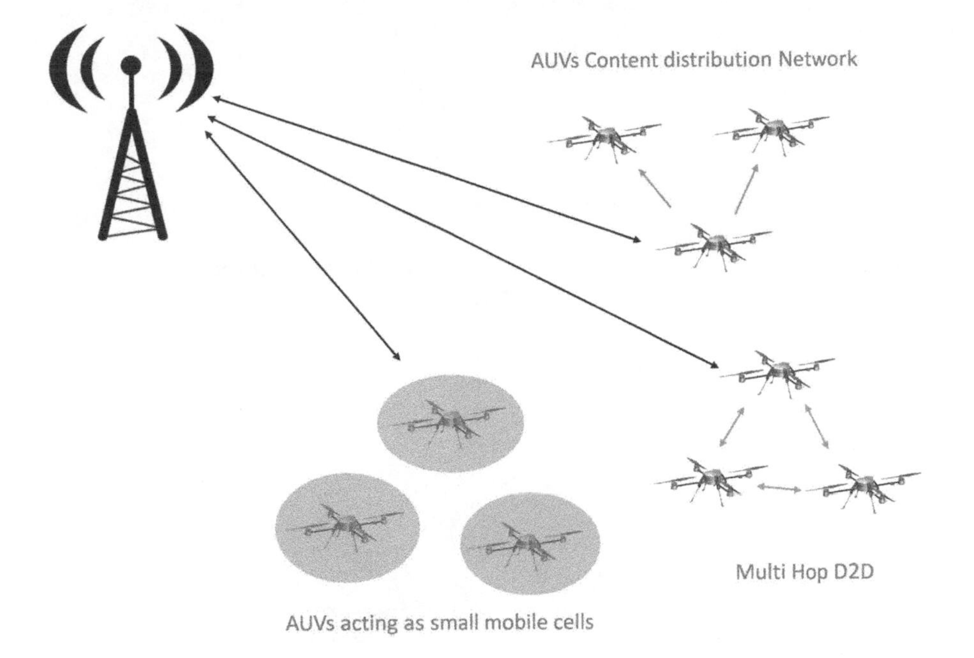

FIGURE 4.2 D2D Communication scenarios.

will be using licensed spectrum, therefore it will have higher communication reliability (i.e., lower packet loss rate) in 5G systems. Figure 4.2 illustrates a multi-hop D2D scenario for V2V communications.

Some studies [35, 37] suggested an alternative approach for D2D communication, in which vehicles can be considered as mobile small cells. Such a model provides a better way to accommodate the traffic, but it will also cause the densification of the network and requires interference management techniques to overcome such a drawback.

In [35, 38–41] the vehicle content distribution network (VCDN) method is proposed for D2D, which enables better bandwidth utilization as illustrated in Figure 4.2.

4.5.1 Multi-hop D2D for UAVs

The multi-hop D2D prototype is a new method of realizing 5G-based UAV communication. In [23, 35, 42] the authors' proposed the concept of an overlay network above the LTE-A system. They called it the first step toward an integrated network by using 5G communication. The three main elements of their methodologies are (1) discovery of network and services, (2) scheduling of users and (3) efficient allocation of the network resources.

Using these three elements we can further define the following three phases used to achieve high-throughput 5G transmissions: (1) the discovery phase of the D2D cluster,

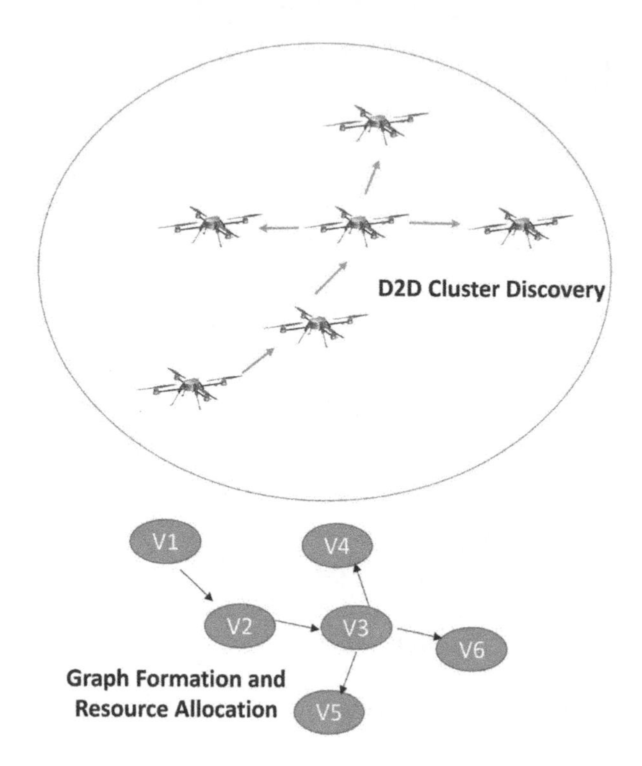

FIGURE 4.3 D2D cluster discovery, graph formation and resource allocation.

(2) the graph information formation phase and (3) the resource allocation phase. Those three phases are illustrated in Figure 4.3.

Mobility is the most important factor in defining the overall connectivity performance. Here it is assumed that all vehicles are enabled with cellular communications, and direct data exchange is performed through mobile D2D, which requires a multi-hop communication scheme and an optimized routing protocol [43, 44]. The discovery and clustering of D2D devices is the key element of multi-hop D2D [42] for UAV-enabled 5G communication. It uses a cellular-evolved home node for D2D cluster formation and discovery. The gNB can choose information provided by incoming vehicles including location, direction, speed and final destination to form the cluster.

After the clustering process, an authentication process is performed to establish a direct communication between vehicles in the 5G communication network. Channel state information (CSI) [35, 45, 46] is used (via direct beaconing) to establish the potential wireless links between cluster members.

After this step, the network topology can be determined depending on the nature of the applications. The network topology must be optimized with respect to the targeted UAV network QoS performance. Tree topology for low-power and lossy networks may be used in this case. After this step we move to the last step, i.e., efficient allocation of the resources. Fair resource allocation is needed to overcome the routing bottleneck problem for the vehicles that are very close to the tree root. A real-time resource distribution policy should consider each vehicle's data priority, and extra resources may be provided to the vehicles near the tree root. An optimal solution for this problem would be the use of the proportional fair scheduling algorithm [35].

Let us denote the scheduling weights as λ_i. For the ith vehicle, assume its estimated rate is σ_i, and the rate of subtree is ξ_i, and ξ represents the overall tree network topology. Then we have the following relationship between those parameters [35]:

$$\lambda_i \doteq \frac{\sigma_i + \sum_{j \in \xi_i} \sigma_j}{\sum_{k \in \xi} \sigma_k + \sum_{j \in \xi_k} \sigma_j} \tag{4.1}$$

4.5.2 Using Game Theory and Matching Theory

Game theory has been extensively deliberated as a promising methodology for modeling the collaboration between various independent, self-organized, decentralized and autonomous network devices. The main element of game theory is known as players. It assumes the existence of collaborations between the individual rational players (i.e., decision makers). Players are considered as selfish in nature, and the sole purpose for each of them is to achieve high node throughput (in network applications).

Various studies [47–50] have exploited the game theory for D2D resource allocation in an effective way. In [35, 51] researchers have used diverse transmission modes, mutual interference and a resource sharing policy to form a coalition formation game that can be used for resource allocation. For example, if we want to increase the capacity of the vehicle network, the signal to interference plus noise ratio (SINR) can be an important performance metric. For the ith vehicle that receives signal from the jth vehicle, we can define the following:

$$\text{SINR} = \rho_i = \frac{\rho_j \xi_{i,j}^{-\alpha} |h_o|^2}{P_{\text{inter},i} + N_0} \tag{4.2}$$

where $P_{\text{inter},i}$ is the interference signal power, $\xi_{i,j}^{-\alpha}$ is the tree network topology, α is the attenuation coefficient, h_o is channel gain and N_0 is total noise power.

The value of P_{inter} for any particular vehicle v paired with d vehicles will can be calculated as

$$P_{\text{inter},\upsilon} = \sum_{d \in D} x_{\upsilon,d} P_d |h_{db}|^2 \tag{4.3}$$

where $x_{v,d}$ is the distance between the vehicle of interest, i.e., v and other pair vehicles $d, |h_{db}|^2 = \xi_{\upsilon,d}^{-\alpha} |h_o|^2$. P_d is the power signal from the d vehicle.

The average uplink rate for a vehicle will be [35]

$$R_{\upsilon} = \log_2\left(1+\frac{P_{\upsilon}\xi_{\upsilon,b}^{-\alpha}|h_o|^2}{\sum_{d\in D} x_{\upsilon,d}P_d\xi_{d,b}^{-\alpha}|h_o|^2 + N_0}\right) \quad (4.4)$$

where P_v is the power signal from the vehicle of interest, i.e., v and P_d are the power signals from other pair vehicles d, and $\xi_{\upsilon,b}^{-\alpha}$ is the tree network topology for vehicle v with α as the attenuation coefficient. Likewise, $\xi_{d,b}^{-\alpha}$ is the tree network topology for vehicle d with α as the attenuation coefficient, h_o is channel gain and N_0 is total noise power.

The average D2D pair rate between two vehicle pairs:vehicle v, vehicle d and vehicle (d') is [51]

$$R_d = \log_2\left(1+\frac{P_d\xi_{d,d'}^{-\alpha}|h_o|^2}{\sum_{\upsilon\in V} x_{\upsilon,d}P_{\upsilon}\xi_{\upsilon,d}^{-\alpha}|h_o|^2 + \sum_{d'\in D} y_{d,d'}P_{d'}\xi_{d',b}^{-\alpha}|h_o|^2 + N_0}\right) \quad (4.5)$$

where P_d and P_v are power transmitted by d and v vehicles, respectively, and $\xi_{d,d'}^{-\alpha}$ and $\xi_{\upsilon,b}^{-\alpha}$ are tree network typologies for vehicles d and v, respectively, with α as the attenuation coefficient. $x_{v,d}$ is the distance between v and d, whereas $y_{d,d'}$ is the distance between d and d', h_o is channel gain and N_0 is total noise power.

A new approach to game theory also is considered for the D2D scenario, which is known as coalitional games [35, 52]. The main feature of the coalition game is the partition formation, which enables the players to form a set by joining or leaving the coalition. This sort of coalition is formed in a systemic way such that every player is able to form and compare its coalition based on its own preferences. With this method a strict form of game theory is applied to make sure that D2D resource allocation is energy efficient.

In UAV communications based on a 5G platform, we can follow this coalition-based game for better utilization of available resources. In spite of their prospective, the game theoretical methods have some inadequacies as well. For example, to find out other players' actions we need to evaluate it from a single player point of view. This is a main drawback when it is used for UAV communication, particularly in cluster formation process. To overcome this shortcoming many solutions such as [35, 53, 54] have proposed new approaches for more efficient vehicular networking, which is called matching theory.

Matching theory offers statistically tractable solutions for the combinatorial problem of matching players in two distinct sets [35], contingent on the information and partiality of the individual player. The main benefits of using matching theory for wireless resource management include the following four aspects:

1. Appropriate models for describing the interfaces among diverse wireless nodes. All of these nodes must be defined by a particular type, objective and data.
2. The capacity to describe universal partialities, i.e., preferences that can be used to manage QoS for various wireless nodes.
3. New system objectives that can provide in-depth understanding and optimal solutions for the stability and other goals.
4. Competent algorithmic enactments that have self- organizing features and can determine other players' actions with low computation complexity.

Table 4.2 shows a list of matching theory types with their particular applicability. Canonical matching theory belongs to class I and can be used in orthogonal spectrum allocation, especially in cognitive radios. Matching with externalities is the feature of class II, which can be used in some applications such as proactive cell association, context-aware allocation, interference management and load balancing. Last, class III uses a matching model with dynamics adaptation. It can be utilized in dynamical environments such as fast fading, mobility and time-varying traffic. Some research, such as [35, 55, 56] and [57], has shown that for

TABLE 4.2
Matching Theory Types for Wireless Communication

Class Types	Matching Types	Preferences	Resource Management	Example Applications
I	Canonical matching games	Based on information accessible at the resource or other resources with which it is looking for to match.	Resource management in a single cell for allotted orthogonal frequency spectrum.	Cognitive radios for spectrum allocation. Suitable for dense networks.
II	Matching with externalities	Required matchings in which the problem shows "externalities"; it translates various dependencies among the players' preferences.	The resource will be contingent on the identity and total strength of other users matched to similar resources.	Proactive cell association, context-aware allocation and interference management and load balancing.
III	Matching with dynamics	The preferences of the players change with respect to time, therefore the time factor must be considered while obtaining a matching solution.	Resources must adjust the matching processes based on the dynamics of the environment, such as fast fading, mobility and time-varying traffic.	Resource management with quick condition deviations such as mobility scenarios.

wireless networks matching theory could solve the clustering problem in a more efficient way than game theory.

For UAV communications that use 5G as the backbone network with D2D support, matching theory can play an important role in terms of defining a fair resource allocation among vehicles. D2D is a significant application field for matching theory because it involves the radio channel allocation issues among neighboring devices.

4.5.3 MILLIMETER WAVE BAND AS A CANDIDATE FOR 5G ENABLED UAV COMMUNICATION

The mmWave frequencies range from 6 to 100 GHz [58], including 6, 15, and 28 GHz [59, 60], 38GHz [59], 60GHz [60], and E-band (71–76 GHz, 81–86GHz) [61, 62]. mmWaves will become an essential component of 5G systems because of its large available spectrum [63, 64]. The mmWave is more efficient for small cells and networks with high data rates and traffic density [65]. mmWave signals have good directionality [66, 67], thus it is suitable for small area coverage.

The two key attributes of mmWaves include (1) small cells and small area coverage and (2) providing high data rates for dense data traffic scenarios. Those two features make mmWaves a promising technology to enable 5G communications [67, 68]. There are various issues to be considered if we use mmWaves for UAV, such as channel modeling, beam tracking using various antenna deployment techniques such as massive MIMO [69] and the MAC collision avoidance issue. Let us first take a look at what types of studies have been done for those issues.

1. *Channel modeling for mmWaves:* The design of MAC/routing schemes should be based on accurate channel models in mmWave communications. Mobile and Wireless Communications Enablers for the Twenty-Twenty Information Society (METIS) has suggested a map-based channel model based on the ray tracing method. The Institute for Electrical and Electronic Engineers (IEEE) has proposed the 802.11ay standard that extends the Saleh-Valenzuela (S-V) model for 60 GHz and considers the distance of communication around 100 m. The 3GPP radio access network (RAN) and International Telecommunication Union Radio communication section (ITU-R) Working Party 5D (WP5D) are both working independently to define the channel model by using spatial channel models (SCMs) for frequency bands of 6 GHz and above. Furthermore, 3GPP has elongated SCM to 3D scenarios. Meanwhile the 3G and 4G cellular systems are already using the same SCM model to perform system-level estimation. Thus, channel model for 5G systems will be an addition to the SCM model.

The mmWave communications require high-gain antenna beams at the BS and user terminal (UE) with the purpose of compensating for the large propagation loss of mmWave signals. For this purpose, we need to use large-size antenna arrays to

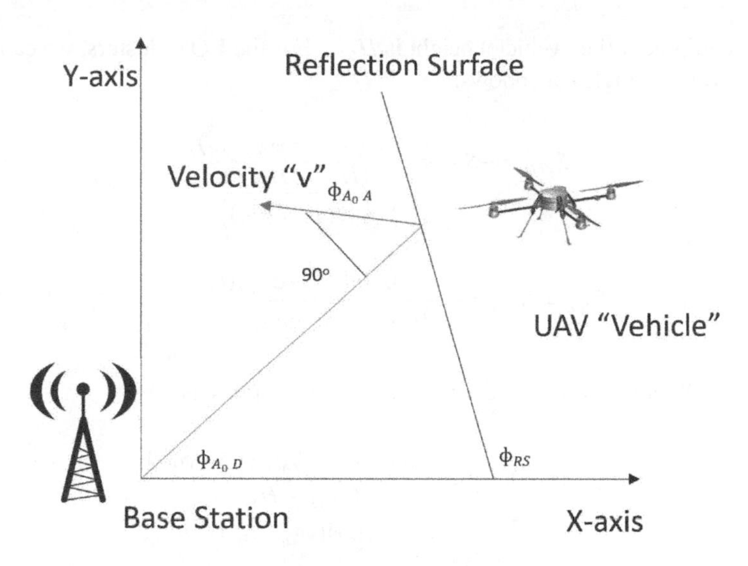

FIGURE 4.4 Variant angle formation in NLOS [70].

achieve high gain. The main drawback of large antennas is that the transmission link needs to tolerate the subtle angle changes.

To account for the subtle angle changes, the concept of variant angles for mmWave channel models is proposed. A derivation of variant angles can be done from the line-of-sight (LOS) channel and random angles at the non–line-of-sight (NLOS) channel in the local coordination systems (LCS), as shown in Figure 4.4 [70]. The main theme of the SCM model with variant angles (VA-SCM) is to update the angles at each time instance, when the receiver is moving within the duration of one drop. A drop duration represented by $T_{\max}$ is around 1000 transmission time intervals (TTIs) [60, 71]. A linear model for variant angles of the rth ray of the cth cluster is given by

$$\theta_{r,c,Z_0A}(t) = \theta_{r,c,Z_0A}(t_0) + S_{r,c,Z_0A} \cdot (t - t_0) \tag{4.6}$$

$$\theta_{r,c,Z_0D}(t) = \theta_{r,c,Z_0D}(t_0) + S_{r,c,Z_0D} \cdot (t - t_0) \tag{4.7}$$

$$\emptyset_{r,c,A_0A}(t) = \theta_{r,c,A_0A}(t_0) + S_{r,c,A_0A} \cdot (t - t_0) \tag{4.8}$$

$$\emptyset_{r,c,A_0A}(t) = \theta_{r,c,A_0A}(t_0) + S_{r,c,A_0A} \cdot (t - t_0) \tag{4.9}$$

$$\emptyset_{r,c,A_0D}(t) = \theta_{r,c,A_0D}(t_0) + S_{r,c,A_0D} \cdot (t - t_0) \tag{4.10}$$

where t belongs to $[t_0,\ t_0 + T_{\max}]$, and S is slope variant angles. Assume that the receiver is moving at velocity v with an angle of $\emptyset_v$, the height of the BS is H_{BS}, and the user equipment (i.e., vehicle) height is H_{VEH}. For the LOS clusters, we can define the slope variant angles as follows:

$$S_{Z_0D} = -S_{Z_0A} = \frac{\dfrac{\upsilon \cos\left(\phi_\upsilon - \phi_{A_0D}(t_0)\right)}{(H_{BS} - H_{VEH})}}{\cos\left(\phi_{Z_0} D(t_0)\right)} \tag{4.11}$$

$$S_{A_0D} = S_{A_0A} = -\frac{\upsilon \sin\left(\phi_\upsilon - \phi_{A_0D}(t_0)\right)}{(H_{BS} - H_{VEH})\tan\left(\phi_{Z_0} D(t_0)\right)} \tag{4.12}$$

For the NLOS clusters we can define slope variant angles as

$$S_{Z_0D} = -S_{Z_0A} = -\frac{\dfrac{\upsilon \cos\left(\phi_\upsilon + \phi_{A_0D}(t_0) - \phi_{RS}\right)}{(H_{BS} - H_{VEH})}}{\cos\left(\phi_{Z_0} D(t_0)\right)} \tag{4.13}$$

$$S_{A_0D} = S_{A_0A} = -\frac{\upsilon \sin\left(\phi_\upsilon + \phi_{A_0D}(t_0) - \phi_{RS}\right)}{(H_{BS} - H_{VEH})\tan\left(\phi_{Z_0} D(t_0)\right)} \tag{4.14}$$

where ϕ_{RS} is an angle of reflection from the surface.

2. *Beamforming and beam tracking in mmWaves:* The mmWave uses an antenna array beamforming and tracking to guarantee the high signal quality. The use of antenna array is to compromise the effect of propagation loss in the link. But the design of high-gain antenna with sharp and narrow beams is very challenging. One solution to this problem is to consider uniform air interface with the following considerations such that the antenna is flexible enough to meet various QoS requirements [5, 72, 73]:
   a. Air interface between various mmWave bands allotted to 5G must be uniform. This can be achieved through various frame format definitions and physical signal shaping techniques.
   b. The backhaul and radio access must share uniform air interface on the request of unified access. In this case the backhaul must work as a special user entity.
   c. Some metrics, such as power efficiency, latency and so forth, must be achieved by using special waveform processing methods based on different QoS requirements for various UAVs applications. This can be done by using software-defined, adaptive air interfaces with various waveform shaping methods.

Based on the previous considerations, in [70] the author has proposed another method to form uniform air interface, which is known as multi-mode beamforming. It assigns array elements, frequency band and corresponding backhaul and radio access channels adaptively. User data and backhaul data are first forwarded for adaptive MIMO mode selections. Here each data stream is processed according to its channel environment, spatial multiplexing techniques and space-time beamforming patterns. For UAV communication scenarios, we can use a multimode beamforming approach as it can be very useful for different mobility modes, but this will add protocol overhead to the 5G system design. To overcome this issue, we can use the compression sensing (CS) method, which utilizes the sparse property of the mmWave channel, and thus can considerably decrease the network overhead.

3. *mmWave network topology and dynamic resource allocation:* mmWave communication follows a network topology that has unified radio access and backhaul network, which makes 5G cellular network support coexist with low-frequency and high-frequency communication techniques. For this very reason mmWave has three main elements: (1) macro BSs (MBs), (2) millimeter cellular BSs (mBs), and (3) UEs, which can be vehicles.

MBs uses lower frequency bands to transmit control messages in macrocells, whereas mBs will operate at higher frequency bands such as 28 GHz in smaller cell sizes (such as picocells). All the UEs (i.e., vehicles) will first connect to the network through mBs via higher frequency bands. But if they fail to do so, they can just connect to MBs through lower frequency bands. Thus, we can see that both the unified radio access and the backhaul are needed to share a common continuous spectrum in mmWave frequency bands.

To avoid the interference between radio access and backhaul, it is necessary to distinguish between them in time or frequency domains in terms of resource allocation blocks. This helps mBs achieve similar resource division ratios without much interference between them. The ratios can be dynamically modified over the various interval times. Assume that the total bandwidth is B, the bandwidth allocated to backhaul is B_{BA} and the one allocated to radio access is B_{RS}. Then the total bandwidth will be

$$B = B^{BA} + B^{BS} \tag{4.15}$$

Denote the backhaul data rate as R_{BA}, and the radio access data rate is R_{RS}, then for ith MB, the total throughput will be

$$T_i^{BA} = R_i^{BA} B^{BA} \tag{4.16}$$

$$T_i^{RS} = R_i^{RS} B^{RS} \tag{4.17}$$

$$T_i^{RS} = R_i^{RS} (B - B^{BA}) \tag{4.18}$$

Now for any MB, T_i can be limited to

$$T_i = \min(T_i^{BA}, T_i^{RS}) \tag{4.19}$$

The total throughput is maximized when [70]

$$B_{i,\max}^{BA} = \frac{R_i^{RS}}{R_i^{BA} + R_i^{RS}} B \tag{4.20}$$

At peak value [70]

$$T_{i,\max} = \frac{R_i^{BA} R_i^{RS}}{R_i^{BA} + R_i^{RS}} B \tag{4.21}$$

Furthermore, various dynamic resource allocation algorithms can be adopted. Typical algorithms include [70] (1) max-min, maximization of the minimum throughput; (2) max-sum, maximization of the sum throughput and (3) quasi-PF, maximization of the satisfactory factor.

4.6 CONCLUSIONS

In this chapter we have provided a brief layout of 5G-enabled UAV communication systems. By utilizing the strengths of 5G such as high capacity, wide coverage, heterogeneous network integration and so forth, we can overcome the drawbacks (such as low scalability, low capacity, intermittent connectivity, etc.) of previously used communication standards for vehicular communications, such as LTE, 802.11p, and so forth. 5G's inherent ability to support UAV communications with much lower latency and higher reliability makes it suitable to UAV applications with high QoS demand. The extensive 5G coverage can be exploited for the dependable broadcasting of event-triggered safety messages in large areas for various critical applications in an effective manner.

REFERENCES

1. M. Nekovee, "Opportunities and enabling technologies for 5G and beyond-5G spectrum sharing," *Handbook of Cognitive Radio*, pp. 1–15, 2018.
2. D. Soldani and A. Manzalini, "Horizon 2020 and beyond: on the 5G operating system for a true digital society," *IEEE Vehicular Technology Magazine*, vol. 10, no. 1, pp. 32–42, 2015.
3. G. A. A. (5GAA), "5GAA white paper: C-V2X use cases methodology, examples and service level requirements," 2019.
4. G. Elia, M. Bargis, M. P. Galante, N. P. Magnani, L. Santilli, G. Romano, and G. Zaffiro, "Connected transports, V2X and 5G: Standard, services and the Tim-Telecom Italia experiences," *2019 AEIT International Conference of Electrical and Electronic Technologies for Automotive (AEIT AUTOMOTIVE)*, IEEE, pp. 1–6, 2019.
5. F. Hu, *Opportunities in 5G networks: A Research and Development Perspective*, Boca Raton, FL: CRC Press, 2016.
6. G. Araniti, C. Campolo, M. Condoluci, A. Iera, and A. Molinaro, "LTE for vehicular networking: a survey," *IEEE Communications Magazine*, vol. 51, no. 5, pp. 148–157, 2013.
7. I. Rasheed, "Basics of 5G," *Opportunities in 5G Networks: A Research and Development Perspective*, p. 1, 2016.

8. V. Vukadinovic, K. Bakowski, P. Marsch, I. D. Garcia, H. Xu, M. Sybis, P. Sroka, K. Wesolowski, D. Lister, and I. Thibault, "3GPP C-V2X and IEEE 802.11p for vehicle-to-vehicle communications in highway platooning scenarios," *Ad Hoc Networks*, vol. 74, pp. 17–29, 2018.
9. ETSI, "5G; service requirements for enhanced V2X scenarios (3GPP TS 22.186 version 15.4.0 release 15)," *3GPP Technical Specification Report*, 2018.
10. H. Onishi, "A survey: why and how automated vehicles should communicate to other road-users," *2018 IEEE 88th Vehicular Technology Conference (VTC-Fall)*, IEEE, pp. 1–6, 2019.
11. F. Camacho, C. Cárdenas, and D. Munoz, "Emerging technologies and research challenges for intelligent transportation systems: 5G, HETNETS, and SDN," *International Journal on Interactive Design and Manufacturing (IJIDeM)*, pp. 1–9, 2017.
12. S. Chen, J. Hu, Y. Shi, Y. Peng, J. Fang, R. Zhao, and L. Zhao, "Vehicle- to-everything (V2X) services supported by LTE-based systems and 5G," *IEEE Communications Standards Magazine*, vol. 1, no. 2, pp. 70–76, 2017.
13. Z. MacHardy, A. Khan, K. Obana, and S. Iwashina, "V2X access technologies: regulation, research, and remaining challenges," *IEEE Communications Surveys & Tutorials*, vol. 20, no. 3, pp. 1858–1877, 2018.
14. C2R Consortium. [Online]. http://www.com2react-project.org/
15. K. Skala, D. Davidovic, E. Afgan, I. Sovic, and Z. Sojat, "Scalable distributed computing hierarchy: Cloud, fog and dew computing," *Open Journal of Cloud Computing (OJCC)*, vol. 2, no. 1, pp. 16–24, 2015.
16. G. Dimitrakopoulos, "Intelligent transportation systems based on internet-connected vehicles: fundamental research areas and challenges," *2011 11th International Conference on ITS Telecommunications (ITST)*, IEEE, pp. 145–151, 2011.
17. Qualcomm. "Accelerating C-V2X commercialization," 2018. [Online]. http://www.qualcomm.com
18. Auto talks.com, "Accelerating global V2X deployment for road safety," *white paper*, 2019.
19. O. Weisman, A. Aharon, and T. Philosof, "Resource allocation for channel access in V2X communication systems," *U.S. Patent App. 15/600,938*, 2018.
20. T. Li, L. Lin, and N. Center, "Byzantine-tolerant v2x communication system," in *2018 INFORMS Annual Meeting Phoenix*, 2018.
21. Ericsson, "More than 50 billion connected devices," *white paper*, 2011.
22. J. G. Andrews, S. Buzzi, W. Choi, S. V. Hanly, A. Lozano, A. C. Soong, and J. C. Zhang, "What will 5G be?" *IEEE Journal on Selected Areas in Communications*, vol. 32, no. 6, pp. 1065–1082, 2014.
23. A. Osseiran, F. Boccardi, V. Braun, K. Kusume, P. Marsch, M. Maternia, O. Queseth, M. Schellmann, H. Schotten, H. Taoka *et al.*, "Scenarios for 5G mobile and wireless communications: the vision of the metis project," *IEEE Communications Magazine*, vol. 52, no. 5, pp. 26–35, 2014.
24. T. E. Bogale and L. B. Le, "Massive Mimo and mmWave for 5G wireless HETNET: potential benefits and challenges," *IEEE Vehicular Technology Magazine*, vol. 11, no. 1, pp. 64–75, 2016.
25. S. E. Elayoubi, M. Fallgren, P. Spapis, G. Zimmermann, D. Martín-Sacristán, C. Yang, S. Jeux, P. Agyapong, L. Campoy, Y. Qi *et al.*, "5G service requirements and operational use cases: analysis and METIS II vision," *2016 European Conference on Networks and Communications (EuCNC)*, IEEE, pp. 158–162, 2016.
26. E. Hossain, M. Rasti, H. Tabassum, and A. Abdelnasser, "Evolution toward 5G multi-tier cellular wireless networks: An interference management perspective," *IEEE Wireless Communications*, vol. 21, no. 3, pp. 118–127, 2014.

27. H. U. H. Takahashi and K. Aoyagi, "Technical overview of the 3GPP release 15 standard," 2019.
28. G. Naik, B. Choudhury *et al.*, "IEEE 802.11 BD & 5G NR V2X: evolution of radio access technologies for V2X communications," arXiv preprint arXiv: 1903.08391, 2019.
29. M. Yao, M. Sohul, V. Marojevic, and J. H. Reed, "Artificial intelligence defined 5G radio access networks," *IEEE Communications Magazine*, vol. 57, no. 3, pp. 14–20, 2019.
30. N. P. Lawrence, B. W.-H. Ng, H. J. Hansen, and D. Abbott, "5G terrestrial networks: mobility and coverage-solution in three dimensions," *IEEE Access*, no. 5, pp. 8064–8093, 2017.
31. S. Eichler, "Performance evaluation of the IEEE 802.11p wave communication standard," *Vehicular Technology Conference, 2007*, IEEE, pp. 2199–2203, 2007.
32. A. Vinel, "3GPP LTE versus IEEE 802.11p/wave: which technology is able to support cooperative vehicular safety applications?" *IEEE Wireless Communications Letters*, vol. 1, no. 2, pp. 125–128, 2012.
33. R. Coppola and M. Morisio, "Connected car: technologies, issues, future trends," *ACM Computing Surveys (CSUR)*, vol. 49, no. 3, p. 46, 2016.
34. R. Trivisonno, R. Guerzoni, I. Vaishnavi, and D. Soldani, "SDN-based 5G mobile networks: architecture, functions, procedures and backward compatibility," *Transactions on Emerging Telecommunications Technologies*, vol. 26, no. 1, pp. 82–92, 2015.
35. F. Chiti, R. Fantacci, D. Giuli, F. Paganelli, and G. Rigazzi, "Communications protocol design for 5g vehicular networks," *5G Mobile Communications*. Berlin: Springer, pp. 625–649, 2017.
36. M. Amoozadeh, A. Raghuramu, C.-N. Chuah, D. Ghosal, H. M. Zhang, J. Rowe, and K. Levitt, "Security vulnerabilities of connected vehicle streams and their impact on cooperative driving," *IEEE Communications Magazine*, vol. 53, no. 6, pp. 126–132, 2015.
37. O. Onireti, J. Qadir, M. A. Imran, and A. Sathiaseelan, "Will 5G see its blind side? Evolving 5G for universal internet access," *Proceedings of the 2016 workshop on Global Access to the Internet for All*, ACM, pp. 1–6, 2016.
38. U. Shevade, Y.-C. Chen, L. Qiu, Y. Zhang, V. Chandar, M. K. Han, H. H. Song, and Y. Seung, "Enabling high-bandwidth vehicular con- tent distribution," *Proceedings of the 6th International Conference*, ACM, p. 23, 2010.
39. J.-B. Seo, T. Kwon, and V. C. Leung, "Social groupcasting algorithm for wireless cellular multicast services," *IEEE Communications Letters*, vol. 17, no. 1, pp. 47–50, 2013.
40. F. Mezghani, R. Dhaou, M. Nogueira, and A.-L. Beylot, "Content dissemination in vehicular social networks: taxonomy and user satisfaction," *IEEE Communications Magazine*, vol. 52, no. 12, pp. 34–40, 2014.
41. Y.-D. Lin and Y.-C. Hsu, "Multihop cellular: A new architecture for wireless communications," *Proceedings of the Nineteenth Annual Joint Conference of the IEEE Computer and Communications Societies*, IEEE, vol. 3, pp. 1273–1282, 2000.
42. G. Rigazzi, F. Chiti, R. Fantacci, and C. Carlini, "Multi-hop D2D networking and resource management scheme for m2m communications over LTE-a systems," *2014 International Wireless Communications and Mobile Computing Conference (IWCMC)*, IEEE, pp. 973–978, 2014.
43. H. Nishiyama, M. Ito, and N. Kato, "Relay-by-smartphone: realizing multihop device-to-device communications," *IEEE Communications Magazine*, vol. 52, no. 4, pp. 56–65, 2014.
44. A. Rasheed, H. Zia, F. Hashmi, U. Hadi, W. Naim, and S. Ajmal, "Fleet & convoy management using VANET," *Journal of Computer Networks*, vol. 1, no. 1, pp. 1–9, 2013.

45. R. Monteiro, S. Sargento, W. Viriyasitavat, and O. K. Tonguz, "Improving VANET protocols via network science," *2012 IEEE Vehicular Networking Conference (VNC)*, pp. 17–24, 2012.
46. C. Liang and F. R. Yu, "Wireless network virtualization: a survey, some research issues and challenges," *IEEE Communications Surveys & Tutorials*, vol. 17, no. 1, pp. 358–380, 2015.
47. M. Cagalj, S. Ganeriwal, I. Aad, and J.-P. Hubaux, "On selfish behavior in CSMA/CA networks," *Proceedings of the IEEE 24th Annual Joint Conference of the IEEE Computer and Communications Societies*, vol. 4, pp. 2513–2524, 2005.
48. M. Felegyhazi, J.-P. Hubaux, and L. Buttyan, "Nash equilibria of packet forwarding strategies in wireless ad hoc networks," *IEEE Transactions on Mobile Computing*, vol. 5, no. 5, pp. 463–476, 2006.
49. M. Gerla and L. Kleinrock, "Vehicular networks and the future of the mobile internet," *Computer Networks*, vol. 55, no. 2, pp. 457–469, 2011.
50. V. Srinivasan, P. Nuggehalli, C.-F. Chiasserini, and R. R. Rao, "Cooperation in wireless ad hoc networks," *IEEE Societies Twenty-Second Annual Joint Conference of the IEEE Computer and Communications*, vol. 2, pp. 808–817, 2003.
51. W. Saad, Z. Han, M. Debbah, A. Hjorungnes, and T. Basar, "Coalitional game theory for communication networks," *IEEE Signal Processing Magazine*, vol. 26, no. 5, pp. 77–97, 2009.
52. R. J. La and V. Anantharam, "A game-theoretic look at the gaussian multiaccess channel," *DIMACS Series in Discrete Mathematics and Theoretical Computer Science*, vol. 66, pp. 87–106, 2004.
53. L. Song, D. Niyato, Z. Han, and E. Hossain, "Game-theoretic resource allocation methods for device-to-device communication," *IEEE Wireless Communications*, vol. 21, no. 3, pp. 136–144, 2014.
54. Y. Li, D. Jin, J. Yuan, and Z. Han, "Coalitional games for resource allocation in the device-to-device uplink underlaying cellular networks," *IEEE Transactions on Wireless Communications*, vol. 13, no. 7, pp. 3965–3977, 2014.
55. Z. Zhou, M. Dong, K. Ota, R. Shi, Z. Liu, and T. Sato, "Game-theoretic approach to energy-efficient resource allocation in device-to- device underlay communications," *IET Communications*, vol. 9, no. 3, pp. 375–385, 2015.
56. E. Yaacoub and O. Kubbar, "Energy-efficient device-to-device communications in LTE public safety networks," *2012 IEEE Globecom Workshops (GC Wkshps)*, pp. 391–395, 2012.
57. B. Wang, Z. Han, and K. R. Liu, "Distributed relay selection and power control for multiuser cooperative communication networks using Stackelberg game," *IEEE Transactions on Mobile Computing*, vol. 8, no. 7, pp. 975–990, 2009.
58. T. Bai and R. W. Heath, "Coverage and rate analysis for millimeter-wave cellular networks," *IEEE Transactions on Wireless Communications*, vol. 14, no. 2, pp. 1100–1114, 2015.
59. A. I. Sulyman, A. T. Nassar, M. K. Samimi, G. R. Maccartney, T. S. Rappaport, and A. Alsanie, "Radio propagation path loss models for 5G cellular networks in the 28 GHz and 38 GHz millimeter-wave bands," *IEEE Communications Magazine*, vol. 52, no. 9, pp. 78–86, 2014.
60. Y. Wang, Z. Shi, M. Du, and W. Tong, "A millimeter wave spatial channel model with variant angles and variant path loss," *2016 IEEE Wireless Communications and Networking Conference (WCNC)*, pp. 1–6, 2016.
61. Y. Wang, L. Huang, Z. Shi, H. Huang, D. Steer, J. Li, G. Wang, and W. Tong, "An introduction to 5G mmWave communications," *Proceedings of IEEE Globecom 2015 Workshop*, 2015.

62. Y. Wang, J. Li, L. Huang, Y. Jing, A. Georgakopoulos, and P. Demestichas, "5G mobile: spectrum broadening to higher-frequency bands to support high data rates," *IEEE Vehicular Technology Magazine*, vol. 9, no. 3, pp. 39–46, 2014.
63. W. Roh, J.-Y. Seol, J. Park, B. Lee, J. Lee, Y. Kim, J. Cho, K. Cheun, and F. Aryanfar, "Millimeter-wave beamforming as an enabling technology for 5G cellular communications: theoretical feasibility and prototype results," *IEEE Communications Magazine*, vol. 52, no. 2, pp. 106–113, 2014.
64. Y.-L. Tseng, "LTE-advanced enhancement for vehicular communication," *IEEE Wireless Communications*, vol. 22, no. 6, pp. 4–7, 2015.
65. H. Li, L. Huang, and Y. Wang, "Scheduling schemes for interference suppression in millimeter-wave cellular network," in *2015 IEEE 26th Annual International Symposium on Personal, Indoor, and Mobile Radio Communications (PIMRC)*, pp. 2244–2248, 2015.
66. J. Zhang, A. Beletchi, Y. Yi, and H. Zhuang, "Capacity performance of millimeter wave heterogeneous networks at 28GHz/73GHz," *2014 IEEE Globecom Workshops (GC Wkshps)*, pp. 405–409, 2014.
67. "Further advancements for E-Utra physical layer aspects," *3GPP Technical Specification Report*, 3GPP TR 36.814 V.9.0.0, 2010.
68. G. R. MacCartney, J. Zhang, S. Nie, and T. S. Rappaport, "Path loss models for 5G millimeter wave propagation channels in urban micro-cells," *2013 IEEE Global Communications Conference (GLOBECOM)*, pp. 3948–3953, 2013.
69. J. He, T. Kim, H. Ghauch, K. Liu, and G. Wang, "Millimeter wave mimo channel tracking systems," *2014 Globecom Workshops (GC Wkshps)*, IEEE, pp. 416–421, 2014.
70. Y. Wang and Z. Shi, "Millimeter-wave mobile communications," *5G Mobile Communications*. Berlin: Springer, pp. 117–134, 2017.
71. Y. Wang, L. Huang, Z. Shi, K. Liu, and X. Zou, "A millimeter wave channel model with variant angles under 3GPP SCM framework," *2015 IEEE 26th Annual International Symposium on Personal, Indoor, and Mobile Radio Communications (PIMRC)*, IEEE, pp. 2249–2254, 2015.
72. H. Huang, K. Liu, R. Wen, Y. Wang, and G. Wang, "Joint channel estimation and beamforming for millimeter wave cellular system," *2015 IEEE Global Communications Conference (GLOBECOM)*, pp. 1–6, 2015.
73. M. A. Nadeem, M. A. Saeed, and I. A. Khan, "A survey on current repertoire for 5G," *International Journal of Information Technology and Computer Science (IJITCS)*, vol. 9, no. 2, pp.18–29, 2017.

Part II

UAV Mobility

5 Machine Learning for Multi-UAV Mobility Prediction

Viprav Raju
Department Electrical and Computer Engineering,
University of Alabama, Tuscaloosa, AL

ShiZe Huang and DongXiu Ou
Department of Transportation Information and
Control Engineering, College of Transportation
Engineering Tongji University, Shanghai. China

CONTENTS

5.1 INTRODUCTION

Unmanned aerial vehicles (UAVs) are becoming popular in the modern world enabling us to monitor and control situations that can be too strenuous to handle manually. UAVs have been used in transportation, traffic control, remote sensing, wild-life monitoring, smart-system agriculture, surveillance, broadband communications and border patrolling [1–7]. Search and rescue (SAR) UAVs are another popular area of growing importance with respect to UAVs. Algorithms are intended to increase the area of coverage by optimizing the network paths to decrease the overlap of two UAVs' coverage areas and increase the chance of exploration of a previously unexplored area.

Drones are one category of UAVs. There were more than 1 million registered drones according to the Federal Aviation Administration (FAA) as of 2018. UAVs are typically used as groups instead of individual-based systems to increase the flight time, payload and range of travel [8]. These networked UAVs offer more flexibility and are especially useful in SAR missions because the networks increase the probability of mission success [9]. The network topology has a mobility nature that poses some challenges for protocol designs [10–12]. The topology consists of a large number of nodes with multiple levels of reliability, sensing, maneuverability, actuation and communication capabilities. Predictive algorithms are beneficial because they can help the network pre-allocate the resources to different nodes to achieve a *smooth* data transmission [13–15].

Predictive routing algorithms can avoid network failures as shown in Figure 5.1. They can predict which link will be broken due to mobility and thus find the backup routes earlier, which is unlike some reactive routing schemes that identify a failure only after it has occurred. Network interruptions have been discussed in [1, 2]. The links' signal quality could be used to predict potential link failure [3]. The system anticipates the possible network paths taken by each UAV.

Predicting motion trajectories of individual UAVs also can enable the autonomy of UAV networks that do not require prior path planning. There are two main categories of prediction algorithms, namely, data driven and model based. Data-driven algorithms require a large amount of data to identify the most certain patterns, in other words, the patterns with the highest probabilities [16]. The system captures the

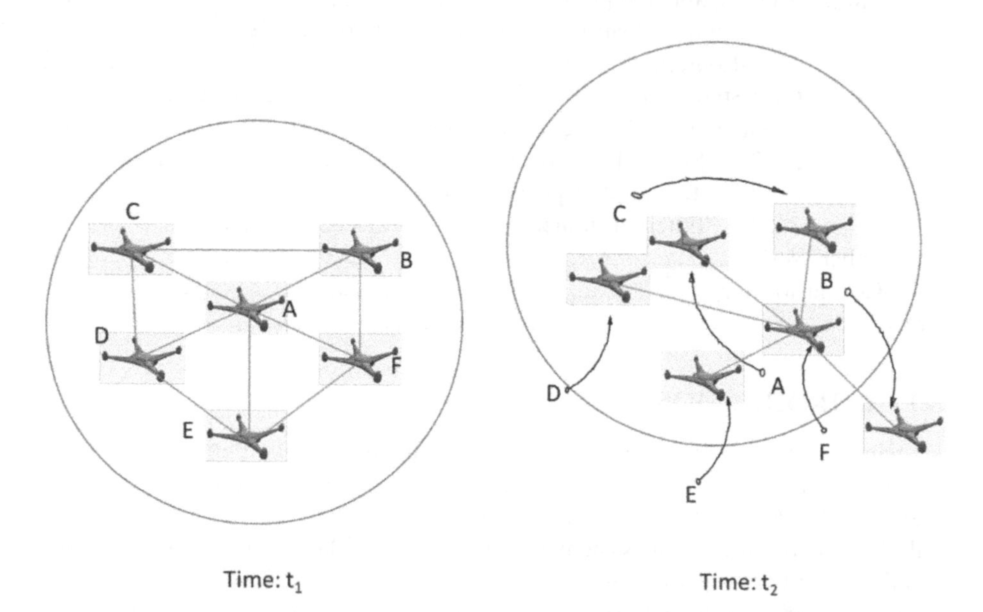

FIGURE 5.1 Depiction of UAV topology at two instances of time. The circle depicts the accessible range of UAV A. At time $t = t_2$, UAV B flies away from UAV A's accessible range and loses network connectivity. Predicted network topology can be used to avoid such connectivity issues by selecting optimal routes that are less likely to cause network disconnections.

dependence of spatial and temporal features of the UAV's mobility. Model-based algorithms try to identify motion trajectories based on the UAV's past trajectories [5]. Previously, several methods were employed in model-based algorithms: hidden Markov models (HMMs), the levy-flight process, Bayesian learning, manifold learning and Gaussian mixture modeling [5–7, 17, 18].

Some of the studies have focused on clustering motion paths by using unsupervised methods, which include tree based, grid, distance-based clustering and kernel methods [13, 14, 19–21]. The limitation of these methods is that they use distance metrics to identify location-optimal paths rather than identify the nature of the paths. In [15], the authors summarized these clustering algorithms. More recent work involves clustering operations based on UAV trajectory/network shapes along with the distances. The methods are multivariate in that they consider more than one parameter when performing prediction or clustering. Some of the popular multivariate models used in the clustering of UAVs include von Mises distribution, non-negative matrix factorization and circular statistics [22, 23]. These methods only focus on identifying the similarities between motion paths.

The lack of accurate predictive models in UAV networking has generated a great deal of interest in deep neural network (DNN)-based or machine learning–based predictive models. The UAV network can be considered to be a special type of spatio-temporal system that can only be solved when we consider both the spatial and temporal features of the UAV network.

DNNs are a possible solution for the spatio-temporal prediction problem. They have been previously used in dynamic applications that involve planning and development, transportation, environment and energy planning [24]. Two popular DNNs are the convolutional neural nets (CNNs) and recurrent neural networks (RNNs). They are alternatively used to solve the spatial or temporal problems, respectively. It has been a challenge to apply these networks to the combination of these two cases, i.e., the spatio-temporal problem.

Time series with internal spatial dependencies are abundant in many fields including meteorology, medicine, biology, ecology, traffic and economics. Their data come from multiple observations in various sources such as satellite imagery, video cameras, global positioning satellites and a variety of different sensors. One major issue in modeling spatio-temporal data is the size/scale of the data coming from the sensors that cover very large spaces. Another critical criterion to be considered is the temporal lags that can evolve due to the complexity of the data sourcing.

Dimensionality reduction results in latent dynamic models, which have been exploited in machine learning [25] and statistics [26]. Different modalities and tasks can be captured using deep learning to extract valuable information. RNNs can handle dynamic data for the purposes of classification and sequence prediction/generation [27]. RNNs operate well with sequential data and the processes associated with such data. Recently, convolutional RNNs [28, 29] have been used to handle both the temporality and the spatiality.

Deep spatio-temporal networks have been extensively used for time series forecasting of spatial processes. Time series forecasting/modeling has received a great deal of attention in the past, both in machine learning and statistics. In statistics, linear models are based on moving average components and autoregression items that assume linear dependencies with a noise factor [9]. In machine learning, non-linear models based on neural networks are used primarily for dynamic state space modeling for time series forecasting [10].

There has been successful implementation of such models in fields like speech recognition [11] and language processing [12, 30]. Spatial dependencies are not addressed in this model. The development of non-parametric generative models has led to different promising methods such as the stochastic gradient variational Bayes algorithm (SGVB) [31], which provides a DNN with stochastic learning for latent variables [4–6].

Spatio-temporal statistics typically uses first- and second-order moments for modeling the spatio-temporal dependencies. Very recently, dynamic state space models have been explored [17] where time and space can either be continuous or discrete. Discrete time modeling can be effective when the space is divided and coded in these cases. The models are represented by linear integral difference equations. When time and space are both discrete, general vectoral autoregressive formulations are employed. However, the curse of dimensionality is a big challenge when there are a large number of sources. Parameter reduction is one possible way to overcome this issue.

Gaussian Markov random fields were used in climatology to map temporal and geographical dependencies [18]. Brain computer interfacing is another domain where spatio-temporal data analysis is desired [20, 21].

In this chapter, we will identify the state-of-the-art implementations of spatio-temporal predictive models in machine learning, and particularly focus on deep spatio-temporal neural networks. Ten studies will be compared using a tabular form that discusses the methods, advantages and some of the possible challenges in the future. Also, a few works are studied in detail to identify the suitability of spatio-temporal neural networks for UAV mobility prediction.

Note that the methods to be summarized here may not directly discuss UAV applications. However, because the UAV network is a special dynamic multi-hop communication system, all the models presented here could potentially be applied to UAV state prediction.

5.2 MULTI-VARIATE SPATIO-TEMPORAL PREDICTIONS

In this section we summarize the results of the multi-variate spatio-temporal prediction models and explore two papers that use DNNs for prediction. Table 5.1 summarizes ten studies that involve spatio-temporal predictions.

5.2.1 Case Study 1: Spatiotemporal Prediction for Mobile Crowdsourcing Services

Mobile crowdsourcing has become a popular method that makes use of the ubiquity of mobile devices. The crowd together forms a powerful source for data services. The spatio-temporal data in this case can be stored on cloud servers.

These crowdsourced services, such as the Wi-Fi hotspots, can provide better mobility than traditional service generators or mediums. As the mobility increases,

TABLE 5.1
Comparison of Ten Spatio-Temporal Prediction Models

Authors	Application	Method	Pros	Challenges
[33]	SST prediction problem	Two-scale modeling for spatio-temporal datasets, using EOFs	• Efficient in capturing variability when standard dimension reduction methods do not perform well • It delivers more balanced results in the presence of outliers	High-dimensional datasets that have complicated dependence structure in both spatial and temporal scales
[34]	Real-time crowd flow forecasting system	DeepST	Works better than four other existing techniques, can be applied to diverse ST applications and can perform well in multi-step ahead predictions	Complex datasets and real-time applications
[35]	Prediction of root zone soil moisture	Multivariate relevance vector Machines	Bootstrapping is used to check over-fitting/ under-fitting and uncertainty in model estimates Good generalization capabilities providing robustness	Other fields of hydrologic science
[36]	Prediction of speciated fine particles	Spatio-temporal linear coregionalization model	Combines different sources of data and accounts for bias and measurement error in each data source	Accurate modeling of fine-resolution particles
[37]	Prediction on latent low-dimensional functional structure with non-stationarity	Represent non-parametrically the linear dependence structure of a multivariate spatio-temporal process in terms of latent common factors	Low-dimensional structure is completely unknown in our setting Method accommodates non-stationarity over space	Algorithm complexity could be high
[38]	Partitioning of large urban networks for travel time prediction	Probabilistic principal component analysis	Versatile, and can be applied for any source of travel time data and multi-variate travel time prediction method	Complexity, heterogeneity, noisy data, unexpected events

(Continued)

TABLE 5.1 (Continued)
Comparison of Ten Spatio-Temporal Prediction Models

Authors	Application	Method	Pros	Challenges
[39]	Taxi demand prediction	DMVST-Net	• Temporal view (modeling correlations between future demand values with near time points via LSTM) • Spatial view (modeling local spatial correlation via local CNN) • Semantic view (modeling correlations among regions sharing similar temporal patterns)	Complex non-linear spatial and temporal relations
[40]	Hierarchical spatio-temporal cortical dynamics of human visual object recognition	Artificial DNN	Comparing DNNs with human brains in space and time provides a spatio-temporally unbiased algorithmic account of visual object recognition in human cortex	If the data source is not an image, a new DNN structure is needed
[41]	Large-scale transportation network speed prediction	CNN	– Can train the model in a reasonable time – Proposed method outperforms OLS, KNN, ANN, RF, SAE, RNN and LSTM NNs with an average accuracy promotion of 42.91%	The large-scale network may require long computation time
[42]	Geo-sensory time series prediction	Multi-level attention-based RNN	• A multi-level attention mechanism to model the dynamic spatio-temporal dependencies • A general fusion module to incorporate the external factors from different domains • Method outperforms nine baseline methods	Complex datasets and real-time applications

ANN: Artificial Neural Networks, CNN: convolutional neural nets; DeepST: deep learning network; DMVST-Net: deep multi-view spatial-temporal network; DNN: deep neural network; EOF: Empirical Orthogonal Functions, KNN: Kernel Neural Networks, LSTM: Long-/short-term memory, NN: Neural Networks, OLS: ordinary least squares, RF: Recurrent Function; RNN: recurrent neural network; SAE: sparse autoencoder; SST: sea surface temperature.

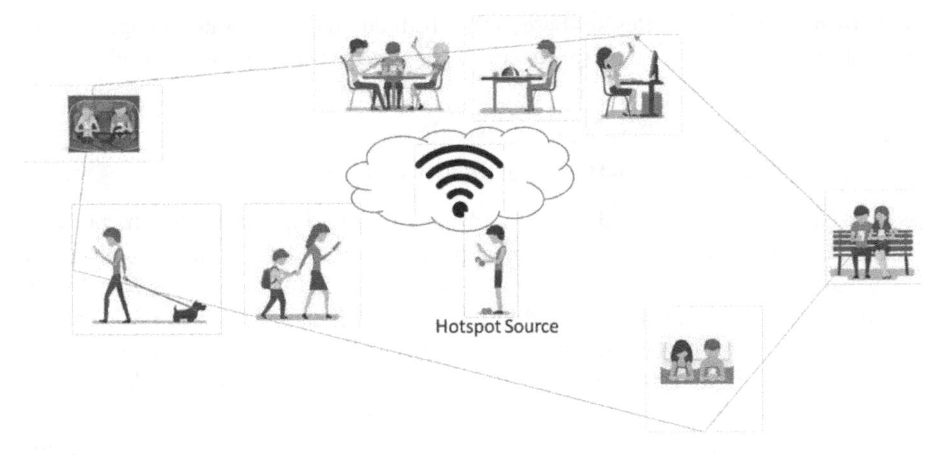

FIGURE 5.2 Mobile crowdsourced service.

there arises a challenge to predict the precise location and availability of such mobile services. The main aspect of crowdsourced hotspots are the spatio-temporal attributes that will be considered while selecting a particular service.

This study [32] involves a deep learning–based predictive model for identifying the availability of crowdsourced services in both the temporal and spatial domains. The model consists of a two-stage structure that uses a historical dataset of mobile crowdsource services. Gramian angular fields are used to predict the availability of services. The services are clustered into regions prior to making the predictions.

A non-deterministic crowdsourced service whose time and availability are not known is considered in [32]. The constraint to this service is that it is dependent on the spatio-temporal attributes of the service itself. As an example, in Figure 5.2 the Wi-Fi hotspot is dependent on the source and all the neighbors within the range of the hotspot. It is essential for these neighbors to predict the location and availability of the source.

The application presents a deep learning–based strategy to determine whether a crowdsourced service (e.g., Wi-Fi hotspot) is available given a particular location and timing. Sensor data collected from smartphones is used to gather information regarding user behaviors and routines (both temporally and spatially). Therefore, one can successfully predict the spatio-temporal aspects of that user.

The network has two stages. First, each geolocation is associated with the source. Using the historical data, deep learning predicts the available service at a specific location for that time t_i. The Gramian angular field is used to determine the availability duration of the service at that location. There is a pre-processing stage in which the services are clustered in space. The service discovery at each location at a particular time can be predicted using a DNN given information about the location and time of the service. The second stage is a time series prediction model that predicts the availability of these services. The time series are transformed into a polar coordinates system and the Gramian angular fields are derived. This offers a new data representation of the time series and enables the application of a deep learning–based forecasting technique. A ResNet-based DNN trained on the generated images is able to predict the presence of the crowdsourced service for the multi-step ahead.

5.2.2 Case Study 2: Predicting Citywide Crowd Flows using Deep Spatio-Temporal Residual Networks

It is essential to predict crowd flows in urban areas (see Figure 5.3) for the purpose of traffic control, risk aversion and general public interest. Risks from massive crowd flows can be analyzed to prevent disasters. As a historical example, the 2015 New Year's Eve celebrations in Shanghai resulted in a catastrophic stampede with 36 casualties. By predicting and anticipating such overwhelming crowd flows, one can avert such tragedies by utilizing emergency mechanisms, such as conducting traffic control, sending out warnings or evacuating people, in advance.

Predicting the flow of crowds is of great importance as well as very challenging given the complexity of the data. The data can consist of temporal, spatial and

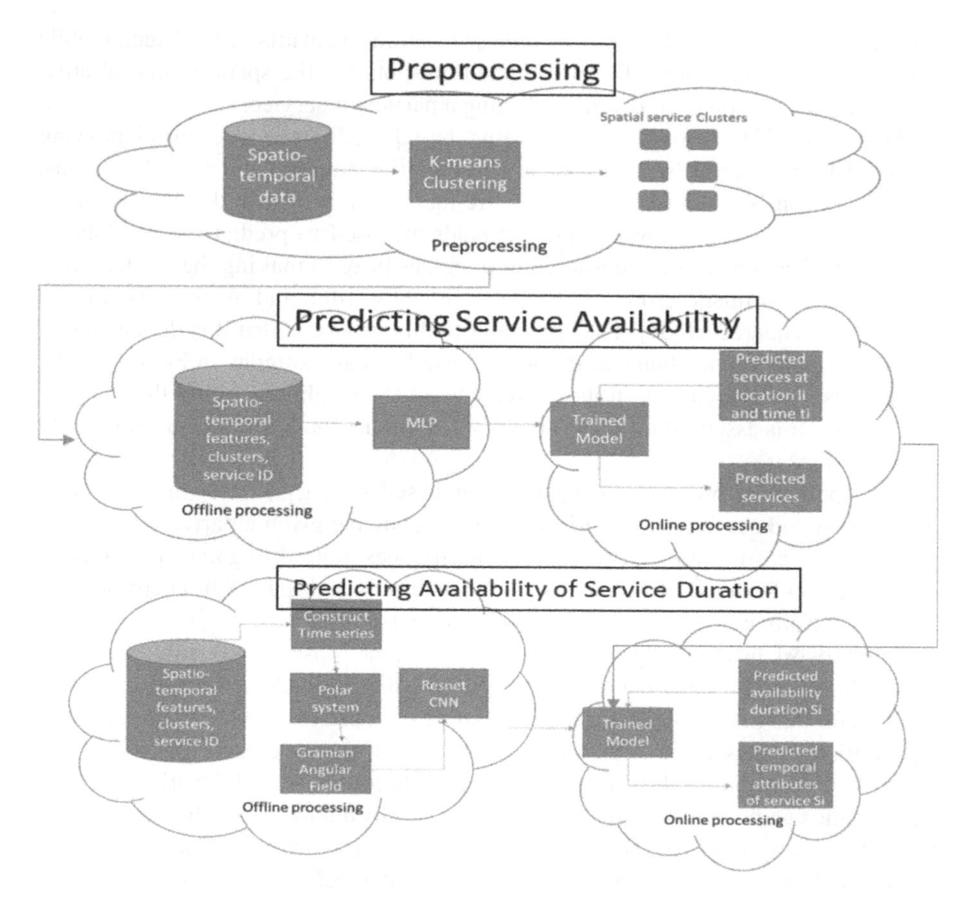

FIGURE 5.3 Two-stage prediction framework to predict crowdsourced service availability spatially and temporally.

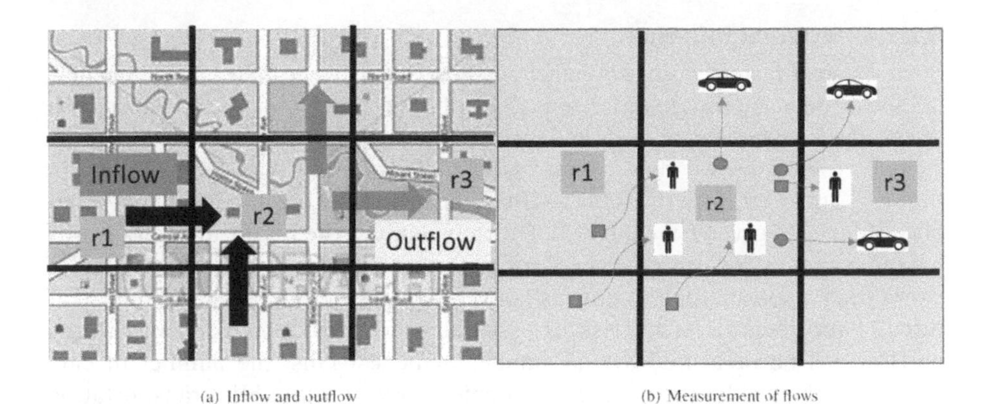

FIGURE 5.4. Crowd flows in a region [34]. (a) Inflow and outflow. (b) Measurement of flows.

external dependencies. The authors [34] proposed a deep learning–based approach called ST-ResNet to collectively forecast two types of crowd flows (i.e., inflow and outflow) in each and every region of a city. They designed an end-to-end structure of ST-ResNet based on the unique properties of spatio-temporal data. The dependencies are defined as follows.

5.2.2.1 Spatial Dependencies

- *Nearby*: The inflow of region r_2 [shown in Figure 5.4(a)] is affected by the outflows of nearby regions (like r_1). Likewise, the outflow of r_2 would affect the inflows of other regions (e.g., r_3). The inflow of region r_2 would affect its own outflow as well.
- *Distant*: The flows can be affected by the flows of distant regions. For instance, people who live far away from the office area always go to work by metro or highway, implying that the outflows of distant regions directly affect the inflow of the office area.

5.2.2.2 Temporal Dependencies

- *Closeness*: The flow of crowds in a region is affected by recent time intervals, both near and far. For instance, traffic congestion occurring at 8:00 a.m. will affect that of 9:00 a.m., and the crowd flows of today's 16th time interval is more similar to that of yesterday's 16th time interval than that of today's 20th time interval.
- *Period*: Traffic conditions during morning rush hours may be similar on consecutive weekdays, repeating every 24 hours.
- *Trend*: Morning rush hours may gradually happen later as winter comes. When the temperature gradually drops and the sun rises later in the day, people get up later and later.

5.2.2.3 External Influence

Some external factors, such as weather conditions and events, may change the flow of crowds tremendously in different regions of a city. For example, a thunderstorm affects the traffic speed on roads and further changes the flows of regions.

Two types of crowd flows were predicted, inflow and outflow, as shown in Figure 5.5(a). Inflow is the total traffic of crowds entering a region from other places during a given time interval. Outflow denotes the total traffic of crowds having a region for other places during a given time interval. Both types of flows track the transition of crowds between regions. Understanding their features is very beneficial for risk assessment and traffic management. Inflow/outflow can be measured by the number of pedestrians, the number of cars driven nearby roads, the number of people traveling on public transportation systems (e.g., metro, bus) or all of them together if the data are available. One can use mobile phone signals to measure the number of pedestrians, showing that the inflow and outflow of r_2 are (3, 1), respectively. Similarly, using the Global Positioning System (GPS) trajectories of vehicles, two types of flows are (0, 3), respectively. Therefore, the total inflow and outflow of r_2 are (3, 4), respectively. Apparently, predicting crowd flows can be viewed as a kind of spatio-temporal prediction problem. To address the aforementioned issues, the authors explored the use of DNNs and proposed a deep spatio-temporal residual network (ST-ResNet) (Figure 5.6) to predict collectively the flow of crowds in every region.

The authors designed individual branches for each property mentioned above. Each branch is a residual convolutional unit that models the spatial properties of crowd traffic. ST-ResNet dynamically aggregates the output of the three residual neural networks. Different weights are assigned to each region and branch.

The aggregation is further combined with external factors, such as weather and day of the week, to predict the final traffic of crowds in each and every region. The authors developed a real-time system based on Microsoft Azure Cloud called Urban

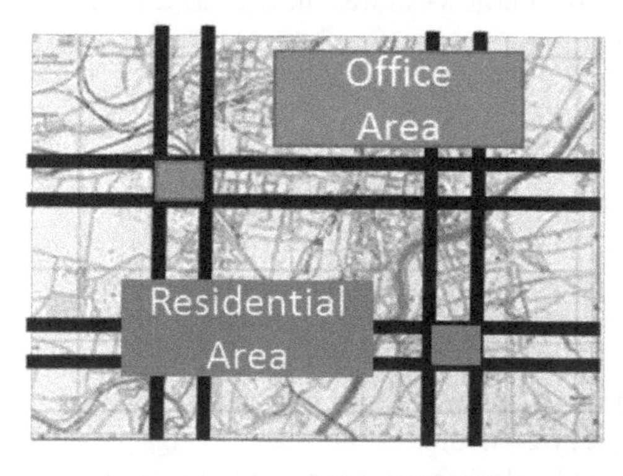

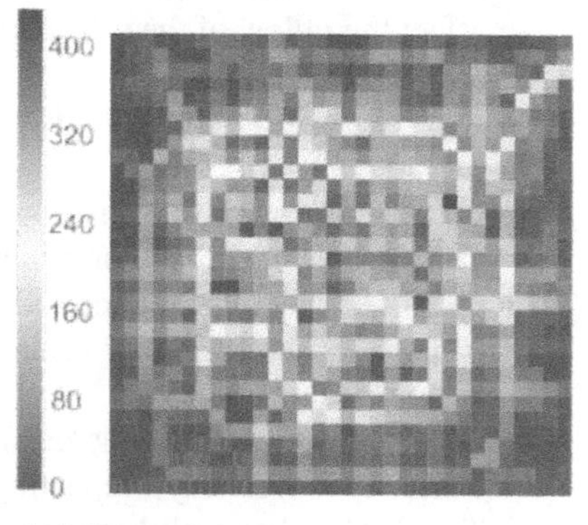

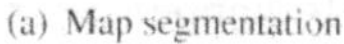
(a) Map segmentation (b) Inflow matrix

FIGURE 5.5. City Inflow [34]. (a) Map segmentation. (b) Inflow matrix.

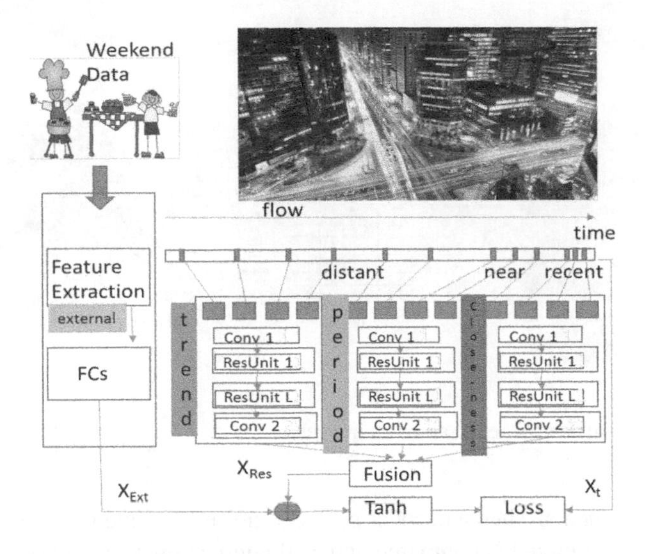

FIGURE 5.6 ST-ResNet architecture. Conv, convolution; ResUnit, residual Unit; FC, fully-connected [34].

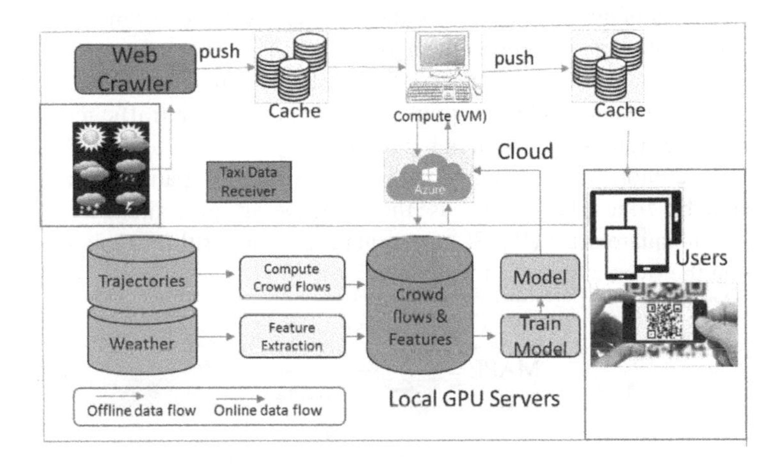

FIGURE 5.7 System framework.

Flow (Figure 5.7), providing the crowd flow monitoring and forecasting in Guiyang City, China. In addition, they presented an extensive experimental evaluation using two types of crowd flows in Beijing and New York City, where ST-ResNet outperforms nine well-known baselines.

5.3 OTHER WORKS

In [43] the authors introduced a dynamic spatio-temporal model that consists of an RNN for time series forecasting. The learning is done through structured latent

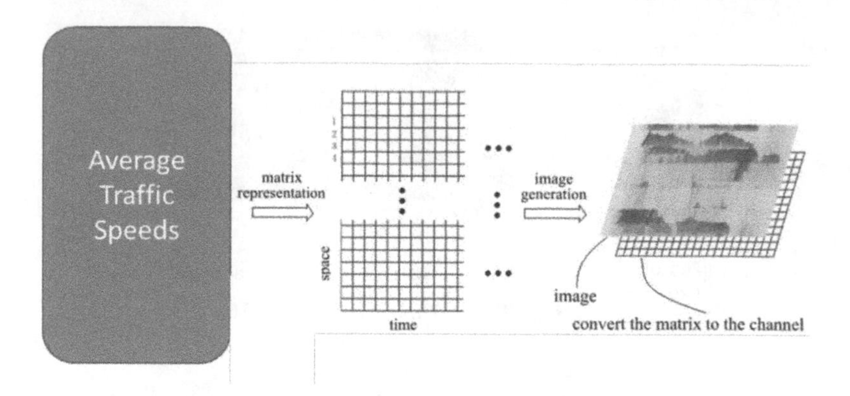

FIGURE 5.8 Traffic flow represented as images.

dynamic components. Then a decoder predicts the future state by using these latent observations. The model is compared and evaluated against several other baselines. The comparison is done using a variety of datasets. We have included this study here to show the robustness of deep spatio-temporal neural nets when considering different datasets.

Another aspect of deep spatio-temporal neural nets is that one can use spatial and temporal information and represent them as images. This enables one to capture the information as a picture. This picture could then be used in a neural net to train the network. In [41], the authors have captured data acquired from traffic monitoring as individual image frames, as shown in Figure 5.8.

In [39], the authors have compared the performance of a spatio-temporal ResNet with different baselines. The authors use the mean average percentage error (MAPE) and root-mean-square error (RMSE) to evaluate their algorithm. Equations (1) and (2) represent the two metrics used for evaluation

$$\text{MAPE} = \frac{1}{n}\sum_{i=1}^{n}\frac{\left|\widehat{y}^{i}_{t+1} - y^{i}_{t+1}\right|}{y^{i}_{t+1}} \tag{5.1}$$

$$\text{RMSE} = \left(\frac{1}{n}\sum_{i=1}^{n}\left(\widehat{y}^{i}_{t+1} - y^{i}_{t+1}\right)^{2}\right)^{\frac{1}{2}} \tag{5.2}$$

where $\hat{y}^{i}_{t+1}$ and y^{i}_{t+1} mean the prediction value and real value of region i for time interval $t + 1$ and n is total number of samples.

The different tools used in some popularly cited literature for state prediction include:

1. *Historical average (HA)*: This predicts the demand using average values of previous demands at the location given in the same relative time interval (i.e., the same time of the day).

2. *Autoregressive integrated moving average (ARIMA)*: This is a well-known model for forecasting time series that combines moving average and autoregressive components for modeling time series.
3. *Linear regression (LR)*: This could be ordinary least-squares regression (OLSR), Ridge regression (i.e., with L2-norm regularization) and Lasso (i.e., with L1-norm regularization).
4. *Multiple layer perceptron (MLP)*: This neural network could have different numbers of fully connected layers (such as 128, 128, 64 and 64).
5. *XGBoost*: This is a powerful boosting tree-based method and is widely used in data mining applications.
6. *ST-ResNet*: This is a deep learning–based approach for traffic prediction. The method constructs a city's traffic density map at different times as images. CNN is used to extract features from historical images.

5.4 CONCLUSIONS

This chapter has summarized several state-of-the-art spatio-temporal prediction models and particularly explored the use of DNNs in a large-scale system state prediction. Those methods also could be employed to solve the multi-UAV mobility prediction problem. Several strategies and evaluation metrics were discussed that can be used to implement a DNN in the UAV domain.

REFERENCES

1. C. A. Thiels, J. M. Aho, S. P. Zietlow, and D. H. Jenkins, "Use of unmanned aerial vehicles for medical product transport," *Air Medical Journal*, vol. 34, no. 2, pp. 104–108, 2015.
2. K. Kanistras, G. Martins, M. J. Rutherford, and K. P. Valavanis, "Survey of unmanned aerial vehicles (UAVs) for traffic monitoring," *Handbook of Unmanned Aerial Vehicles*, pp. 2643–2666, 2014.
3. J. Xu, G. Solmaz, R. Rahmatizadeh, L. Boloni, and D. Turgut, "Providing distribution estimation for animal tracking with unmanned aerial vehicles," *2018 IEEE Global Communications Conference (GLOBECOM)*, 2018.
4. P. K. Freeman and R. S. Freeland, "Agricultural UAVs in the U.S.: potential, policy, and hype," *Remote Sensing Applications: Society and Environment*, vol. 2, pp. 35–43, 2015.
5. T. Wall and T. Monahan, "Surveillance and violence from afar: the politics of drones and liminal security-scapes," *Theoretical Criminology*, vol. 15, no. 3, pp. 239–254, 2011.
6. C. Joo and J. Choi, "Low-delay broadband satellite communications with high-altitude unmanned aerial vehicles," *Journal of Communications and Networks*, vol. 20, no. 1, pp. 102–108, 2018.
7. A. R. Girard, A. S. Howell, and J. K. Hedrick, "Border patrol and surveillance missions using multiple unmanned air vehicles," *2004 43rd IEEE Conference on Decision and Control (CDC)*, 2004.
8. M. L. Stein, *Interpolation of Spatial Data: Some Theory for Kriging*, New York: Springer, 2012.
9. J. G. De Gooijer and R. J. Hyndman, "25 years of time series forecasting," *International Journal of Forecasting*, vol. 22, no. 3, pp. 443–473, 2006.
10. J. T. Connor, R. D. Martin, and L. E. Atlas, "Recurrent neural networks and robust time series prediction," *IEEE Transactions on Neural Networks*, vol. 5, no. 2, pp. 240–254, 1994.

11. A. Graves, A. Mohamed, and G. Hinton, "Speech recognition with deep recurrent neural networks," *2013 IEEE International Conference on Acoustics, Speech and Signal Processing*, Vancouver, BC, Canada, 2013, pp. 6645–6649.
12. I. Sutskever, J. Martens, and G. Hinton, "Generating Text with Recurrent Neural Networks," p. 8. In Proceedings of the 28th International Conference on International Conference on Machine Learning (ICML'11). Omnipress, Madison, WI, USA, pp. 1017–1024.
13. V. Sharma, K. Kar, R. La, and L. Tassiulas, "Dynamic network provisioning for time-varying traffic," *Journal of Communications and Networks*, vol. 9, no. 4, pp. 408–418, 2007.
14. A. Urra, E. Calle, J. L. Marzo, and P. Vila, "An enhanced dynamic multilayer routing for networks with protection requirements," *Journal of Communications and Networks*, vol. 9, no. 4, pp. 377–382, 2007.
15. Y. Zhang, X. Zhang, W. Fu, Z. Wang, and H. Liu, "HDRE: Coverage hole detection with residual energy in wireless sensor networks," *Journal of Communications and Networks*, vol. 16, no. 5, pp. 493–501, 2014.
16. M. Hess, F. Buther, and K. P. Schafers, "Data-driven methods for the determination of anterior-posterior motion in PET," *IEEE Transactions on Medical Imaging*, vol. 36, no. 2, pp. 422–432, 2017.
17. L. Gupta, R. Jain, and G. Vaszkun, "Survey of important issues in UAV communication networks," *IEEE Communications Surveys & Tutorials*, vol. 18, no. 2, pp. 1123–1152, 2016.
18. M. Quaritsch, K. Kruggl, D. Wischounig-Strucl, S. Bhattacharya, M. Shah, and B. Rinner, "Networked UAVs as aerial sensor network for disaster management applications," *e & i Elektrotechnik und Informationstechnik*, vol. 127, no. 3, pp. 56–63, 2010.
19. S. Rosati, K. Kruzelecki, G. Heitz, D. Floreano, and B. Rimoldi, "Dynamic routing for flying ad hoc networks," *IEEE Transactions on Vehicular Technology*, vol. 65, no. 3, pp. 1690–1700, 2016.
20. Z. Kaleem and M. H. Rehmani, "Amateur drone monitoring: state-of-the-art architectures, key enabling technologies, and future research directions," *IEEE Wireless Communications*, vol. 25, no. 2, pp. 150–159, 2018.
21. F. Afghah, M. Zaeri-Amirani, A. Razi, J. Chakareski, and E. Bentley, "A coalition formation approach to coordinated task allocation in heterogeneous UAV networks," *2018 Annual American Control Conference (ACC)*, 2018.
22. J. H. Sarker and R. Jantti, "Connectivity modeling of wireless multihop networks with correlated and independent factors," *6th International Conference on Advanced Communication Technology*, 2004.
23. M. Khaledi, A. Rovira-Sugranes, F. Afghah, and A. Razi, "On greedy routing in dynamic UAV networks," *2018 IEEE International Conference on Sensing, Communication and Networking (SECON Workshops)*, 2018.
24. Y. Zheng, L. Capra, O. Wolfson, and H. Yang, "Urban computing: concepts, methodologies, and applications," *ACM Transactions on Intelligent Systems and Technology*, vol. 5, no. 3, pp. 1–55, 2014.
25. M. T. Bahadori, Q. (Rose) Yu, and Y. Liu, "Fast multivariate spatio-temporal analysis via low rank tensor learning." In: Z. Ghahramani, M. Welling, C. Cortes, N. D. Lawrence, and K. Q. Weinberger, Eds. *Advances in Neural Information Processing Systems 27*, Red Hook, NY: Curran Associates, 2014, pp. 3491–3499.
26. N. A. C. Cressie and C. K. Wikle, *Statistics for Spatio-Temporal Data Series. Wiley Series in Probability and Statistics*, Hoboken. NJ: Wiley, 2011.
27. Y. Bengio, "Neural net language models," *Scholarpedia*, 2008.

28. X. Shi, Z. Chen, H. Wang, D.-Y. Yeung, W. Wong, and W. WOO, "Convolutional LSTM network: a machine learning approach for precipitation nowcasting." In: C. Cortes, N. D. Lawrence, D. D. Lee, M. Sugiyama, and R. Garnett, Eds. *Advances in Neural Information Processing Systems 28*, Red Hook, NY: Curran Associates, 2015, pp. 802–810.
29. N. Srivastava, E. Mansimov, and R. Salakhutdinov, "Unsupervised learning of video representations using LSTMs," *arXiv: 1502.04681*, 2015.
30. K. Cho et al., "Learning phrase representations using RNN encoder-decoder for statistical machine translation," *arXiv:1406.1078 [cs, stat]*, 2014.
31. D. P. Kingma and M. Welling, "Auto-encoding variational Bayes," *arXiv:1312.6114 [cs, stat]*, 2013.
32. A. B. Said, A. Erradi, A. G. Neiat, and A. Bouguettaya, "A deep learning spatiotemporal prediction framework for mobile crowdsourced services," *Mobile Networks and Applications*, vol. 24, pp. 1120–1133, 2019.
33. L. Chen, "Predictive modeling of spatio-temporal datasets in high dimensions," Ph.D. dissertation, The Ohio State University, 2015.
34. C. Zhang and P. Patras, "Long-term mobile traffic forecasting using deep spatio-temporal neural networks," *Proceedings of the Eighteenth ACM International Symposium on Mobile Ad Hoc Networking and Computing–Mobihoc '18*, pp. 231–240, 2018.
35. B. Zaman and M. McKee, "Spatio-temporal prediction of root zone soil moisture using multivariate relevance vector machines," *Open Journal of Modern Hydrology*, vol. 4, no. 3, pp. 80–90, 2014.
36. J. Choi, B. J. Reich, M. Fuentes, and J. M. Davis, "Multivariate spatial-temporal modeling and prediction of speciated fine particles," *Journal of Statistical Theory and Practice*, vol. 3, no. 2, pp. 407–418, 2009.
37. E. Y. Chen, Q. Yao, and R. Chen, "Multivariate spatial-temporal prediction on latent low-dimensional functional structure with non-stationarity," *arXiv:1710.06351 [stat]*, 2017.
38. M. Cebecauer, E. Jenelius, and W. Burghout, "Spatio-temporal partitioning of large urban networks for travel time prediction," *2018 21st International Conference on Intelligent Transportation Systems (ITSC)*, pp. 1390–1395.
39. H. Yao, J. Ke, X. Tang, et al., *"Deep multi-view spatial-temporal network for taxi demand prediction,"* p. 8. https://arxiv.org/abs/1802.08714
40. R. M. Cichy, A. Khosla, D. Pantazis, A. Torralba, and A. Oliva, *"Deep neural networks predict hierarchical spatio-temporal cortical dynamics of human visual object recognition,"* p. 15. eprint arXiv:1601.02970
41. X. Ma, Z. Dai, Z. He, J. Ma, Y. Wang, and Y. Wang, "Learning traffic as images: a deep convolutional neural network for large-scale transportation network speed prediction," *Sensors*, vol. 17, no. 4, p. 818, 2017.
42. Y. Liang, S. Ke, J. Zhang, X. Yi, and Y. Zheng, "GeoMAN: multi-level attention networks for geo-sensory time series prediction," *Proceedings of the Twenty-Seventh International Joint Conference on Artificial Intelligence*, pp. 3428–3434.
43. A. Ziat, E. Delasalles, L. Denoyer, and P. Gallinari, "Spatio-temporal neural networks for space-time series forecasting and relations discovery," *2017 IEEE International Conference on Data Mining (ICDM)*, pp. 705–714, 2017.

6 Routing and Mobility Models for Ad Hoc UAV Networks

Jared Helton[1], Qian Mao[1], Immanuel Manohar[1], Fei Hu[1] and Haijun Wang[2]*
[1]Electrical and Computer Engineering, The University of Alabama, Tuscaloosa, AL
[2]Computer Applications, Shanghai Dianji University, Shanghai, China {*: Corresponding author}

CONTENTS

6.1 INTRODUCTION

Along with the increasing use of unmanned aerial vehicles (UAVs) and other mobile units, there has been an increase in designing mobility models to describe the behavior of mobile nodes. UAVs are typically used for reconnaissance missions as well as search and rescue operations. Many of the models that have been developed in the last decade have been focused on specific applications and used the information that is present in the system to dictate intelligent mobility strategies. Recently, there has been a push to create more mobility models that are application independent, so that the model can be used in a wide variety of applications without being chained to one application or another.

In this chapter, three types of applications for the mobility models are identified: reconnaissance, search and rescue. Sections 6.2 and 6.3 describe those three applications, and Section 6.4 delves into a more flexible mobility model strategy that allows for a wide range of manipulations so that users can tailor any application using the model (we call them application-independent models). In each section, the effectiveness of the proposed models compared with other mobility models is discussed.

The first UAV application is reconnaissance swarms. For many applications of mobility strategies, reconnaissance is a prime inspiration. Many of the reconnaissance operations are concerned with military strategies seeking to identify areas of interest (AoIs) in which either a general overview or specific target is the subject of the investigation using UAVs. The missions that are carried out often use numerous UAVs, or a "swarm." Reconnaissance missions are instructed in a way to cover large distances and survey the area within an acceptable time frame, such that the UAVs could collect the data necessary for the mission and transmit their information to a command and control center (C&C). A common example analyzed in this chapter is having UAVs navigate an AoI. The UAVs scan an AoI, move through a tunnel and then exit on the other side of the tunnel that is connected to another AoI, all based on predetermined force equations. Some published studies provided details of reconnaissance applications and investigated the communication issues among UAVs in swarms, and some studied the inherent limitations that come with attempting to maintain continuous communications throughout the mission.

The second and third UAV applications are search and rescue operations. In many of the search and rescue operations, UAVs or multi-robot teams, are used with the knowledge of the network topology that helps to create effective routing procedures. These systems are usually composed of either a unicast or multicast system. The unicast system focuses primarily on coordination-oriented communications between two robots in the multi-robot swarm. The multicast system focuses more on group communications in which close and constant communications among the surrounding devices are needed to successfully complete the mission tasks.

The two main types of data forwarding are coordinate based and data based. Unicast protocols are tuned toward coordinated-based communications with one-to-one commanding and requesting of information, while multicast protocols are more diverse and can occupy both coordinate-based and data-based communications. This flexibility allows multicast systems to command many mobile robots/swarms of UAVs at once, while being proficient in transmitting and receiving random or periodic data.

The final topic we will cover in this chapter is application-independent mobility modes, which are suitable to heterogenous (not homogenous) applications. To adapt to heterogenous communications, the mobility model needs to be able to generate different mobility patterns. Those models achieve this by capturing UAV speed, angular velocity and pitch angles as the Gauss-Markov (GM) process, which uses memory state to determine the next state. The GM process can accurately model real mobility behaviors in airborne networks such as smooth turn and continuous movement. Models that are more flexible and application independent face the challenge of being able to prove that they can be used in a wide variety of situations. The model analyzed here is shown to demonstrate traceable behaviors in predictable cases as well as erratic behaviors in random cases.

The sections are organized as follows. Section 6.2 looks into the reconnaissance-specific applications of mobility models. Section 6.3 summarizes search and rescue applications within mobility models. Section 6.4 delves into an application-independent mobility model, and Section 6.5 concludes the chapter.

6.2 RECONNAISSANCE SWARMS

In this section, a brief overview is first given to introduce the basic mobility models within reconnaissance operations. Within the individual subsections, comparisons are drawn between the different mobility models to highlight similarities and differences between models used for specific reconnaissance missions.

6.2.1 Overview

Different mobility strategies are for specific purposes such as mobile sensor networks, human-centered mobile networks and swarms of unmanned vehicles. For many applications of mobility strategies, reconnaissance is a prime inspiration. Reconnaissance operations can identify AoIs in a large UAV network. The missions that are carried out often require the cooperation among numerous UAVs, called a "swarm." Reconnaissance missions often cover a big area for environment surveillance. They can collect environment data and transmit the information back to a C&C in real time.

Figure 6.1 shows a configuration proposed in [1] that gives a topographical overview of how the UAVs were oriented throughout the simulation. The UAVs scan an AoI by using different sensors, move through a tunnel with a small range and then exit on the other side of the tunnel, which is connected to another AoI, all based on predetermined force equations. The swarm of UAVs is experimentally shown to decrease the amount of time needed to surveil large AoIs compared with a single UAV case.

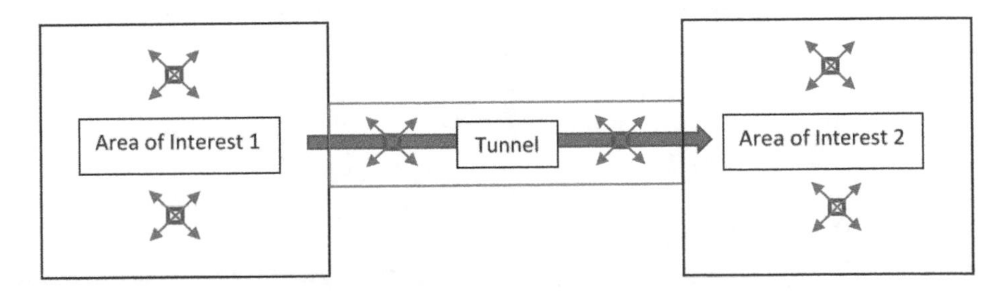

FIGURE 6.1 Example of reconnaissance map used.

6.2.2 Virtual Forces–Based Mobility Model

Falomir et al. [1] focused on the problem of creating virtual forces among sets of UAVs and boundaries within the environment to create an efficient mobility model. Their model was able to perform reconnaissance on a single AoI, move through a tunnel connecting the first AoI to another and move all used UAVs safely to the other side of the tunnel into the second AoI. The mobility model created attempted to seek a balance between attractive and repulsive forces, first among UAVs and then among objects in the environment. The distributions used were based on the safe distances between UAVs as well as necessary forces for UAVs to maintain a swarming shape. Each section of the attractive-repulsive force distribution was either a constant or linear force. Because an object could interact with multiple forces in any given environment, a weighted average of forces was used to calculate the next direction that the UAV would move. The weight for each of the forces was the exponential of its magnitude, similar to spring forces. Figure 6.2 shows how the different forces, highlighted as blue arrows, all interact with the UAV and guide it in a given AoI. The orange arrows show a link to the nearest UAV, the unfilled brown arrow shows the net force guiding the single UAV and the brown circle at the edge of the figure is the target destination. Each UAV would have forces surrounding them determined by their environment.

After all of the distributive rules were established to guide the UAV interactions in the target areas, strategies were established for mobility control. The *surveillance mobility* strategy links two UAVs together, which travel in an "S-shape," to efficiently survey the AoI. Connectivity within the swarm is maintained by each UAV waiting for the 1- and 2-hop neighbors to complete their own missions. After surveillance is completed, the swarm comes together to form a more compact version to align itself with the entrance of the tunnel. Communication links are established by spacing the UAVs within a distance that complies with both transmission and safety limitations. The UAVs then establish a *passage crossing* strategy to move through the tunnel. The passage strategy is in concurrence with attractive-repulsive forces between the surrounding UAVs, the passage itself and objects encountered in the passage.

Falomir et al. [1] then simulated the proposed mobility model in JBotSim between variations of 1–6 UAVs in a swarm and a total of five tunnels. The simulations showed a significant increase in speed for the reconnaissance missions when using six UAVs compared with the 1–5 UAVs. The different types of tunnels were established in

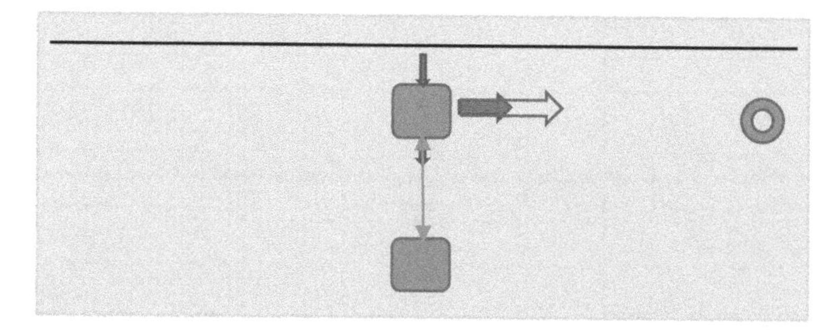

FIGURE 6.2 Attractive-repulsive force interactions on a single UAV in its environment.

a way that some were unique, whereas others were created to show similar trends during tunnel crossings. The only significant performance variations came when increasing the number of UAVs in a way that created additional radio links in the overall swarm, during the sections with linear obstructions, which had to be broken when passing over a corner. Some of the limitations found in the simulations were the lack of ability for the UAVs to communicate between each other to identify obstacles. This obstacle avoidance system was also identified with the pheromone model described in Kuiper and Nadjm-Tehrani [2], in which each drone communicates through pheromones to prevent over searching of the AoI. This is detailed in the following section.

6.2.3 Pheromone Model

In [2], how 10 centrally located UAVs randomly proliferate into a wide area to effectively scan the whole region is described. Two different mobility models were analyzed: the first model sends out each UAV with a random distribution for each movement after the UAV had reached an area for surveillance, and the second model sends out each UAV from a centrally located area but with pheromones attracting and repulsing the subsequent movement of each of the 10 UAVs. They have used the idea of "pheromones" (which are actually special network messages) exchanged among UAVs. Such an idea originates from the behaviors of certain insects, like ants, which interact while searching for food.

In [2], the first system is a fairly straightforward one that implements a Markov process to determine the probability of a UAV turning left, right or proceeding forward toward the boundary of the AoI. The motivation of using a Markov process is to use a historical state to estimate the future state. For example, if we know which direction a UAV is already headed in one time window after reaching a flight goal, then the direction that the UAV will go next should follow a similar trend. The Markov process assigns discrete probabilities to each movement state. The next state can be determined by using the state transition matrix.

In [2] the second system uses an attractive-repulsive pheromone model that could direct a UAV toward or away from a target area. This model is used throughout the AoI originating from boundaries and individual UAVs to prevent any overlap between UAVs' areas of the region that had already been covered. The mobility model requires that each UAV looks left, right and center of its location and summarizes a total of the pheromones to determine the next movement. The movement can be determined by comparing the individual pheromones from the cardinal directions with the total sum of the pheromones. The direction that had the least number of pheromones compared with the sum should have the highest probability of movement.

Both the pheromone and random mobility models were simulated for 2 hours, and they compared the results of the coverage of the area and the probability distribution of the time between scans. It showed that the pheromone model had a coverage level significantly better than the random mobility model, with the pheromone mobility model covering 90% of the area over an hour compared with the 80% coverage over 2 hours performed by the random mobility model. The paper also concluded that the pheromone mobility model was more uniformly distributed than the random case.

The Markov model could enable the UAVs to scan the AoI efficiently. The limitations from the pheromone model, and the testing process as a whole, were in the form of rigidity. The model was assumptively simplistic to the point that the movement and communication were simplified to demonstrate a phenomenon but not emulate a real UAV mission. They claimed to use more dynamic movement models in the next step of research.

6.3 SEARCH AND RESCUE

This section will first provide an overview of mobility models used in search and rescue applications. It outlines a primary distinction between two types of models used for efficient routing protocols, *unicast* and *multicast* routing. Both systems have advantages and disadvantages, and both are built to be used in multi-robot teams to successfully carry out missions specifically related to search and rescue operations. The first section gives a brief overview of the material, while the following two sections detail the unicast and multicast routing procedures and their abilities.

6.3.1 Overview

In many search and rescue operations, UAVs or multi-robot teams are used with a foray of knowledge that helps to create effective routing procedures. In Das et al. [3], two main procedures were considered, built and tested—a unicast and a multicast system. The unicast system focuses primarily on coordination-oriented communications between two robots within the multi-robot swarm. The multicast system focuses more on group communication applications where high-quality communication among robots are needed to successfully complete the operations.

Figure 6.3 illustrates a taxonomy of the different roles that are played within unicast and multicast routing systems. The two main types of communications are

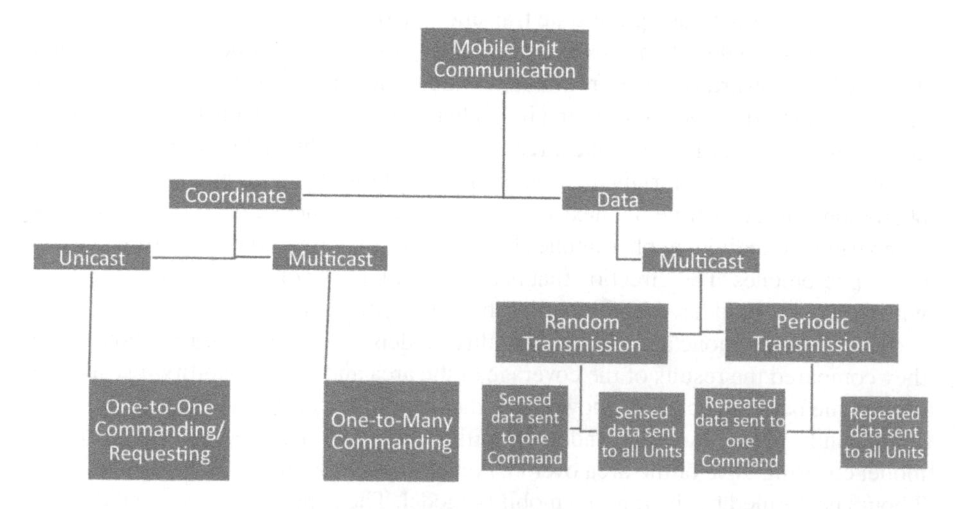

FIGURE 6.3 Breakdown of UAV communications.

coordinate based and data based. Unicast protocols are tuned toward coordinated-based communications with one-to-one commanding and requesting of information, whereas multicast protocols are more diverse and include both coordinate and data-based communications. This flexibility allows multicast systems to be able to command mobile robots/swarms of UAVs at the same time, while being proficient in transmitting and receiving both randomly generated and periodic packets. These two models were created and tested to see how effectively they could be used for routing procedures in communications in search and rescue operations of multi-robot swarms, and the goal is to achieve low-overhead and energy-efficient communications.

6.3.2 Unicast Routing

Most applications with search and rescue operations are dependent on reliably sending coordinates to the UAVs or robots within the swarm. The coordinates can be most effectively relayed from one UAV to another within the swarm without overall group communication to reduce overhead and delays for effective area coverage in a given mission. The communication does not have to be limited to the relay of coordinates. Other forms of effective communication for unicast routing include sending data in the form of images for other robots in the swarm to process and use, as well as asking for assistance between robots to navigate the AoI most effectively.

Unicast routing is extremely effective for the coordinate-based communication. But it has a disadvantage when compared with multicast routing, it only allows communications from one node to another; thus, when trying to relay information between multiple nodes, unicast routing is not an optimal system.

The effectiveness of a unicast routing system in a search and rescue application was tested in [3]. They focused on two different routing protocols, mobile robot distance vector (MRDV) and mobile robot source routing (MRSR). Both models are described in the following section.

6.3.2.1 Mobile Robot Source Routing (MRSR)

The mobility strategy is concerned with three major aspects: route discovery, route construction and route maintenance. Figure 6.4 breaks down the major aspects involved in completing a connection. Route construction is mostly handled at the source or destination node, so most of the connection and maintenance operations are managed by exchanging control messages between the source and destination. The two main control messages are labeled as ROUTE REQUEST (RREQ) and ROUTE REPLY (RRER).

The routing protocol is first concerned with route discovery, which aims to find an end-to-end data delivery path. A packet labeled RREQ, which contains the address of the initiator of the request, the target of the request and a route record, is sent out across the network in a controlled flooding process.

To determine how frequently the request should be sent out, a probability variable p_r is defined. A lower p_r helps to reduce energy consumption and routing overhead. Its calculation depends on the distance weighting factor γ, which is formulated based on the wireless transmission range R and a value known as the distance to rest d_{rest}, which

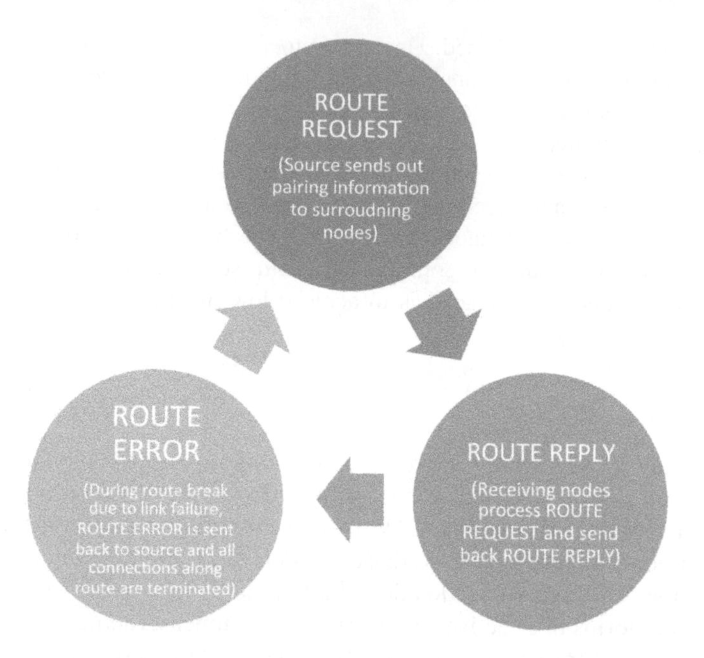

FIGURE 6.4 MRSR routing procedure.

is calculated as $(D - d)$. Here the value of D is the total distance that the mobile robot will travel, and the value of d is the distance that a robot already traveled. The formula governing the weight factor γ is shown below:

$$\gamma = \begin{cases} \dfrac{d_{\text{rest}}}{R/2} \; if \; d_{\text{rest}} \geq R/2 \\ 0 \text{ otherwise} \end{cases}$$

With γ defined, p_r is calculated to be inversely proportional to the the velocity of the mobile robot in such a way that a highly mobile robot will have a lower probability of rebroadcasting the routing package. The distance weighting factor γ helps to drive the probability value in a relation such that as it is increased, p_r decreases exponentially. In this way, p_r is set to 1 when γ equals 0, as described by

$$p_r = \min\left(1, \left(\frac{1}{v}\right)^{\gamma}\right)$$

After the probability value is determined and the request is received, a subsequent RRER packet is sent back. This packet can come from either the node where the mobile robot will reach its destination or an intermediate node that serves as a stepping stone for the robot to eventually reach its destination where the route is stored in a cache. The MRSR technique requires that each mobile robot encode its mobility information into the reply packet, which consists of an estimated timeout value to determine when the device will be out of range and a single bit to tell whether the robot is moving or paused. The time-out value is calculated from the flag bit and the current speed of the robot so that the time-out can be used to efficiently manage and utilize routes.

The second step in the routing protocol is route construction. The way that MRSR achieves this aspect of the routing protocol is similar to the dynamic source routing (DSR) protocol, in that it uses aggressive caching to lower the frequency of the routes as well as the propagation of their discoveries.

The MRSR takes the distinct links of the routes to create an overall graph of the network. To build this topological graph, each mobile robot connected to the network maintains its graph cache based on its discovered link information as well as the forwarded link information from other mobile robots.

A simple, single-source shortest path algorithm like Dijkstra's algorithm is used by the mobile robots to compute source routes. To determine whether a link is broken, MRSR checks the time-out values from the RRER packets that were generated in the route discovery phase of the protocol. The broken links are removed in such a way that all other links tied back to an end point of origin could find out that their time-out values are reduced. MRSR attempts to isolate a path with the longest lifetime that has been provided from other mobile robots, and such a path specifically corresponds to the highest time-out of the links.

The final process used in MRSR is called route maintenance. Once the routes have been discovered and constructed, maintenance is used to monitor the operation and functionality of the routes in use. A ROUTE ERROR packet is sent from a detecting host that identifies a route break or a link failure. When this error occurs, the host eliminates all of the routes in its cache that might use that link.

6.3.2.2 Mobile Robot Distance Vector (MRDV)

The second type of *unicast* routing scheme is the MRDV routing protocol [3]. This protocol is similar to MRSR with distinctions in the route discovery aspect. It is primarily based on a routing protocol known as ad hoc on-demand distance vector routing (AODV), and it is broken down into two processes known as route discovery and route maintenance.

The first process of the routing protocol is route discovery. To perform this function, MRDV proceeds in a similar way to MRSR. The main difference between the two is that MRDV only stores its routing information as one entry with every destination, whereas MRSR uses a cache of information to determine the best route to a destination. ROUTE REQUEST and ROUTE REPLY packets are also used, but reverse paths are established for MRDV to determine whether or not to rebroadcast the request packets.

Time-out values are set to indicate how long each node will remain a part of the route, and MRDV only uses one time-out value in the request packet instead of performing source routing as in MRSR. The link with the lowest time-out decides the circumstance of link breakage, and with each check along the nodes the weakest link determines the final time-out value encoded into the ROUTE REPLY packet sent back to the source node for calculating a routing table.

The final process in the routing protocol MRDV is route maintenance. The main focus of this procedure is to use timers to eliminate routes that have not been used for a long time. Because MRDV only encodes one route, the protocol avoids route errors by allowing the route to expire at appropriate times. In the same way that MRSR maintains the routes, if a discovery fails, a packet is sent out again to ensure connectivity.

6.3.2.3 Comparison of MRSR and MRDV

To test and validate the system, Das et al. [3] used a four-metric system aimed at isolating points of interest in a nominal system that these protocols would be used for in a search and rescue mission. The four main metrics used were *routing overhead*, *packet delivery ratio*, *average delay* and *energy consumed.* All four mobility strategies were compared with each other, but specifically the MRDV protocol system was compared with a common routing strategy called AODV, and MRSR was compared with DSR.

When the network was static, the pause time equals to the time needed to complete a task and the performance of MRDV and AODV was the same. In the cases where the network is mobile, MRDV outperforms the AODV model significantly. There is a 50% reduction in *routing overhead* without any loss of performance in *packet delivery ratio.* The other point of comparison comes in *energy consumption*, where MRDV has an energy saving of 30% in high-mobility scenarios and 18% in low-mobility ones.

When looking at the performance of MRSR against DRR, the trend of static network performance arises similarly to the comparison of MRDV and AODV. The higher mobility causes big differences. In these cases, MRDV achieved a 30% reduction in the *routing overhead* along with a 12% reduction in medium mobility scenarios. The delay between the two strategies was similar, but MRSR outperformed DSR in most scenarios when pause time was increased. In terms of energy consumption, MRSR achieved a 12% energy savings in high mobility and a 5% energy saving in medium mobility, compared with DSR.

For the comparison between MRSR and MRDV, two network sizes are used, 50 and 100 nodes. In the 50-node case, the performance is similar. MRSR outperforms MRDV with a lower overhead, but MRDV has a lower delay than MRSR.

TABLE 6.1
Comparison of MRSR to DSR, and MRDV to AODV [3]

	Routing Overhead	Packet Delivery Ratio	Average Delay	Energy Consumed
MRSR (DSR)	30% reduction in high-mobility models	Similar performance to DSR	Similar performance to DSR	12% savings in high mobility/5% in medium
MRDV (AODV)	50% reduction in mobile models	Similar performance to AODV	Similar performance to AODV	30% savings in high mobility/18% in low mobility

The main differences come when the network size is increased to 100 nodes. In this case, MRDV outperforms all other mobility strategies by significantly increasing energy efficiency while having much more manageable *average delay* and *packet delivery ratio* values. Overall, both MRSR and MRDV outperform DSR and AODV, and they have similar results to each other in smaller to medium-sized networks, while MRDV provides a significantly better option as a routing strategy in large networks.

6.3.3 Multicast Routing

For swarms of multi-robot teams, multicast routing creates an efficient system that allows group communication for multiple purposes. Multicast routing is most effectively used for sending out information to multiple nodes. This information could be the data used to navigate the AoI, or the specific instructions from a centrally located command center detailing specific flight plans.

There are three primary scenarios in which multicast routing is effective in multi-robot operations. The first scenario is to utilize the resources of a few mobile robots for the benefit of the entire swarm. In certain applications, instead of loading every mobile robot with bulky or cost-ineffective sensors, a few mobile robots in the swarm can be used to collect very specific data that can be sent to the rest of the swarm to more effectively cover an AoI. A good example is to put sophisticated Global Positioning System (GPS) hardware on a few nodes and broadcast landmarks that the rest of the swarm should avoid. The second scenario concerns having human operators managing the swarm. Human operators can bring in a source of information outside the scope and ability of a multi-robot group and help target specific sections of an AoI to complete a search within a tighter restraint of time. The third scenario is a sensor network with specific controllers to guide operations. This can be best envisioned by considering a small team of robots with specific tasks relaying information to a commanding mobile robot, which allocates subsequent tasks to other teams of robots that depend on the success of the original small team of robots. Overall, multicast routing presents a large upside for sensor-like communication between multiple nodes but with the disadvantage of higher overhead and slower routing procedures between the communication nodes.

To understand the effectiveness of multicast routing protocols, Das et al. [3] tested a system called mobile robot mesh multicast (MRMM) for deployment in mobile robot networks. This strategy is based on the On-Demand Multicast Routing Protocol (ODMRP), which is specifically made for mobile ad hoc networks (MANETs). MRMM follows closely to ODMRP, but the significant design differences come in creating a strategy to be more reliable and efficient in creating and maintaining meshes for sending/receiving data. The strategy is built on two major phases, *mesh construction/maintenance* and *data delivery*.

The mesh construction aims to create a subset of robots within a larger mesh network that can efficiently forward packets with as little necessary redundancy as possible. When a source wants to send out information without a mesh setup, it floods a JOIN QUERY control packet that proliferates throughout the network. Each node that receives the packet stores the source ID for detecting duplicates, and a routing table is updated with the source node ID to set up a reverse path back to the source. Once the control packet is received, a JOIN REPLY is sent from each node to neighboring nodes in the mesh. IDs are checked in each node and when they align, paths are established back to the source by setting a forwarding group (FG) flag, which creates a mesh of nodes known as a forwarding group. Each node in the mesh exists with a soft state, so the JOIN QUERY control packet has to be continuously transmitted to maintain an efficient network.

The control process for MRMM is the same as for ODMRP, but the difference lies in the fact that MRMM has mobility knowledge that is used to create a mesh pruning algorithm, which can more effectively select nodes to be in the FG, based on the mobility information of each node. Similar to the unicast systems, MRMM uses a probability value, p_i, to determine whether or not each node will rebroadcast the JOIN QUERY packet sent from the source. The variables are similar to the unicast system above, except a new variable, ω, known as the mesh sparsness factor, is used to control how aggressively nodes are removed from the forwarding group. When ω is large, p_i is small and the mesh will be sparse. As a precaution, MRMM has nodes that want to leave or join a group and can flood JOIN/LEAVE messages so that information for group membership is known throughout the network. The source node uses this information to collect all available nodes in the network before running the mesh pruning algorithm.

To ensure that the data are delivered successfully, MRMM operates similarly to ODMRP in that each node that has not expired from time-out sends out a received non-duplicated packet. Where MRMM is unique is that it also checks whether it is within an acceptable distance from the mesh to be able to send out the information effectively, which helps with data forwarding efficiency. As described previously, MRMM takes advantage of reverse multicast, or many-to-one communications, to create a mesh that can report all of its readings back to a central node, which is the source node in this case.

The main processes of creating efficient routing protocols for MRMM are labeled in Figure 6.5. A similar packet transmission style to the unicast routing mobility models is used in MRMM, but the main focus lies more on creating a mesh of nodes instead of individualized communications. The main packets transmitted for communication are the JOIN QUERRY and JOIN REPLY transmissions for the establishment of the meshes. A mesh pruning algorithm running at the core of MRMM is used to establish the most efficient communication channels between the mesh nodes and the source.

6.3.4 Comparison of MRMM with ODMRP

To evaluate how successful MRMM is compared with its predecessor ODMRP, four criteria are established similar to the evaluation techniques created for the unicast communication systems. The four criteria are *control overhead, multicast delivery ratio (MDR), average delivery latency (delay)* and *forwarding efficiency*. Because the two forms of communication operate distinctly, MRMM was compared only with ODMRP and not the unicast procedures outlined in the previous sections. With the four criteria, three different factors are tested: mobility, group size and number of traffic sources.

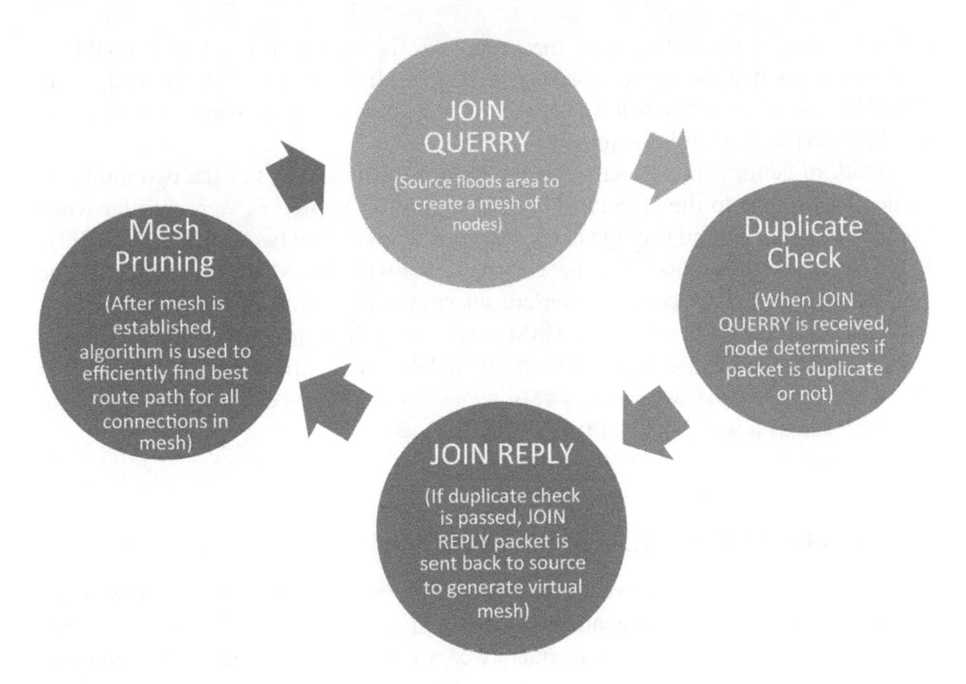

FIGURE 6.5 MRMM routing procedure.

The first factor is on the effect of mobility. ODMRP and MRMM are primarily compared, but another special case of MRMM with random pruning was evaluated where the nodes were identified to randomly rebroadcast the JOIN QUERY packets with a fixed probability of 0.5. Under the criteria of control overhead, both MRMM methods were shown to be significantly better than the ODMRP strategy.

In scenarios where there was continuous movement, MRMM had an overhead that was 44% lower than ODMRP. This is highlighted even more when MDR is taken into consideration. Even though the control overhead was significantly reduced in MRMM compared with ODMRP, the proposed strategy delivered just as many multicast packets as needed. When looking at the delay comparison between MRMM and ODMRP, both had comparable delays around 0.01 seconds for pause times ranging from 0 to 500 ms. When comparing forwarding efficiency, MRMM outperformed ODMRP: MRMM has 27% better efficiency at 100-ms pause time and 42% better efficiency at 0 pause time. The main comparison between the two MRMM strategies, one with mobility information and one with random pruning, is performed in terms of control overhead and forwarding efficiency. In these two criteria, the mobility-based MRMM outperformed random pruning because random pruning had a larger tendency to make the mobility mesh sparser.

The second factor is group size and its impacts on MRMM and ODMRP. The two mobility strategies were similar when compared in the criteria of MDR and average delay. The MDR was around 100%, and the average delay was around 0.01 seconds for both strategies. The primary difference came in the criteria of control overhead and forwarding efficiency. The MRMM strategy consistently had lower control overhead no matter what the group size is used compared with ODMRP. Both strategies had the higher control overhead with the increase of group size, but MRMM started at a much lower value. When considering forwarding efficiency, MRMM was significantly better than ODMRP.

The third factor is the effect of the number of traffic sources of the two multicast strategies. Similar to the effect of group size, the two strategies were similar when the criteria of MDR and average delay were considered. The two criteria were 100% for MDR and 0.01 seconds for the average delay with MRMM and ODMRP. The significant comparison comes when considering control overhead and forwarding efficiency. For control overhead, MRMM was once again significantly better, having a 40% reduced overhead compared with ODMRP. When comparing the forwarding efficiency between the two multicast strategies, MRMM was once again superior to ODMRP across a wide array of traffic sources.

6.4 APPLICATION-INDEPENDENT MODELS

This section first gives an overview of a specific mobility model that provides a resource for application-independent airborne networks. The mobility model is generalized using GM processes to adequately capture realistic movement for UAVs by using old memory to calculate new states.

6.4.1 Overview

Unlike previous sections, the mobility model covered in this section is outlined as heterogenous (not homogenous) in UAV applications. To be heterogenous, the mobility model needs to be able to generate multiple mobility patterns. One such model that has found success in creating a heterogenous mobility model is described in Anjum and Wang [4]. This model does so by capturing speed, angular velocity and pitch angle as a GM process, which allows memory to be a factor in determining the next state. The use of a GM process allows this method to accurately model real behaviors in airborne networks like smooth turn and continuous movement.

Other models are either limited to some specific application, or they do not have the capability of performing smooth operations in movements. These different mobility models include, but are not limited to, the random waypoint mobility model, smooth-turn mobility model, semi-random circular movement mobility, and three-way random mobility. By altering a selection of variables we could generate a set of mobility patterns that span numerous applications. Those models are shown to demonstrate traceable behaviors in predictable cases as well as erratic behavior in random cases.

6.4.2 Mobility Model

To create a comprehensive mobility model for an airborne network, it is essential to be capable of understanding movement in 3D space, be able to consider angular velocity for smooth-turning, be able to link past states and current states to avoid sharp changes in speed or direction, be able to deal with boundary effects and be flexible and not application specific. The two main drawbacks of 3D MANET mobility models are (1) sudden movement and change of directions and (2) that the models possess inappropriate aerodynamics. The first problem can be solved by using GM models, while the second issue can be solved by adequately assessing aerodynamic properties, such as angular velocity, turn angle, turn radius, turn center, pitch angle, vertical velocity and vertical displacement. The GM model is applied to the velocity, angular velocity and pitch angle of the mobility model. The high-level specifics are in the form of GM memory coefficients, GM variances and state intervals.

The main variables of interest are the velocity, angular velocity and pitch angle. Each of these variables are quantified with the GM process with a variance σ and degree of memory ρ. The first main interest, velocity, is detailed as follows:

$$v(t) = \rho v(t-1) + (1-\rho)\overline{v} + \sqrt{1-\rho^2}\,\Gamma_{v(t-1)}$$

In this equation $v(t)$ is the velocity with respect to time, $\overline{v}$ is the mean velocity and $\Gamma_{v(t-1)}$ is the Gaussian random variable. To illustrate this, when $\rho = 1$, the movement of the current state follows exactly to the movement of the previous state, and when $\rho = 0$ the movement is completely independent and memoryless. The velocity equation gives a way to create a variety of controlled or random cases. The probability coefficient is an important parameter. Another important variable is the angular velocity. This variable is similar to the velocity by being modeled as a GM process as follows:

$$\omega(t) = \rho\omega(t-1) + (1-\rho)\overline{\omega} + \sqrt{1-\rho^2}\,\Gamma_{\omega(t-1)}$$

This equation follows the velocity equation, with the difference that the variable of interest now is ω, which is the angular velocity. To get a feeling for the equation, to achieve sharper turns, $|\overline{\omega}|$ must be large, and to achieve straighter trajectories, $|\overline{\omega}|$ must be reduced. Concerning direction, a positive value for ω indicates a counterclockwise turn, whereas a negative value indicates a clockwise turn. The angular velocity equation is similarly a function of the probability coefficient, which allows a wide range of applications to be tested. The final variable determined by a GM process is the pitch angle. This variable, similar to the velocity and angular velocity above, can be derived in a similar process as follows:

$$\theta(t) = \rho\theta(t-1) + (1-\rho)\overline{\theta} + \sqrt{1-\rho^2}\,\Gamma_{\theta(t-1)}$$

The familiar equation is now focused on the azimuthal angle θ. The pitch angle does cause the velocity to split into two major components, which are the vertical direction along the Z-axis as well as the horizontal direction along the Z-axis. This final equation for the angle of the UAV allows a similar variability as the velocity and angular velocity equations.

The velocity, angular velocity and pitch angle are not the only variables of interest, and other variables are detailed below. The turn radius, r, is equated by taking the division of the vertical velocity against the angular velocity. The turn duration, τ, is determined by taking the time between each state when updating the dynamic variables. It depends on a variable called the *mean turn duration*, λ. As an illustration, a larger value of λ indicates a longer turn and circular motion, whereas a smaller value means a very quick change in dynamic variables. The turn angle, $\alpha(t)$, is the derivative of the angular velocity. Finally, the direction of motion $\phi(t)$ is evaluated as the subtraction of the turn angle from the previous direction of motions.

A common point of interest in mobility models is in considering what happens at a boundary. In an airborne network, a UAV may be unable to enter a location due to some reasons, such as a physical landmark or man-made landmark like a no-fly zone. Because of this restriction, it is necessary to consider boundary effects in this mobility model. The method that deals with the boundary effect creates an effective boundary area that lists a physical distance from which a UAV would need to slow down. As the airborne network approaches the boundary, it either turns clockwise or counterclockwise, depending on the angle when crossing the border before the redirection.

Stochastic variables are commonly used in mobility models to evaluate its performance. Anjum and Wang [4] detail the necessary stochastic variables needed to evaluate the performance of a network.

6.4.3 Simulations

After defining the above variables, the mobility strategy was evaluated in a single environment with nine different cases that explored a variety of situations. They showcased the flexibility of the mobility strategy by altering various variables described above. The changes made in each of the cases are described in Table 6.2. As an example, the second simulation differs from the first by altering parameters concerned with the angular velocity and its variance, while simulation 9 differs from simulation 7 to highlight the changes within the mobility model that the GM coefficient can specifically change. The cases that are related back to the original test case just have the subsequent minor changes. Most of the tests have a velocity of 100 m/sec, angular velocity around 0.001 rad/sec, pitch around 0, variance of velocity of 1, variance of angular velocity around 0.01, mean turn duration between 1 and 10 seconds and GM parameter around 0.7.

Each of the test cases were simulated to demonstrate a particular mobility model and the characteristics associated.

The first simulation demonstrated a fairly straightforward flight plan but had a significant variation in the vertical axis, due to the variance of the pitch angle being equal to 0.01 rad.

The second simulation varied from the first with respect to the angular velocity, which changed from 0.0002 to 0.001 rad, and the angular velocity variance, which changed from 0.00001 to 0.01 rad. These alterations simulated a flight path that had sharper turns and less straightforward movement.

TABLE 6.2
Summary of Application-Independent Test Cases

Case Number	Summary	Results
1	Tests base case for general model	Trajectory is fairly straight with slight curves at turning points
2	Angular velocity is increased from case 1	Trajectory is more erratic with sharper turns
3	Angular velocity is increased from case 2	Trajectory does not deviate as much from center path but resembles case 2
4	Mean turn duration is increased from previous test cases	Trajectory has smoother turns and is more circular
5	Mean turn duration decreased, angular velocity decreased and pitch angle increased	Similar trajectory to case 3 but with larger, non-circular loops
6	Mean turn duration decreased from case 5	Trajectory is straighter than case 5 with less loops
7	Tests what happens when the probability coefficient goes to one (controlled case)	Trajectory is a series of symmetric loops descending proportional to the pitch angle
8	Mean and variance of angular velocity are negligible	Trajectory is dominated by straight paths descending proportional to the pitch angle
9	Tests what happens when the probability coefficient goes to zero (random case)	Trajectory is random in nature and follows no distinct path

The third simulation increased the angular velocity to 0.01 rad/sec as a change from the second simulation, and the most significant change is that the loops of the UAV are smaller in size.

The fourth simulation increased the mean turn duration to 100 seconds compared with the 5 seconds simulated in the third simulation. This increase caused the model to turn for a much longer time period with a constant radius that resulted in a much more circular flight path.

The fifth simulation carried many changes from the previous simulation set as characterized: the mean turn duration decreased to 10 seconds, the angular velocity decreased to 0.001 rad/sec, and the pitch angle increased 0.1 rad. The impact of all of the changes created bigger loops in the flight path with the loops following less of a circular path.

The sixth simulation varied the mean turn duration to 1 second from the previous set, and the primary change was detailed by a straighter movement path.

The seventh path evaluated how the GM coefficient effected the mobility by increasing the coefficient to 1, which eliminates randomness. With the coefficient being set to 1, the velocity, angular velocity, and pitch angle are all constant, and the motion is similar to a spiral decreased in the *Z*-direction.

The eighth simulation reduced the angular velocity and its variance to negligible values, while setting the mean turn duration to 10 seconds and the GM coefficient back to 0.7.

The final simulation set the angular velocity, its variance, and the mean turn duration to the seventh simulation parameters, but it altered the GM coefficient to equal 0. Setting the coefficient equal to 0 creates an utterly random system, and the result in the mobility created neither a straight nor circular path but a random natured path.

6.5 CONCLUSIONS

In this chapter, three different scenarios were identified and evaluated, i.e., reconnaissance, search and rescue and application independent. Within the first two applications, multiple routing methods were identified, their performance was compared with each other and the mobility strategies are described. The application-independent case targets a generic, flexible mobility model strategy that generated a large number of variables so that a list of different applications could be modeled. Each section created its own justified criteria for evaluation and showed significant improvement when compared with earlier strategies on which the proposed mobility models were based. The application-independent mobility model introduces flexibility in tailoring specific applications from a large variance of possible applications as shown in the testing of nine different cases.

REFERENCES

1. E. Falomir, S. Chaumette, and G. Guerrini, "Mobility strategies based on virtual forces for swarms of autonomous UAVs in constrained environments," *14th International Conference on Informatics in Control, Automation and Robotics*, 2017.
2. E. Kuiper and S. Nadjm-Tehrani, "Mobility models for UAV group reconnaissance applications," *2006 International Conference on Wireless Mobile Communications*, IEEE, pp. 33, 2006.
3. S. Das, Y. C. Hu, C. S. G. Lee, and Y.-H. Lu, "Mobility-aware ad hoc routing protocols for networking mobile robot teams," *Journal of Communications and Networks*, vol. 7, pp. 296–311, 2007.
4. N. Anjum and H. Wang, "Mobility modeling and stochastic property analysis of airborne network," *IEEE Transactions on Wireless Communication*, vol. 7, no. 3, pp. 1282–1294, 2020.

Part III

Communication Protocols

7 Skeleton Extraction of Routing Topology in UAV Networks

Zhe Chu[1], *Lin Zhang*[1], *Zhijing Ye*[1], *Fei Hu*[1], *Bao Ke*[2]
[1]Electrical and Computer Engineering, The University of Alabama, Tuscaloosa, AL
[2]Electrical Engineering, Milligan College, TN

CONTENTS

7.1 INTRODUCTION

Unmanned aerial vehicle (UAV) networks need an efficient routing scheme to form any swarming shape. Such a routing scheme exchanges the node profile information [moving speeds, locations, quality-of-service (QoS) requirements, etc.] among UAVs to speed up the swarming process. To simplify UAV group control and decrease communication complexity in a UAV network, dynamic hierarchical routing schemes are needed. Hierarchical routing is a classic routing with the goal to manage large-scale UAV networks and decrease routing table size in each node. It first separates the nodes into different groups based on some type of criteria such as node proximity and task synchronizations. The routing process will find the group IDs to traverse each time instead of going through each individual node.

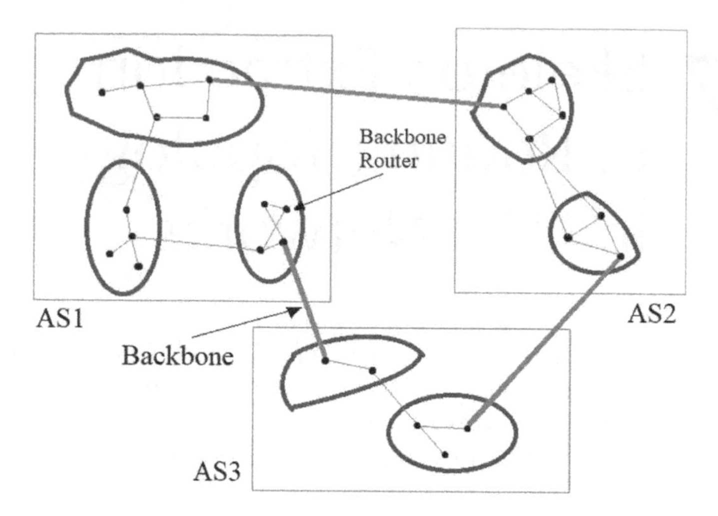

FIGURE 7.1 Internet with hierarchical structure with ASes and backbones.

Internet routing was built in a hierarchical style. Figure 7.1 shows the multilevel Internet structure. Users or customers are first separated into different areas [called autonomous systems (AS)] based on their physical locations or network link states. Several network areas are connected with an Internet backbone. Those areas form one AS. Several ASes can share the same upper-level backbone, which are shown as bold lines in Figure 7.1. Specific routing protocols can operate in different ASes. In Figure 7.1 there are actually three route levels, i.e., intra-area, inter-area, and inter-AS. They are responsible for transferring packets within the same area, from an area to the backbone or between different ASes (via backbone), respectively. This hierarchical structure works well in wired Internet, which use cables or optical fiber links. However, it cannot be directly used for UAV networks because the backbone is not easy to identify.

To build such a hierarchical structure in UAV networks, the skeleton of a UAV network needs to be identified first. Here the skeleton is defined as the contour that reflects the approximate shape of the whole UAV swarm. From a geometry viewpoint, such a skeleton often represents the median axis of the entire shape. It is typically located in the core area of the network so it has the most stable routing topology. In other words, the nodes located in the skeleton do not move as much as the nodes in the marginal areas during the swarming process. Hence, a virtual backbone of the UAV network can be established by using the skeleton nodes, and a hierarchical routing topology can be formed.

Figure 7.2 shows a general ideal of a UAV network by utilizing hierarchical routing structure. In this figure, the first-level routers (which are special UAVs located in the main skeleton sections) are located in the "trunk" (red). The second-level routers are located in the branches of the trunk (brown). Other UAVs (blue) use the second-level routers to reach the first-level routers. The benefits for such a multilevel routing structure are straightforward: it is very easy to determine the communication routes by just searching the closest skeleton UAVs.

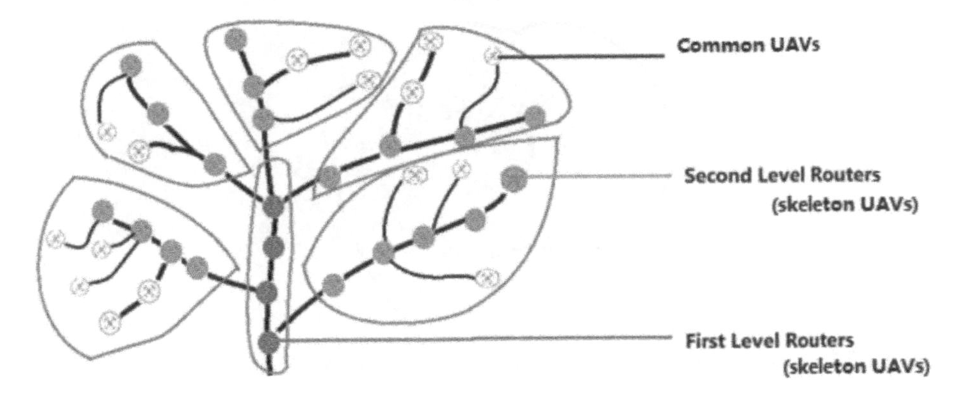

FIGURE 7.2 UAV network with hierarchical routing structure.

This chapter aims to establish such a UAV virtual backbone, given a group of UAVs. Our approach is based on the principle of computer graphics. We will first go through the process of generating a trunk/branch skeleton structure. Then, more details will be given about the hierarchical routing establishment. We will also compare with previous ad hoc routing protocol like optimized link state routing (OLSR). Details about physical and media access control (MAC) layer protocols will also be discussed.

7.2 UAV SKELETON AND SEGMENTATION

This section will first describe how we use computer graphics algorithms to figure out network boundaries. Then we will use boundary information to find the joint positions that are located in the transition areas between neighboring swarm regions. Those joints are often the locations of traffic bottlenecks during data transmission. Those boundary and joint locations can be used to find the skeletons of the entire network. Then, virtual backbones can be created in those skeleton positions, and trunk/branch structure is thus formed. Different levels of routers can be assigned to trunk/branch nodes.

7.2.1 Boundary Detection Issues

To find the framework of the entire network, the coordinates of boundary nodes in 3D networks need to be found first. This is because the joint and skeleton detection algorithms need the information of 3D coordinates to find the median axis of the whole shape based on the boundary nodes in the 3D surface. The boundary nodes are the nodes located at the network surface or the perimeters of holes [1]. To detect the boundary, surrounding nodes' density of each node can be used. For node *n*, a broadcast message will be sent to detect the hop IDs of each neighbor. Then the nodes that received the message return the responses. Figure 7.3 illustrates such an ideal. Node A with enough neighbors around it is not considered a boundary node. In contrast, node B is located at the boundary. The radio range and average distance between nodes could affect the boundary detection result [2] shows a small radio

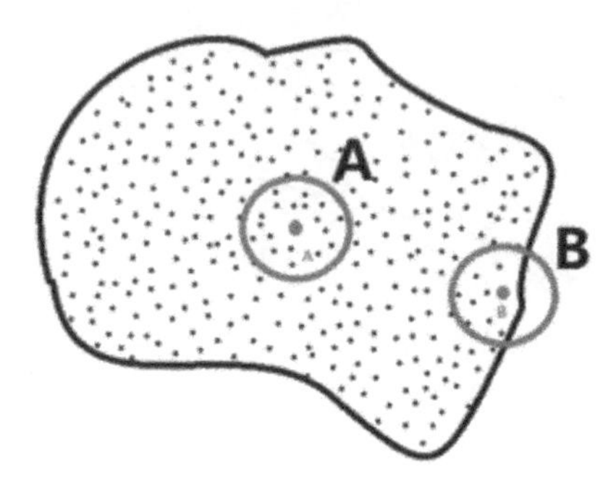

FIGURE 7.3 Boundary detection in wireless sensor networks.

range that may make the boundary detection more difficult than the long radio range case.

7.2.2 Nodes' Virtual Coordination in The Surface Network

The second step is to find the coordination of each surface node. Note that the UAV network we target here is GPS free, which means that we can only find the approximate, relative coordinates of each boundary node. The real Euclidean distance between two adjacent sensors can be approximately measured by received signal strength (RSS) or time difference of arrival (TDOA), whereas the real distance between two remote nodes is often estimated by their shortest path [3]. In 2D networks, it is sufficient to estimate the approximate relative coordinates of each node via mathematical geometry, such as trilateration theory. GPS also uses such a theory to compute the coordinate, but it has accurate distance measurement via satellite signals. Triangulation is a popular location measurement method by using either angles or distances.

However, utilizing trilateration in 3D space may generate multiple solutions. Figure 7.4(a) shows three distances between nodes, and two results on the location

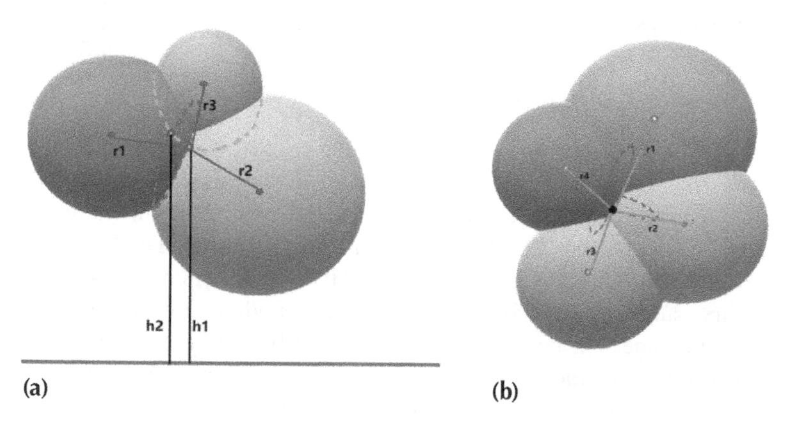

FIGURE 7.4 (a) Trilateration in 3D space generates two solutions. (b) Coordination calculation with four known nodes' distances.

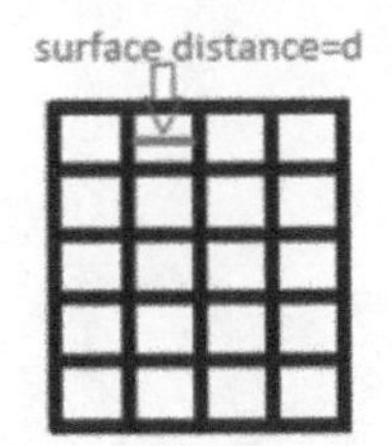

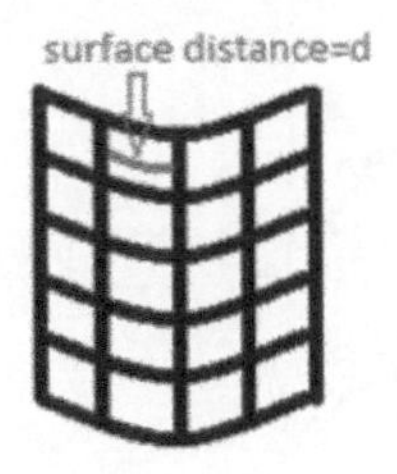

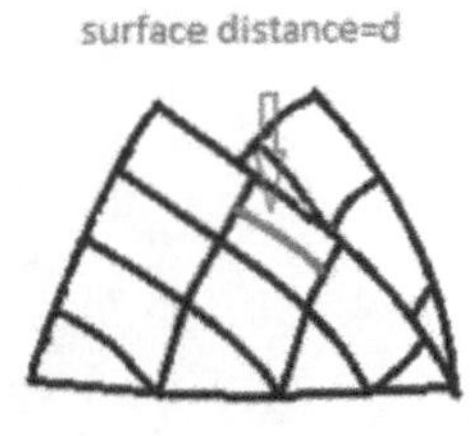

FIGURE 7.5 3D surface with surface distance constraints.

of the node are generated (violet dots). Hence, the trilateration method should be extended via the quad-points calculation method, as illustrated in Figure 7.4(b). Four neighbors' distances need to be known to calculate the unknown node's position. Only one solution can be identified in most cases.

However, the real world is complex. Both RSS and TDOA could have large measurement errors. The measured distance between two nodes on a 3D surface may not be Euclidean distance; instead, they may be surface distances. A 3D space with only surface distances between given nodes are unsolvable because there will be non-single solutions for the system. Figure 7.5 depicts a system with surface distance information given, and there are multiple solutions for this system's shape.

To overcome the above issues and minimize the final localization errors, one solution is to use the cut-and-sew theory [3], which adds height information of each UAV by using barometers. The surface network can be cut into pieces, and we can calculate nodes' coordinates in each piece. After that, combine and stick each small part together can obtain a result with a relatively low error rate. Results and details for such a surface network localization scheme can be find from here [3].

7.2.3 Skelton and Trunk Identification

By using the abovementioned cut-and-sew scheme, the boundary of the UAV network can be detected. All boundary UAVs generate a surface network. After that, the virtual coordinates of the surface nodes can be found. Figure 7.6 shows that some

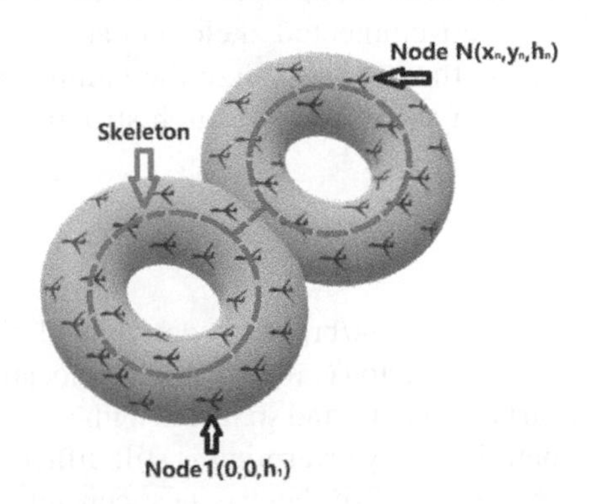

FIGURE 7.6 UAV networks with surface nodes' coordinates determined.

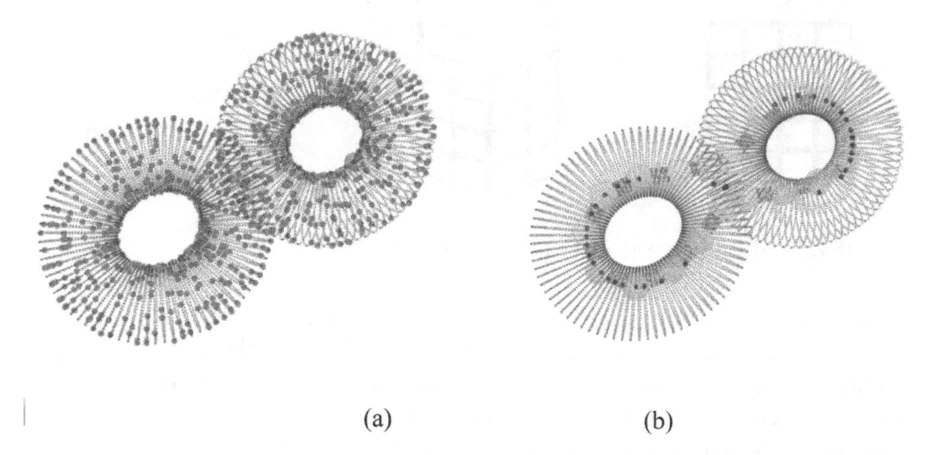

FIGURE 7.7 L_1-medial skeleton of point cloud. (a) Random choice points from raw data. (b) Final skeleton established.

nodes' locations have been determined. There is a virtual origin node (i.e., node 1) with coordinates $(0,0, h_1)$. Node N has a relative location (x_n, y_n, h_n) when taking node 1 as a reference.

With the coordinate information from Figure 7.6, the skeleton of the UAV network can be determined. For example, in Figure 7.6, the two connected cycles (red) form a skeleton of this 3D network. For a mobile ad hoc network (MANET) such as a UAV network, the calculation time is a critical point. A fast skeleton processing algorithm should be used. L_1-medial skeleton of point cloud [4] can tolerate location measurement noise, outliers and large areas of missing data, and it has fast calculation speed. It randomly samples a set of points from the given raw data, which is the surface nodes' positions in our case. Then, each point is iteratively projected and redistributed to the center of the input points within its local neighborhood, and the size of the neighborhood is gradually increased. Finally, a well-connected skeleton can be generated. Figure 7.7 shows how we apply this method to recognize the number 8 skeleton (shown in Figure 7.6). With the median skeleton, the rough structure of routing topology is thus found.

7.2.4 Bottleneck

After the rough structure of the trunk/branch is determined, the next step is to distinguish which part is the core skeleton (called trunk), the location of the joint, which part is branch 1, which part is branch 2 and so forth. Without the joint (or bottleneck) information, mobility control of UAV swarming is still difficult.

Here we use two strategies to solve these issues: concrete skeleton localization and using the ideal of injectivity radius to find the bottlenecks. From the perspective

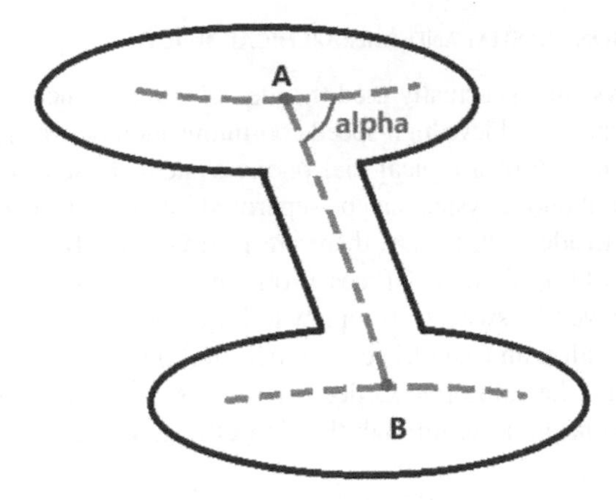

FIGURE 7.8 Bottleneck detection: use skeleton angle change gradient.

of the skeleton, we can trace the skeleton angle change gradient. Figure 7.8 shows a basic ideal. At point A, the angle change rate increases from 0 to around 90 degrees per sample distance. From such a large angle change gradient we can see that point A should be a bottleneck/joint point in general cases. Similarly, point B is also a joint.

However, the method of using angle only is restricted, and it only provides rough joints' information. A good example is shown in Figure 7.9. The angle of the skeleton stays on zero forever; hence, the gradient of angle changes and stays on zero. But, there is obviously a bottleneck between the left and right side; that is, there should be a joint at point A.

The other bottleneck detection method does not have this issue. This method uses injectivity radius to find joints [5]. It is calculated by each individual sensor on the network boundary, and it helps to measure the narrowness of the corresponding boundary area. Hence, the method helps to dedicate the location of bottlenecks. There are some other methods that can be applied to find the bottle necks (joints).

7.3 ROUTING BASED ON SKELETON

After trunk/branches are detected, we can then use the skeleton information to facilitate the communications in a UAV network. The routing scheme aims to find an end-to-end path to deliver the packets from one end to the other.

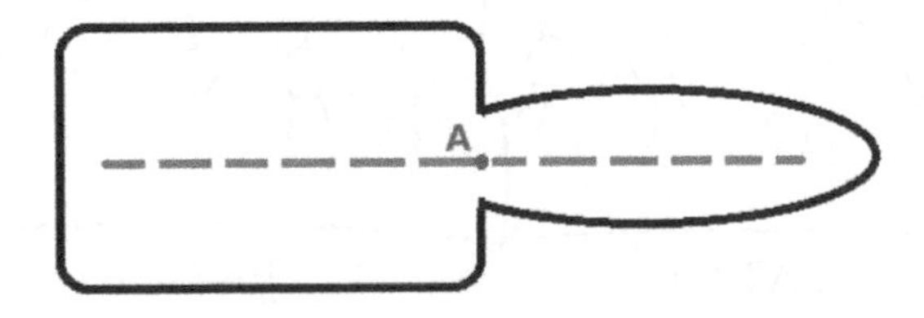

FIGURE 7.9 Limitation of skeleton angle change gradient.

7.3.1 Dial Phone System and Hierarchical Routing

The dial phone system has actually used the idea of trunk/branches. It utilizes different levels of switches to achieve high-speed communication between pairs. Figure 7.10 illustrates the structure of a typical dial phone system. Based on the locations of those areas, the dial phone system can be separated into Local Access and Transport Areas (LATAs). Inside each LATA, there are End Offices, IntereXchange Point of Presences (IXC POP), Tandem Offices (Toll Office) and IntereXchange carriers (IXCs). They behave like switches from low to high levels.

For example, within an End Office area, two dial phones can communicate with each other through the same End Office. In contrast, two dial phones in different LATAs need to communicate through the End Office, Tandem Office, IXC #1 and IXC #2 on both sides.

Extending this ideal into UAV networks is straightforward. Figure 7.11 shows the basic idea. Like the dial phone system, a leveled Access Point (AP) routing protocol can be applied here. Among the skeleton nodes of the network, the first level of APs can be established. After that, joints are selected as the second-level APs. Within each first-level AP area, surrounding UAVs can communicate with each other through the same first-level AP. Within one area (or section), first-level APs construct a loop or continuous line. Hence, the connected communication skeleton can become the backbone of the system. Packets go through the chain of first-level APs. Furthermore, packets in first-level APs can be uploaded/downloaded to/from the second-level APs when UAVs belonging to different areas try to communicate with each other.

As we mentioned in Section 7.1, our UAV routing algorithm is a type of hierarchical routing scheme that separates nodes into groups. The idea is originally used in broadcast routing. Figure 7.15 illustrates a hierarchical routing scenario that fits general UAV networks. Nodes are separated into groups, as seen in Figure 7.12(a).

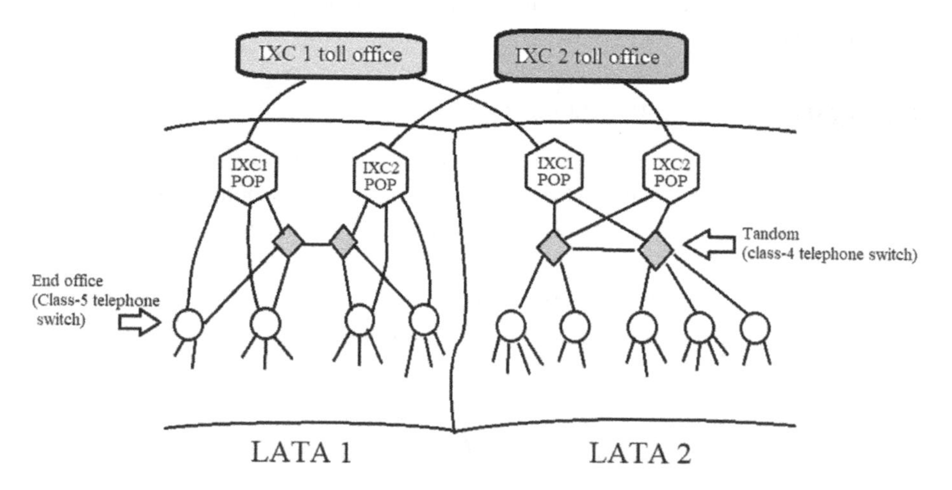

FIGURE 7.10 Dial phone system.

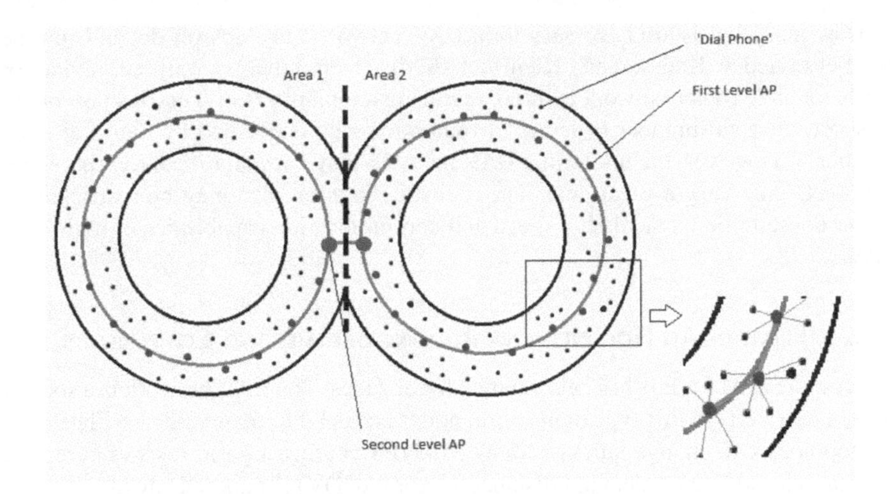

FIGURE 7.11 UAV communication routing.

Without a hierarchical routing scheme, the routing table for each node will be long and oversized. Figure 7.12(b) shows the long routing table of node 1A.

In contrast, the routing table only needs to have a group number when using the idea of hierarchical routing. It is critical for large UAV networks that the nodes are separated into groups and their addresses contain area codes. For example, all nodes on left area 1 have addresses that start with 1, and all nodes' addresses on area 2 start with 2. Furthermore, without the need for detailed routing path in each node, we just need to know which area the packet needs to send to. Within a subgroup, the packet finds its final destination via the details in the address field. Hence, it decreases the size of the routing table and increases routing table calculation speed.

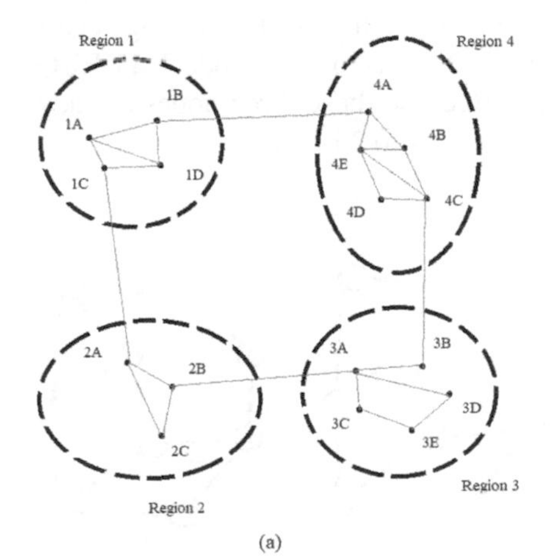

Full Table for 1A		
Destination	Lines	Hops
1A	-	-
1B	1B	1
1C	1C	1
1D	1D	1
2A	1C	2
2B	1C	3
2C	1C	3
3A	1C	4
3B	1C	5
3C	1C	5
3D	1C	6
3E	1C	5
4A	1B	2
4B	1B	3
4C	1B	4
4D	1B	4
4E	1B	3

(b)

Hierarchical Table for 1A		
Destination	Lines	Hops
1A	-	-
1B	1B	1
1C	1C	1
1D	1D	1
2	1C	2
3	1C	4
4	1B	2

(c)

FIGURE 7.12 Hierarchical routing. (a) Whole system overview. (b) Full table for 1A. (c) Hierarchical table for 1A.

How many levels are necessary for a UAV network? The answer depends on the number of nodes. Kamoun and Kleinrock [6] discovered that the optimal number of levels for an N-router network is $\ln(N)$, requiring a total of $e*\ln(N)$ entries per router. In short, the total number of levels of routers or APs is decided by the total node number N. However, the nodes in a UAV network may vary significantly; therefore, the levels may vary a great deal. The realistic situation also may be restricted by power consumption, calculation speed and communication capabilities of high-level UAVs.

7.3.2 Previous Ad Hoc Network Routing Schemes and Comparisons

The routing in MANETs has been studied for decades. The UAV network is a special type of MANET. In this type of network, nodes act as both router and host. The routing topology can change quickly. Many MANET routing algorithms were created, for example, ad hoc on-demand distance vector (AODV) [7] and OLSR [8].

In the AODV routing algorithm, every node is assumed to be able to communicate with its neighbors within its radio range. Figure 7.13(a) shows a simple example for AODV routing. Suppose node 1 wants to communicate with node 9. Node 1 will broadcast a ROUTE REQUEST (RREQ) packet to its neighbors 2, 3, and 4 [Figure 7.13(b)]. With the help of a sequence number in RREQ, nodes can avoid duplicated request packets and continue to broadcast the request [Figure 7.13(c)] until it reaches the destination [Figure 7.13(d)]. After that, node 9 will send back a ROUTE REPLY (RREP) packet back to node 1 through its previous coming path. The hop count in RREP will increase by one whenever the reply packet goes through an intermediate node. Finally, the paths are compared, and a routing path will be found for the sender. For example, in our case, the path will be one of the four cases: 1-3-5-9 or 1-3-6-9 or 1-2-5-9 or 1-4-7-9. All four possible cases use four hops to finish end-to-end routing. To maintain its routing table and make sure the routing path is still available, HELLO packets are sent periodically to make sure neighbors are still active for routing. Moreover, because all the information is stored in each node, the routing vector can be reused by other nodes. For example, if node 2 wants to reach node 9, a RREQ packet will be sent by 2, and reach 1, 3, and 5. Because these three nodes already know how to reach 9, a reply message will be sent back to 2 immediately.

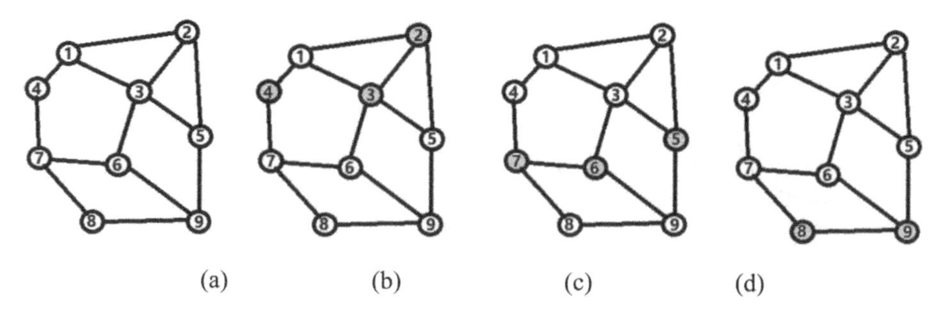

FIGURE 7.13 Basic AODV routing scheme RREQ Flooding in Sequence of Time.

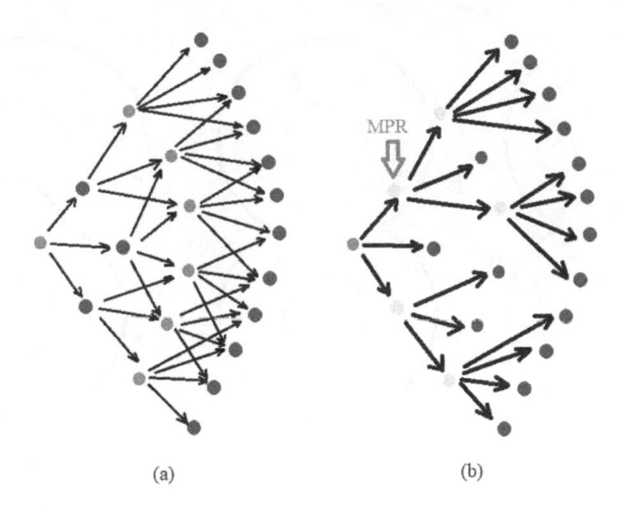

FIGURE 7.14 Pure message broadcast and OLSR. (a) Pure broadcast style. (b) OLSR with MPR messages (yellow circles).

The AODV and many other routing algorithms may create many flooding messages such as the HELLO and RREQ packets mentioned before. These flooding messages decrease the overall efficiency of the routing (longer delay and higher protocol overhead). Other key elements of algorithms we need to consider include convergence speed, routing table size and so forth. OLSR [9] is a good example of what improves the routing efficiency and speed.

The OLSR protocol is a link state protocol, and it selects a specific router on every link to perform flooding. The key ideal of OLSR is to utilize multi-point relay (MPR). In the previous AODV case, node 1 broadcasts REQUEST to every neighbor. In contrast, OLSR only allows node 1 to send a REQUEST to the selected MPRs. Hence, it decreases the overall flooding time and duplicated packets. Figure 7.14 shows the broadcast idea in the OLSR case. Applying this improved routing protocol can decrease the time needed to find the routing path. OLSR also considers the total number of nodes to be covered or reached in the network, but it ignores the quality of the routing path. For example, two nodes could previously directly communicate with each other, but now they have to use an intermediate node when using OLSR.

7.3.3 Dynamic Mobility-Adaptive Hierarchical Routing in a UAV Network

The previous section discussed how to create a hierarchical routing. However, this hierarchical routing may not adapt to mobility very well. To implement such protocols in a UAV network with frequent mobility events, a quick calculation of skeleton positions is needed. Dynamic mobility hierarchical routing is our innovation here. In this case, a quick flooding process will periodically occur based on mobility events. When a UAV node changes its position significantly, or a timer has expired, the new link state and cost information will be sent to nearby nodes. In addition, this flooding packet only reaches the node within this area. The *hierarchical routing* table

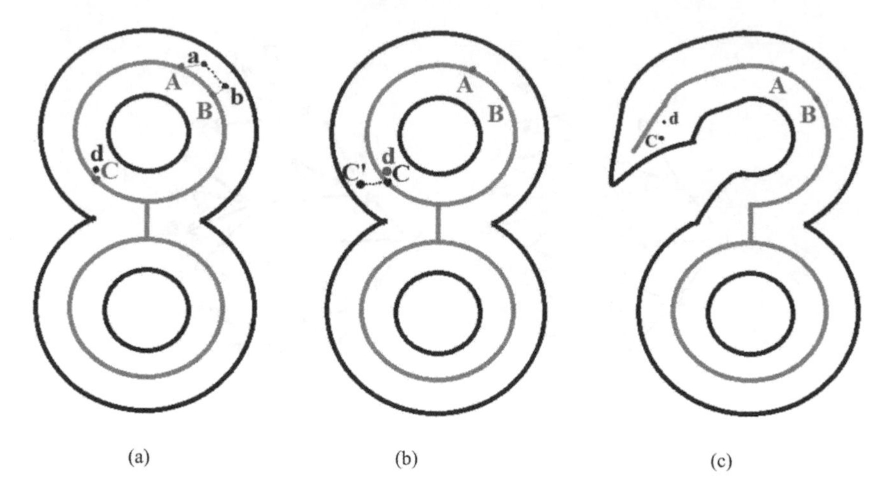

FIGURE 7.15 Dynamic hierarchical routing with mobility events. (a) Normal node mobility with unchanged backbone. (b) Backbone mobility with recalculated backbone. (c) Continuous change of nodes' positions with the entire backbone's renewal.

does not change in the backbone routers. Figure 7.15 describes some cases in dynamic hierarchical routing. Figure 7.15(a) indicates a normal node movement from point a to point b. The backbone does not change. The original router A is now switched to router B. In Figure 7.15(b), backbone router C moves to the C′ position. Hence, a recalculation of the backbone and route flooding are performed. Figure 7.15(c) illustrates continuous node position changes, finally causing the backbone structure to be changed.

More event trigger events are discussed in Table 7.1. Notice that not all activities need to trigger message flooding. More trigger events can be added here, but they should avoid flooding the area as much as possible. Hence, such a UAV communication system can save energy consumed in the message flooding process.

In summary, each routing protocol used for ad hoc networks has drawbacks. When we use AODV for hundreds of UAVs, it may cause many flooding messages and occupies a large amount of link bandwidth. OLSR solves such an issue, but it may still decrease the quality of route path due to the use of specific relay nodes each time. The dial phone system is a kind of modified hierarchical routing and may not be performed very well for superfast mobility speed cases.

7.4 LOWER LAYER DESIGN

After the discussions about the UAV network routing layer, we briefly describe the lower layers, such as the physical and MAC layers, in this section.

7.4.1 MU-MIMO

New antennas or other physical components should be considered in lower layers. In general, there are other four types of antennas [12]: single-input single-output (SISO), single-input multi-output (SIMO), multi-input single-output (MISO) and

TABLE 7.1
Dynamic Hierarchical Routing: Trigger Events

Triger Event	Sign	Activity	Result
Local Area Shared Timer	Time out	Control Message Flooding in Local Area	Local Area Routing Table Renew
General Node Movement	> Threshold	Control Message Flooding in Local Area	Local Area Routing Table Renew
Level 1 Backbone Timer	Time out	Control Message Flooding in Backbone	Backbone Nodes Routing Table Renew
Level 1 Backbone Movement	> Threshold	Control Message Flooding for Whole Network	Rebuild Routing Table for Whole Network
Level 2 Backbone Timer	Time out	Control Message Flooding for Whole Network	Rebuild Routing Table for Whole Network
Level 2 Backbone Movement	> Threshold	Control Message Flooding for Whole Network	Rebuild Routing Table for Whole Network
Level 1 Backbone Beacon ACK	Lost	Control Message Flooding for Whole Network	Rebuild Routing Table for Whole Network
Level 2 Backbone Beacon	Lost	Control Message Flooding for Whole Network	Rebuild Routing Table for Whole Network
Level 1 Backbone Beacon	Lost	Control Message Flooding in Local Area	Rebuild Routing Table for Whole Network

multi-input multi-output (MIMO). The backbone UAVs may need to use MIMOs for multipath propagation. This enables UAVs to send and receive more than one packet simultaneously over the same radio channel. MIMO is often affiliated with orthogonal frequency division multiplexing (OFDM) or orthogonal frequency division multiple access (OFDMA) for high-throughput transmissions.

Although many routers utilize MIMOs, they often communicate with one device at a time. These routers are single-user MIMO devices. Each user takes turns receiving data from an AP, and this will create a long packet waiting time. To increase the overall data rate and decrease device waiting time, multi-user, multiple-input multiple-output (MU-MIMO) may be used. Decreasing packet waiting time in a complex network with a large number of receivers is critical, especially in our hierarchical routing structure. The standards IEEE 802.11ax and 802.11ay with newer designs in MIMO beamforming should be used in the future.

7.4.2 Upcoming Physical and Data Link Layer protocol: IEEE 802.11ax

The last-generation IEEE 802.11ac devices were in the market by Quantenna in 2011 [10]. With MIMO-OFDM modulation and 256-QAM, its theoretical speed can reach 2.34 Gbps. The next-generation protocol 802.11ax [11] will come out this year with 1024-QAM and 37% faster speed compared with 802.11ac.

With 1024-QAM, channel contention increases, which will result in a decrease of network throughput, 802.11ax uses an improved MIMO-OFDM, i.e., MU-MIMO-OFDM.

In 802.11ac, users' signals are separated by space and time. IEEE 802.11ax adds frequency dimension to separate channels into more units. In short, both protocol divide time slots by using *enhanced distributed channel access* (EDCA). In the *spatial* domain, space is divided into units by MIMO. In the *frequency* domain, OFDMA uses frequency division to generate more bandwidth "slots."

7.5 CONCLUSIONS

Ad hoc networks have been used for military and emergency applications, and they typically do not have a clear communication structure such as backbone-based multilevel routing. In this chapter, a virtual backbone is built based on the skeleton extraction of the UAV swarm shape. The routing table of a hierarchical routing scheme gets renewed in major mobility events. It does not require a routing message flooding process for the whole network. As a result, the dynamic hierarchical routing not only decreases routing table size and increases destination address lookup speed, but it also decreases protocol overhead.

REFERENCES

1. I. M. Khan, N. Jabeur, and S. Zeadally, "Hop-based approach for holes and boundary detection in wireless sensor networks," *IET Wireless Sensor Systems*, vol. 2, pp. 328–337, 2012.
2. G. Tyagi, S. Shukla, and R. Matam, "Local connectivity based boundary detection in wireless sensor networks," *2016 IEEE International Conference on Internet of Things (iThings)*, 2016.
3. Y. Zhao, H. Wu, M. Jin, Y. Yang, H. Zhou, and S. Xia. Cut-and-sew: a distributed autonomous localization algorithm for 3D surface wireless sensor networks, *Proceedings of the Fourteenth ACM International Symposium on Mobile Ad Hoc Networking and Computing (MobiHoc '13)*, ACM, pp. 69–78, 2013.
4. H. Huang, S. Wu, D. Cohen-Or, M. Gong, H. Zhang, G. Li, and B. Chen, "L_1-medial skeleton of point cloud," *ACM Transactions on Graphics*, vol. 32, no. 4, Article 65, 2013.
5. H. Zhou, N. Ding, M. Jin, S. Xia, and H. Wu, "Distributed algorithms for bottleneck identification and segmentation in 3D wireless sensor networks," *2011 8th Annual IEEE Communications Society Conference on Sensor, Mesh and Ad Hoc Communications and Networks*, pp. 494–502, 2011.
6. F. Kamoun and L. Kleinrock. "Stochastic performance evaluation of hierarchical routing for large networks," *Computer Networks*, vol. 3, pp. 337–353, 1979.
7. C. E. Perkins and E. M. Royer, "Ad-hoc on-demand distance vector routing," *Proceedings WMCSA'99. Second IEEE Workshop on Mobile Computing Systems and Applications*, pp. 90–100, 1999.
8. P. Jacquet, P. Muhlethaler, T. Clausen, A. Laouiti, A. Qayyum, and L. Viennot, "Optimized link state routing protocol for ad hoc networks," *Proceedings. IEEE International Multi Topic Conference, 2001. IEEE INMIC 2001. Technology for the 21st Century*, pp. 62–68, 2001.
9. Linksys, "What is MU-MIMO and why do you need it," https://www.linksys.com/us/r/resource-center/what-is-mu-mimo/.
10. Wikipedia, "IEEE 802.11ac," https://zh.wikipedia.org/wiki/IEEE_802.11ac.
11. Wikipedia, "IEEE 802.11ax," https://en.wikipedia.org/wiki/IEEE_802.11ax.

8 Routing in 3D UAV Swarm Networks

Katelyn Isbell[1], *Yang-Ki Hong*[1], *Fei Hu*[1]
[1]Electrical and Computer Engineering,
The University of Alabama, Tuscaloosa, AL, USA

CONTENTS

8.1 INTRODUCTION

Due to the recent growth of environment surveillance applications, there is an increasing need for developing efficient communication methods for 3D unmanned aerial vehicle (UAV) networks. Network management methods originally developed for 2D networks, such as topology control, node localization, power management and network routing, must be redesigned to accommodate the complexities of 3D networks. Ad hoc routing protocols, often applied to UAV networks and other mobile systems, can be used in environments lacking fixed network infrastructure. Likewise, sensor networks, as a special type of ad hoc network, are often applied to large, complex spaces, such as the sky, ocean, and space, for some critical applications such as disaster recovery, topographical mapping, space exploration, the Internet of Things and undersea monitoring [1]. Two examples of 3D networks with mobile nodes, geographic obstacles, acoustic links and so forth are shown in Figure 8.1. The complexity, sparsity and size of the networks in these applications demand advanced 3D network protocols. This article will focus on the development of routing in 3D networks.

In mobile ad hoc networks, there typically exist two types of routing protocols: topology based or position based [2]. Topology-based routing schemes, which use information about the radio-frequency (RF) links in the network to forward packets,

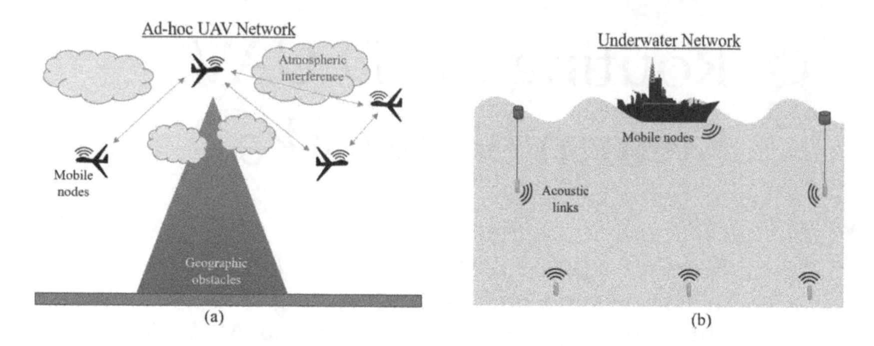

FIGURE 8.1 3D networks. (a) Ad hoc UAV network. (b) Underwater network.

can be classified as proactive or reactive [2]. In proactive routing, a routing table is created and stored with the record(s) of the shortest or most efficient routes. Creation/storage of such routing tables becomes unfeasible in large and unstable networks, although efforts are made with compact routing schemes. Alternatively, reactive routing uses message flooding to determine the route, which occurs only when a new routing request comes. Reactive routing schemes do not maintain the paths, reducing protocol overhead. However, because it establishes the path in an on-demand style, it could have a long delay before the first end-to-end routing path is established.

A popularly used routing protocol developed for ad hoc networks is a position-based scheme, also known as geographic or location-based routing. In geographic routing, Routing REQuest (RREQ) messages are sent to the nodes that are geographically close to the destination. Each node knows its own geographic location, which may be obtained through the Global Positioning System (GPS), as well as the locations of its closest neighbors. In this way, a message can be sent without knowing the layout of the network. Geographic routing is memoryless and local, meaning that the messages do not store information about their states, reducing protocol overhead, and the messages traverse through the network based solely on the nodes' local knowledge [3]. Messages are first transmitted using a forwarding pattern, such as greedy forwarding [4], in which the message is forwarded from one node to a neighboring node that has the shortest distance to the destination [Figure 8.2(a)].

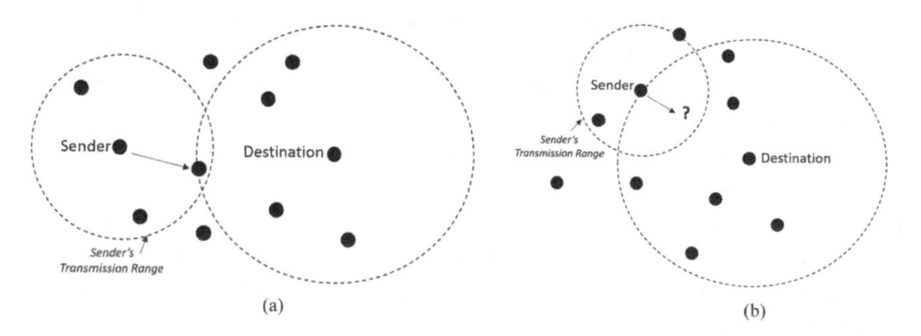

FIGURE 8.2 (a) Greedy forwarding. A sender forwards RREQ to a node within its transmission range that is nearest to destination node. (b) Local minima. The sender is closer to destination node than its neighboring nodes so it cannot perform greedy forwarding.

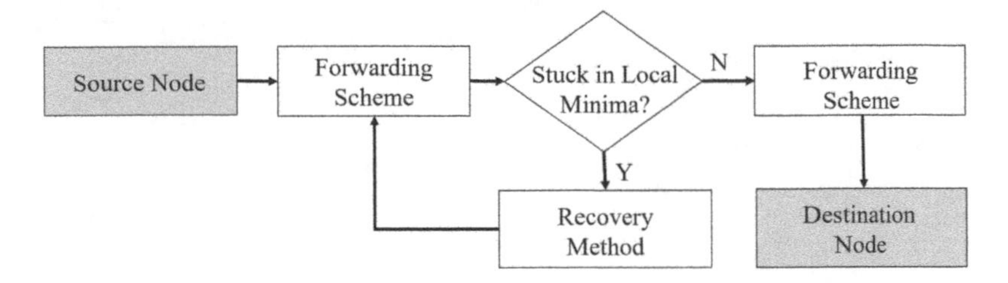

FIGURE 8.3 Flowchart of 3D geographic routing schemes.

Using this scheme, routing search messages can often get stuck in what are called "local minima" or "voids". In this case the closest node it can travel to is farther away from the destination than that node itself, as shown in Figure 8.2(b). Local minima are especially common in sparse networks; therefore, they need to be addressed for large-scale UAV networks. Thus, one of the greatest challenges of 3D geographic routing is avoiding or escaping the local minima by using route recovery methods.

In 2D routing, techniques such as face routing [5] can be used to reliably escape the local minima, but the techniques are not directly transferable to the 3D case because the faces that surround a local minimum are now 2D surfaces. Additionally, local minima are more likely to occur in 3D networks than 2D networks. To travel through a network from the sender to destination node, a forwarding scheme is used to forward the packet until a local minimum is met. In that case, a recovery method may be used until the packet can return to the normal forwarding status, as shown in Figure 8.3.

The routing protocols to address the problem of local minima and ensure packet delivery need to be further explored. First, the coverage of 3D networks should be investigated. These include the analysis of unit ball graphs (UBGs) and k-local routing [6], truncated octahedral tessellation [7], combinatorics [8] and nesting wireless rings [9] to achieve the coverage of the entire network area. Then, forwarding methods based on distance progress can be used [10–12], followed by forwarding methods based on congestion status [13–15]. Then the protocols, with recovery methods [3, 16, 17], can finally be adopted.

8.2 COVERAGE OF 3D UAV NETWORKS

To perform routing through a UAV network, the nodes must be placed properly so that a route can be found between two points. The route cannot be too sparse. With large, dynamic 3D networks, the development of a node placement strategy is crucial. Additionally, controlling the geometry of the network is a common approach to improve the performance of 3D routing [17].

8.2.1 Graphs of 3D Networks

Several graphing techniques conventionally employed in 2D networks have been modified for the 3D case. For instance, UBGs in 3D space have been developed as an

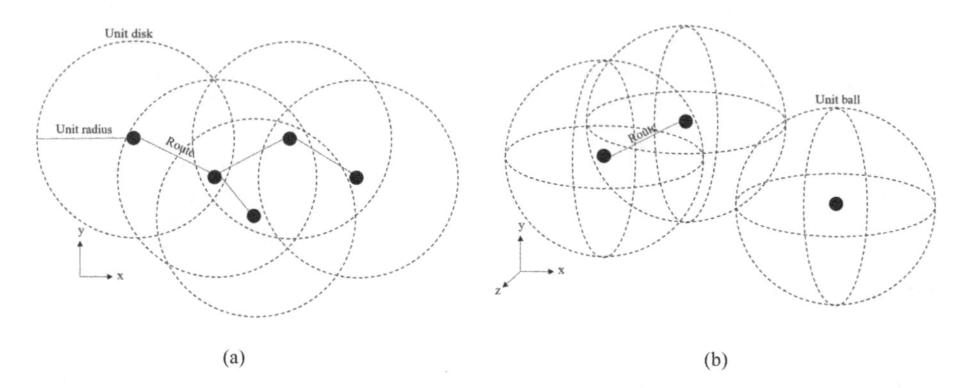

FIGURE 8.4 (a) Unit disk graph for 2D space. (b) Unit ball graph for 3D space.

alternative to unit disk graphs (UDGs) in 2D space. UDGs, which are the intersection graphs of equal sized circles in a plane, are often used to model 2D ad hoc networks [6], as shown in Figure 8.4(a). In a UDG, a node v on the plane is connected to node u, if u lies within the disk of unit radius 1 that is centered at v. A natural conversion to 3D space is to use the UBG, as shown in Figure 8.4(b). In a UBG, node v on the plane is connected via an *edge* to node u, if u lies within the ball of unit radius 1 that is centered at v. Additional graphs of 3D networks include the Delaunay triangulation [10, 11, 17] and hull trees [16], which will be further explored in Sections 8.3 and 8.4. However, the UDG is more suitable for initial basic analysis of 3D networks.

In [6] Durocher investigates k-local routing algorithms on the UBG in 3D space. In geographic routing, each node contains only k-local information, where k is the degree of neighbors stored by the node (for $k = 1$, the node stores only the location of the neighbors that are less than one hop away). Durocher first proves that a UBG contained between parallel planes P_1 and P_2 in R^3 can be transformed into a quasi-unit disk graph (QUDG), which is used to model nodes with irregular transmission ranges, by projecting the points in the UBG to a parallel plane. Additionally, for all finite sets of points P in R^3 contained between parallel planes P_1 and P_2 with a separation of $1\sqrt{2}$, there exists a 2-local routing algorithm for UBG(P). However, if the distance between P_1 and P_2 is greater than $1\sqrt{2}$, there exists no k-local routing path that can guarantee the data delivery on all UBGs. Durocher concludes that no k-local routing algorithms can guarantee 100% data delivery for 3D ad hoc networks modeled by UBG.

8.2.2 Strategic Partition of 3D Networks

While UBGs can be used to model a simple 3D network, partitioning of 3D networks is required to determine the ideal placement of nodes and possible routing techniques. The partitioning requires exploration into the complex geometry of 3D shapes. In [7], the authors proposed a Voronoi tessellation of 3D space as a node placement problem to achieve 100% sensing coverage with the minimal number of nodes required. The Voronoi cell of a point c is a convex polyhedron, which contains

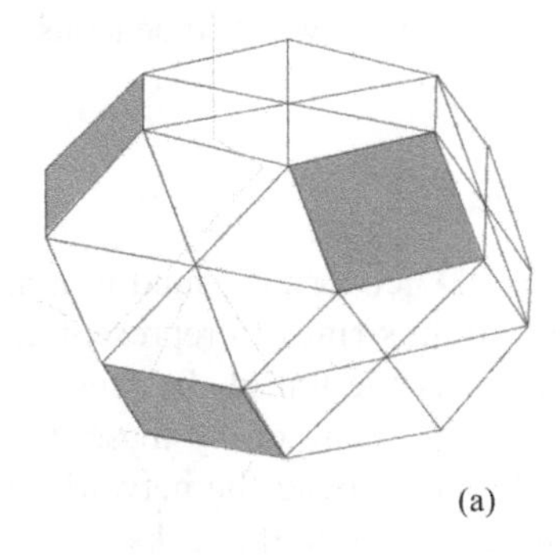

(a)

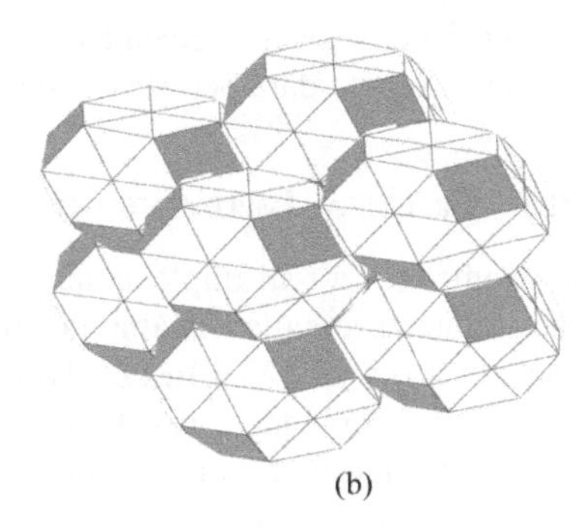

(b)

FIGURE 8.5 (a) Truncated octahedron. (b) Lattice of truncated octahedrons.

the set of points that are closer to c than other points in the space. The partitioning of a space by the polyhedrons is the Voronoi tessellation, which is used to find the optimal shape of the cells that can efficiently fill in a 3D space. The space-filling structure is defined by its isoperimetric quotient (IQ), which is calculated by the following:

$$IQ = 36\pi V^2 / S^3$$

where V is the volume and S is the surface of the structure. The value of IQ ranges from 0 to 1, where the IQ of a sphere has the highest value at 1. The result of applying a Voronoi tessellation to a body-centered cubic (BCC) lattice is a truncated octahedron [Figure 8.5(a)], which was proved to have a volumetric quotient (VQ) of 0.68329 [7], where VQ is the ratio of the volume of a polyhedron to the volume of its circumsphere. To place the network nodes, an arbitrary point (node) is chosen as the input, and other nodes are placed around this node, which acts as the seed for a growing lattice of truncated octahedrons [Figure 8.5(b)]. Using this network coverage, 3D routing can make use of the known locations of all nodes.

In practice, 3D networks require a large number of nodes to achieve full network coverage. In [8], the authors investigated the combinatorics of beacon-based routing in 3D space to determine the upper and lower bounds of required nodes, or beacons. A beacon is a point-like object that attracts all points (or nodes) within a polyhedron P. A point will move in the direction that reduces the distance between itself and the beacon. To analyze the upper and lower bounds of beacons required to route packets between two points in 3D, the polyhedron P is decomposed into m tetrahedra, where $\Sigma = \{\sigma_1, \ldots, \sigma_m\}$ is the tetrahedral decomposition. For an arbitrary n, there is a P with $\Theta(n)$ vertices that require at least $\Omega(n^2)$ convex parts to decompose it. The P can be triangulated with $O(n^2)$ tetrahedra by using $O(n^2)$ Steiner points. They have proved the following result:

Theorem: *Given a polyhedron P with a tetrahedral decomposition Σ with $m = |\Sigma|$ tetrahedra, it is always sufficient and sometimes necessary to place $\left\lceil \frac{m+1}{3} \right\rceil$ beacons to route data between any pair of points in P.*

A special case occurs given a c-corner spiral polyhedron P, where c beacons are necessary to route data between two points.

8.2.3 Practical Implementation of 3D Networks

One example of a real implementation of a complex 3D network is found in a network architecture called Diamond [9], which uses wireless rings to represent data center networks (DCNs). In DCNs, radios are located at the top of the racks and operate in a 2D planar network. To reduce the interference caused by these dense networks, Diamond uses ring reflection spaces (RRSs) to expand the network to a third dimension. Wireless rings and reflectors running parallel to the racks are used to create the 3D network. The RRSs create a hybrid network that makes use of wired or wireless links. The physical distance between two neighboring rings, R_i and R_{i+1}, are referred to as the RRS width Δ_i, which has the following property:

$$\lim_{t \to \infty} \Delta_i = 2L/\pi$$

where L is the reflector length, which equals the height of the racks. As the ring number increases, RRS width becomes stable, demonstrating Diamond's scalability. The servers in the DCNs have three options for routing: (1) wired, (2) wireless or (3) a hybrid of wired and wireless. The proposed architecture was implemented on a 60-GHz test bed and shown to be functional [9].

8.3 DATA FORWARDING METHODS

Once the network has been partitioned and maximum node placement has been determined, the routing scheme can be executed to establish the end-to-end path for data forwarding. In geographic routing, packets are forwarded in a hop-to-hop style based on location information. The node is transmitted to the destination through the network via a series of relay nodes. Selection of the next node is decided at each node by the data forwarding methods. Forwarding methods in geographic routing methods can be classified as (1) progress based and (2) randomized. In *progress-based* forwarding methods, the routing scheme uses only geographic information, which may include the knowledge of neighboring nodes and some limited geographic information stored in the node's data forwarding table. In *randomized* forwarding methods, the routing scheme uses more network information such as traffic congestion and energy consumption status in addition to geographic information. Important parameters to consider for data forwarding are packet delivery success rate, overhead, and path length.

8.3.1 Progress-Based Forwarding Methods in 3D Networks

In progress-based forwarding, the packet is forwarded from the sender to the node that *progresses* it closer to the destination node. In greedy forwarding, a special type of progress-based forwarding, the packet is forwarded to the neighbor that is closest to the destination node. In [10], a geographic routing protocol for 2D, 3D,

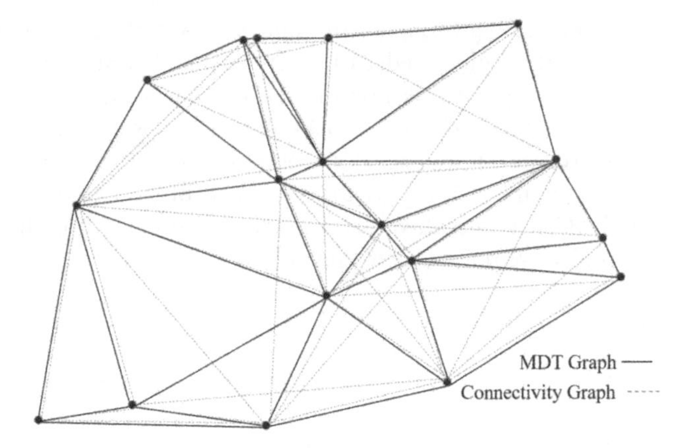

FIGURE 8.6 Delaunay triangulation graph (generated in MATLAB with random coordinates).

and higher dimensions is proposed with guaranteed packet delivery and low routing path stretch by using multi-hop Delaunay triangulation (MDT). In this protocol, nodes are virtually connected through the use of a Delaunay triangulation graph, which is a triangulation of points created so that no point is within a circumcircle of a triangle, allowing for the maximum angles of the triangles (Figure 8.6). The virtual connections generated by Delaunay triangulation and the physical connections between the nodes are mapped onto the MDT graph. To travel through the virtual connections that guarantee data delivery, the message is greedily forwarded through the multi-hop connections by using the soft-state forwarding tables. The soft-state entries consist of the source and destination nodes of a path, and the node's predecessor/successor nodes in the path. The accuracy of MDT can be defined as

$$A = \frac{N_c(\text{MDT}(S)) - N_w(\text{MDT}(S)) - 2 \cdot N_{np}(\text{MDT}(S))}{2 \cdot N_{\text{edges}}(DT(S))}$$

where $N_c(\text{MDT}(S))$ is the number of correct neighbor entries, $N_w(\text{MDT}(S))$ is the number of wrong neighbor entries, $N_{\text{edges}}(DT(S))$ is the number of edges in $DT(S)$ and $N_{np}(\text{MDT}(S))$ is the number of edges in $DT(S)$ without forwarding paths in the multi-hop DT of S. This equation shows that with correct node information it can achieve 100% accuracy. The MDT protocol includes the methods for route maintenance/changes to account for the node/link churn. It also requires node leaving and link failure information to deal with a departing node's multi-hop forwarding issues.

It was shown that MDT provides guaranteed delivery in 3D with a routing stretch close to 1 [10]. However, MDT incurs high per-node storage cost when multiple hops are enabled, because each node stores 1- and 2-hop neighbors' information. MDT's main advantage is its resilience to churn and its ability to handle dynamic topology changes. The protocol also can be applied to nodes with arbitrary coordinates.

Alternatively, in [11], greedy surface routing greedy (GSG) for geographic routing in 3D environments was developed. Using 3D restricted Delaunay triangulation (3D RDT), a 3D network is partitioned into triangles and edges, where the edges longer than one unit and intersecting triangles are removed. A surface is a region that is enclosed by triangles and isolated edges that are defined as 3D components. The messages are routed among the nodes on the surface unless the message meets the local-minimum-destination (*lmd*) line. A 3D component has a central point whose coordinates are the average coordinates of the nodes in the 3D component. The coordinate of a local-minimum L is (x_1, y_1, z_1), the coordinate of destination D is (x_d, y_d, z_d) and the central point C's coordinate of 3D component F is (x_f, y_f, z_f). The coordinate $(x_{f'}, y_{f'}, z_{f'})$ of F's projective point C is obtained by

$$(x_d - x_l)\cdot(x_f - x_{f'}) + (y_d - y_l)\cdot(y_f - y_{f'}) + (z_d - z_l)\cdot(z_f - z_{f'}) = 0$$

and

$$(x_d - x_l)\cdot(x_f - x_{f'}) = (y_d - y_l)\cdot(y_f - y_{f'}) = (z_d - z_l)\cdot(z_f - z_{f'})$$

The distance from C to the *lmd* line is $L_{ff'}$, where

$$L_{ff'} = \sqrt{(x_f - x_{f'})^2 + (y_f - y_{f'})^2 + (z_f - z_{f'})^2}$$

The distance $L_{df'}$ from C to D is calculated as

$$L_{df'} = \sqrt{(x_d - x_{f'})^2 + (y_d - y_{f'})^2 + (z_d - z_{f'})^2}$$

Routing on the surface operates in four modes: (1) FA, where the message is transmitted from the *lmd* line until it reaches the farthest 3D component; (2) FW, where the message routes from the projective point of a 3D component to the destination; (3) CT, where the message routes toward the destination until it crosses the *lmd* line and (4) RD, where the message finally jumps out of the surface by going forward according to the depth search process among 3D components. Additionally, due to the difficulty of forming triangles and dominance of isolated edges in sparse networks, a 3D relative neighborhood graph (RNG)-based depth-first search (DFS) is used to ensure the connectivity of the entire network and achieve high efficiency.

In [11], a simulation of n nodes was conducted in a 3D region, where $200 < n < 1000$. Path lengths from the local minima to the destination are compared for simulations of greedy-random-greedy (GRG) [3], greedy-hull-greedy (GHG) [17] and 3D RNG. The path length of GSG remains stable with increasing node density and hole size. It performs comparably to GHG and better than other techniques at low node densities and better than GHG at high densities. In terms of performance-to-price ratio (here price is the average number of messages exchanged by each node), the 3D RNG outperforms other algorithms, whereas GSG and GHG are similar in sparse networks.

In [12], ellipsoid geographical greedy-face 3D routing (EGF3D) was developed. In EGF3D, the messages are greedily forwarded inside a subminimal ellipsoid. The ellipsoid is constructed by using the source and destination node as the long axis end points. The width of the ellipse is determined by the neighboring nodes that form the subminimal (second least) angular distance. The message is forwarded to the node m with subminimal angle

$$\angle m_i xd \mid m_i \in N(x), i = 1,2,\ldots k$$

TABLE 8.1
Geographic 3D Routing Schemes with Greedy-based Forwarding Methods

Scheme	Forwarding Method	Recovery Method	Performance
Su et al.'s GSG [11]	Greedy surface routing	Local-minimum-destination line	Path length is stable with increasing node density
Lam and Qian's MDT [10]	Greedy forwarding through virtual connections made by Delaunay triangulation stored in the node's soft-state forwarding table	None	Guarantees 100% delivery rate with correct node accuracy, also resilient to churn, topology change and inaccurate coordinates
Wang et al.'s EGF3D [12]	Greedy forwarding inside a subminimal ellipsoid	None	Delivery rate approaches 100% for large networks, with low protocol overhead

where $N(x)$ is the set of neighboring nodes of node x. If the message reaches a local minimum, the face algorithm is used to escape it.

To evaluate the performance of EGF3D, simulations were performed in [12], and it was found that the packet delivery success rate of EGF3D approaches 1.0 as the average node degree increases. It was also found that EGF3D has significantly less overhead, approximately only 50–100 bytes, compared with GRG [3] (approximately 200 bytes at average node degrees above 6). Alternatives to greedy forwarding have been proposed in [10–12], and are summarized in Table 8.1.

8.3.2 Randomized Forwarding Methods in 3D Networks

Randomized forwarding methods have been developed as alternatives to progress-based routing to address other issues of the network, such as high energy consumption and traffic congestion. This is important in UAV networks, which suffer from low computing resources and require energy efficiency and collision avoidance. Those methods are summarized in Table 8.2.

In [13], an energy-aware dual-path geographic routing (EDGR) scheme was proposed that can bypass the routing holes. It is suitable to energy-constrained networks that contain battery-powered sensors and routing holes caused by geographical obstacles. The energy consumption $c(u,v)$ of a node u transmitting data to neighbor v is defined as

$$c(u,v) = c_1 \cdot d(u,v)^{\alpha} + c_2 + c_3 \cdot d(u,v)^2$$

where $d(u,v)$ is the Euclidean distance between nodes u and v; α is the path loss constant depending on the transmission environment and c_1, c_2, and c_3 are the constraints that are dependent on wireless device characteristics. In EDGR, packets are forwarded around local minima based on the knowledge of its neighbors' residual

TABLE 8.2
Geographic 3D Routing Schemes with Randomized Forwarding Methods

Scheme	Application Scenario	Forwarding Method	Recovery Method	Performance
Huang et al.'s EDGR [13]	Resource-constrained networks	Greedy forwarding along two different routes that pass through the routing holes	Random shift to the location of subdestination	High energy efficiency and network lifetime, <90% delivery rate
Rubeaai et al.'s 3DRTGP [14]	Time-sensitive applications	Adaptive packet forwarding region (PFR) limits the forwarding of nodes in the direction of destination to reduce congestion	Retransmission of message with PFR adjustment and backpressure	Low end-to-end delay and miss ratio
Gharajeh and Khanmohammadi's DFTRP [15]	Resource constrained networks	Forwarding decided by fuzzy logic inference machine based on traffic probability	None	High network lifetime and delivery rate

energy, location and energy consumption through the use of two node-disjoint anchor lists. EDGR was simulated and found to have low energy consumption and a long network lifetime. It exhibits a packet delivery success rate higher than 90% for dense networks with moderate routing hole sizes [13]. UAV swarming networks meet such requirements due to their high density.

Furthermore, in [14], a real-time geographic routing protocol for 3D networks (3DRTGP) was proposed based on a *packet forwarding region* (PFR). It aims to reduce the number of redundant packet transmissions, collisions and congestion by checking the network density around each forwarding node. The 3DRTGP performs four functions: (1) *location management*, (2) *forwarding management*, (3) *local minima escape* and (4) *flooding reduction*. In the first function, the message is unicast to all nodes, with the information about the PFR with respect to the sender and destination. The receiving nodes determine whether they are in the PFR by determining whether θ satisfies the following condition:

$$\theta \leq \beta/2$$

where β is the conical forwarding angle and θ can be obtained by solving the following equation:

$$\theta = \cos^{-1} \frac{SN \times SD}{\|SN\| \times \|SD\|}$$

where SN is the vector from the sender to the node itself, SD is the vector from the sender to the destination, $\|SN\|$ and $\|SD\|$ are the Euclidean vector norms and $SN \cdot SD$ is the dot product. If a node fulfills this requirement, it is a potential forwarding node; otherwise, the node stores the information in *VoidNodePacketList* for the use by local minima escape protocol. The initial value of β is set to be $360°/\eta_s$, where η_s is given by $\rho \times V_s$ with V_s being the spherical region around the sender, since β is inversely proportional to the network density.

A forwarding node is chosen from the nodes based on the delay estimation and forwarding probability. The delay estimation is performed based on the number of expected hops and the time required in each hop. If the PFR with the lowest delay estimation contains more than one node, forwarding probability is used to choose the forwarding node from those nodes. The average delay between the sender and the neighboring nodes determines the forwarding probability.

If the initial PFR does not have a forwarding node, due to the presence of a local minimum or traffic congestion, the sender retransmits the message after a timeout period. Nearby nodes identify the retransmission. If the message is stored in its *VoidNodePacketList*, then the node doubles its β, which expands the PFR.

To reduce flooding caused by retransmissions, the sender node drops the message if it receives it again. Simulations were performed to evaluate 3DRTGP, and it was found that it is scalable to network densities. When the packet deadline is ignored, the algorithm achieves a 100% packet delivery success rate. It also outperforms comparable algorithms in terms of end-to-end delay and packet loss ratio, which are important parameters in time-sensitive UAV networks.

In [15], the authors proposed dynamic 3D fuzzy routing based on traffic probability (DFTRP) to increase network lifetime and improve success rate of packet delivery. DFTRP uses hop-to-hop delivery in which the message is transmitted to a neighboring node based on fuzzy logic and local information until it reaches the destination. Fuzzy logic differs from Boolean logic in that it has degrees of "truth" rather than "false" or "true". It uses the concept of human intuition to make decisions. An inference engine takes inputs, applies a fuzzy rule base, and produces outputs. A fuzzy set represents the relationship between an uncertain quantity x and a membership function (MF) μ in the range [0,1]. The fuzzy set A in the universe of discourse U can be represented by a set of ordered pairs such that

$$A = \{(x, \mu_A(x)) \mid x \in U\}$$

where $\mu_A(x)$ is the MF of x in set A. The proposed fuzzy system takes the number of and distance to all neighboring nodes and calculates the traffic probability.

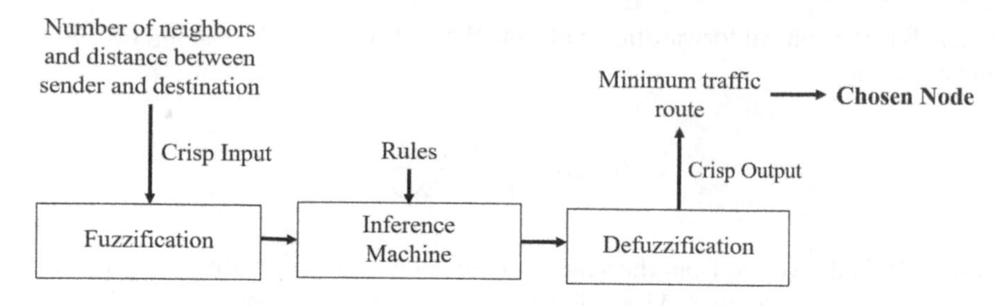

FIGURE 8.7 Process of using fuzzy logic to select the relay node.

The inference engine uses the Mamdani-type fuzzy process with the following rule base:

$$
\begin{aligned}
C' &= (A' + B') \circ R \\
&= (A' + B') \circ (U_{j=1}^{m} R_j) \\
&= (A' + B') \circ (U_{j=1}^{m} A_j \times B_j \times C_j)
\end{aligned}
$$

where C' is the membership grade of the output parameter; A' and B' are the membership grades of the input parameters; R is the total grade of the fuzzy rules; R_j is the fuzzy rule of the rule base; m is the total grade of the fuzzy rules and A_j, B_j, and C_j are the MFs of the input and output parameters. Once the traffic probability is calculated by the proposed fuzzy system, the message is then forwarded to the node with the lowest traffic probability. This allows the packet to travel along the link least likely to cause congestion. This process is illustrated in Figure 8.7.

To evaluate the proposed routing scheme, a mobile node network was simulated, and the results of the network lifetime and packet delivery success rate were compared against the number of transmission packets and nodes for the DFTRP. From the simulations of [15] it was found that the network lifetime and success rate of packet delivery of the DFTRP were high.

8.4 ROUTING RECOVERY METHODS IN 3D NETWORKS

In a routing scheme, the forwarding method is implemented until the message is forwarded to a node in a local minimum. As depicted in Figure 8.3, the scheme must switch to the *recovery* method to route the message out of the local minima, at which point it can resume the forwarding method.

In 2D networks, the problem of local minima is addressed through some techniques such as face routing, where the packet is forwarded along the *face* of a hole, or local minima, until it can be forwarded to a node that is closer to the destination. However, this is only applicable to planar graphs.

In 3D, holes can have 3D structure. Therefore, recovery methods for escaping local minima must be addressed for 3D networks.

Comparison of the recovery methods can be found in Table 8.3.

TABLE 8.3
Geographic 3D Routing Schemes with Recovery Methods

Scheme	Application Scenario	Recovery Method	Performance
Flury and Wattenhofer's GRG [3]	Large mobile ad hoc and sensor networks	Random walks	Recovery cost is bound to $O(k^6)$ hops, no guaranteed delivery
Liu and Wu's GHG[17]	Networks deployed in the sky, underground, and underwater	Hull routing	Low cost, more efficient than GRG [3]
Zhou et al.'s GDSTR-3D [16]	Sensor networks	Tree traversal forwarding mode with two convex hull trees	Delivery rate approaches 100% as average node degree increases

In [3], the authors analyzed randomized geographic routing algorithms for 3D network topologies that use greedy-face routing until the message reaches a local minimum, in which case it reverts to "random walks" to escape it (GRG routing). In [6], the authors argue that there are no routing protocols that can guarantee delivery or escape from a local minimum, and they found that a random walk can be used to efficiently recover from it.

For a connected graph $G = (V, E)$ with the number of nodes $n{:=}|V|$ and number of edges $m{:=}|E|$, a random walk on the graph can be captured by using the Markov chain, which is a sequence of random variables used to model random processes that are memoryless. Another method to capture the random walk is to apply electrical resistance formulas, by replacing the edge of the graph with a resistance of 1 Ω and analyzing the flow of the electric network. For an arbitrary graph, the cover time, C_G, which is the time required to visit all nodes of G at least once and determines the upper bound on the random walk model, is described by

$$C_G = 0(n \cdot m) = 0(n^3)$$

Additionally, for any randomize routing algorithm in 3D, the message requires $\Omega(r^3)$ hops, which means that the expected routing stretch is at least cubic. In [3] they proved this by analyzing a 3D graph in which nodes are placed on the surface of a sphere of radius r with distances between nodes of at least 2. These surface nodes are determined as

$$S_1 := \{(r \cdot \sin(2i \cdot \arcsin(1/r)), 0, r \cdot \cos(2i \cdot \arcsin(1/r))) \mid i \in [0, [0.5\pi / \arcsin(1/r)]]\}$$

Then, for each $(x, y, z) \in S_1$ additional nodes are added to S_2 such that

$$S_2 := \{(x \cdot \sin(2i \cdot \arcsin(1/x)), x \cdot \cos(2i \cdot \arcsin(1/x)), z) \mid i \in [1, [\pi / \arcsin(1/x)]]\}$$

Intermediate nodes are also added to the surface to connect nearby surface nodes. For each surface node, a line of $[(r - l)/2]$ nodes are added so the distance between nodes on the line is 1 and the line is directed toward the node at the center of the sphere (called node t). An arbitrary surface node w is chosen and further nodes are appended to its line so that it can reach t. The number of points per line is denoted as $\Theta(r)$. The number of surface nodes is determined by

$$|S_2| \geq 2 \cdot \sum_{i=1}^{\frac{1}{2}\left[\frac{\pi}{2\arcsin(1/r)}\right]} r \cdot \sin(2i \cdot \arcsin(1/r)) \geq \sum_{i=1}^{r/2} i = \Theta\left(r^2\right)$$

and the total number of nodes in the graph is determined by

$$\left(|S_1| + |S_1|\right) \cdot \Theta(r) = \Theta\left(r^3\right)$$

For an optimal algorithm, the message is routed on the surface until it hits w, then travels down the w line until it reaches t. Because the line contains at most r nodes to traverse, $O(r)$ hops are required to reach the destination. However, for a geographic routing algorithm, because only local information is known, the message will be routed down the lines that are directed toward but do not reach node t before finding node w, which would require it to explore $\Omega(r^2)$ lines and complete $\Omega(r^3)$ hops until it finds t.

The authors compare four variations on a random walk model:

1. For a random walk limited by a ball of radius r, the recovery cost is bound to $O(k^6)$ hops, where k is the length of the shortest path connecting the local minimum and the closest node to the target.
2. For a random walk that is restricted to nodes surrounding the "hole" of the local minimum, such as the face routing in 2D graphs, the expected routing stretch is at least cubic.
3. For a random walk on a sparse subgraph by using a dual graph, the cover time is limited to $O(r^6)$, where r is the maximum distance between two randomly placed nodes in the unit square.
4. For a random walk that does not return to the previous node, the cover time is improved.

From simulations of [3] it was observed that GRG with variation 2 did not reduce the number of nodes visited before the escape operation, because the holes were found to be interconnected and not closed, creating a large face over most areas of the network. In addition, variation 4 could not be implemented with 2. Variation 3 was found to have lower overhead than variation 1. It was found that variations 2 and 3 performed better than flooding techniques for dense networks. Although GRG provides an adequate method for escaping minima, it does not guarantee data delivery. GRG also relies on the assumption of a UBG, which does not accurately model a real wireless network.

To address this, the authors in [17] proposed a memoryless hull-routing technique to escape local minima. Similar to the commonly used greedy-face-greedy (GFG) routing technique of confining the recovery forwarding to the "face" of the local minima, a GHG routing technique confines the recovery to the "hull" of the local minima. The hull of a subspace is defined as a structure containing Delaunay triangles enclosing the subspace and single edges. A partial unit Delaunay triangulation (PUDT) algorithm is applied to remove intersecting edges and triangles, similar to a 2D planar graph construction. When a local minimum is met, the packet is forwarded to valid edges and triangles within the subspace hull, which reduces overhead and guarantees success in searching the hull. The packet is forwarded to at most twice the number of nodes belonging to a hull. Simulations in [17] showed that GHG has a lower path length than GRG [3], especially at higher network density. GHG, on average, has a 20% longer path length than the optimal path, which is determined through message flooding, and is 50% longer than the worst-case path.

In [16], Zhou et al. developed the greedy distributed spanning tree routing for 3D networks (GDSTR-3D). GDSTR is a routing protocol originally developed for 2D networks that uses greedy forwarding until the message reaches a local minimum, where it then switches to tree traversal forwarding mode. The tree traversal mode makes use of hull trees, which are spanning trees in which each node is associated with a convex hull that contains the location information of its descendants (Figure 8.8). The convex hulls reduce the size of the spanning tree and remove unproductive paths. GDSTR uses two convex hull trees, each one located at opposite ends of the network.

To convert GDSTR method to 3D (GDSTR-3D), two 2D convex hulls are used to approximate a 3D convex hull, which would require a large amount of storage and computation. Additionally, GDSTR-3D uses 2-hop neighbor information in greedy forwarding to escape local minima. In this case, each node must store the locations of its closest neighbors (1 hop) and its neighbor's closest neighbors (2 hop). It was found in [16] that any 3D geographic routing can be significantly improved through the use of 2-hop neighbor information.

Evaluation of GDSTR-3D was performed with a wireless sensor network test bed, Indriya, which is composed of 120 TelosB sensor motes deployed on three levels of a building. Using simple greedy forwarding, it was found that the packet success delivery rate dropped from 90% to about 40%, as the number of levels of the building increased from 1 to 3. However, applying the 1- and 2-hop neighbor information

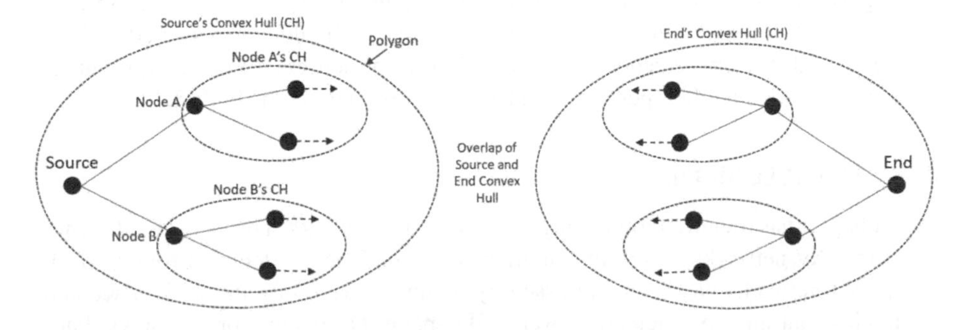

FIGURE 8.8 Two 2D convex hull trees in which the convex hull polygons are approximated with ellipses.

increased the success rate to approximately 95%. Additionally, GDSTR-3D achieved a hop stretch close to 1.

To further evaluate GDSTR-3D, in [16] simulation was performed with TOSSIM, a simulator for TinyOS. Results showed that GDSTR-3D asymptotically approaches a success rate of approximately 100% as average node degree increases. Simulations were also performed for GDSTR-3D with three different tree storage methods: two 2D convex hulls, three 2D convex hulls and a sphere. It was found that two 2D convex hulls had the lowest storage cost and comparable hop stretch to the three 2D convex hulls. Because of its lower overhead cost, two 2D convex hulls are preferable to three 2D convex hulls. Simulations also revealed that greedy forwarding performs well in high-density networks, but they cannot achieve 100% success rate in low-density networks even when using 2-hop neighbor information.

Additionally, the forwarding algorithms discussed in Section 8.1 include route recovery techniques. For instance, in [14] the 3DRTGP algorithm utilizes a back-pressure mechanism if a message fails to reach its destination, due to local minima or traffic congestion. In this mode, the message is broadcast with its packet flag set to 1. The previous sender overhears the packet, sets the packet flag to 2, and retransmits it to its neighbors, who reset the packet flag to 0 and retransmit. This allows the packet to route around the local minima. In [13], the authors that their forwarding method could send messages around local minima based on the knowledge of its neighbors' residual energy, location and energy consumption by using two node-disjoint anchor lists.

8.5 DISCUSSIONS

The development of 3D routing algorithms is inherently more complex than that of 2D routing algorithms. On one hand, this challenge is due to the non-trivial division of 3D space into smaller subspaces through the use of UBGs [3, 6], truncated octahedral tessellation [7], beacons [8], Delaunay triangulation [10, 11, 17], and hull trees [16]. The subspaces are then used to aid in local minima escape or forwarding decisions. Because 3D networks are important in UAV applications, the routing algorithms should address not only local minima escape and forwarding choices, but also consider the sparsity, scalability, and resource-poor qualities of the networks. Therefore, the algorithms developed for churn-resilience [10], high network-lifetime [13–15], low overhead [12, 16], and minimal traffic congestion [15], should be considered and further explored. The abovementioned routing algorithms address different requirements of a 3D network, and their performance has been summarized in Table 8.4.

8.6 CONCLUSIONS

This chapter has reviewed routing algorithms in 3D networks. They can be efficiently used in UAV networks if certain conditions are met. The challenge of routing packets in 3D networks has been addressed by using connectivity and node placement strategies that aim to efficiently cover a 3D space. Durocher's proof shows that a k-local memoryless routing algorithm cannot deterministically guarantee successful delivery in 3D networks. Therefore, geographic routing scheme, the most attractive

TABLE 8.4
Comprehensive Comparison of All Routing Algorithms

	Partition of Space/ Network	Delivery Performance	Network-Conscious Performance
MDT [10]	Multi-hop Delaunay triangulation graph	Guaranteed delivery with correct node information	Resilient to churn, handles topology change
GSG [11]	3D restricted Delaunay triangulation graph	No guaranteed delivery, stable path length with increasing network size	None
EGF3D [12]	Subminimal ellipsoid	Approaches to 100% delivery rate with increasing network size	Low protocol overhead in comparison to GRG [3]
EDGR [13]	Two node-disjoint anchor lists	90% delivery rate for dense networks with moderate hole sizes	Low energy consumption and high network lifetime
3DTRGP [14]	Packet forwarding region	100% delivery rate when packet deadline is ignored	Reduces redundant packet transmissions, collisions, congestion
DFTRP [15]	None (fuzzy probability routing)	High delivery rate	High network lifetime
GRG [3]	UBG	No guaranteed delivery	None
GHG [16]	Partial unit Delaunay triangulation graph	Guaranteed delivery, low path length compared to GRG [3]	None
GDSTR-3D [17]	Two spanning hull trees	Approaches 100% delivery with increasing network size	None

method for 3D routing, cannot reach 100% success delivery. There have been developments of progress-based forwarding methods, which depend on location information of neighboring nodes to forward the packet closer to the destination node with techniques to avoid local minima. Additional methods include randomized forwarding methods, which utilize knowledge of traffic probability and node energy to route the packet to the destination with minimal collisions and low energy consumption. Last, because local minima are unavoidable, route recovery methods are needed to efficiently route packets out of local minima despite the complexities of 3D holes. The network partitioning, packet forwarding and recovery methods discussed in this chapter can be used to meet the needs of 3D UAV networks.

REFERENCES

1. B. Shah and K. Kim, "A survey on three-dimensional wireless ad hoc and sensor networks," *International Journal of Distributed Sensor Networks*, vol. 2014, 2014.
2. A. E. Abdallah, T. Fevens, and J. Opatrny, "High delivery rate position-based routing algorithms for 3D ad hoc networks," *Computer Communications*, vol. 31, pp. 807–817, 2008.

3. R. Flury and R.R. Wattenhofer, "Randomized 3D geographic routing," *IEEE Infocom 2008*, 2008.
4. M. Jouhari, K. Ibrahimi, M. Benattou, and A. Kobbane, "New greedy forwarding strategy for UWSNs geographic routing protocols," *2016 International Wireless Communications and Mobile Computing Conference (IWCMC)*, pp. 388–393, 2016.
5. M. Narasawa, M. Ono, and H. Higaki, "NB-FACE: no-beacon face ad-hoc routing protocol for reduction of location acquisition overhead," *7th International Conference on Mobile Data Management (MDM'06)*, Nara, Japan, pp. 102–102, 2006.
6. S. Durocher, D. Kirkpatrick, and L. Narayanan, "On routing with guaranteed delivery in three-dimensional ad hoc wireless networks," *Proceeding of ICDCN*, 2008.
7. S. M. N. Alam and Z. J. Haas, "Coverage and connectivity in three-dimensional networks," *Ad Hoc Networks*, vol. 34, pp. 157–169, 2015.
8. J. Cleve and W. Mulzer, "Combinatorics of beacon-based routing in three dimensions." In: M. Bender, M. Farach-Colton, M. Mosteiro, Eds. *LATIN 2018: Theoretical Informatics. 2018. Lecture Notes in Computer Science*. Cham, Switzerland: Springer, vol. 10807, 2015.
9. Y. Cui, S. Xiao, et al., "Diamond: nesting the data center network with wireless rings in 3D space," *Proceedings of the 13th USENIX Symposium on Networked Systems Design and Implementation 2016*, 2016.
10. S. Lam and C. Qian, "Geographic Routing in *d*-dimensional spaces with guaranteed delivery and low stretch," *IEEE/ACM Transactions on Networking*, vol. 21, no. 2, 2013.
11. H. Su, Y. Wang, and D. Fang, "An efficient geographic surface routing algorithm in 3D ad hoc networks," *Proc. of the 5th International Conference on Pervasive Computing and Applications (ICPCA)*, pp. 138–144, 2010.
12. Z. Wang, D. Zhang, O. Alfandi, and D. Hogrefe, "Efficient geographical 3D routing for wireless sensor networks in smart space," *Baltic Congress on Future Internet and Communications*, 2011.
13. H. Huang, H. Yin, et al., "Energy-aware dual-path geographic routing to bypass routing holes in wireless sensor networks," *IEEE Trans. on Mobile Computing*, vol. 17, no. 6, pp. 1339–1352, 2018.
14. S. F. Al Rubeaai, M. A. Abd, et. al., "3D real-time routing protocol with tunable parameters for wireless sensor networks," *IEEE Sensors Journal*, vol. 16, no. 3, pp. 843–853, 2016.
15. M. S. Gharajeh and S. Khanmohammadi, "DFRTP: dynamic 3D fuzzy routing based on traffic probability in wireless sensor networks," *IET Wireless Sensor Systems*, vol. 6, no. 6, pp. 211–219, 2016.
16. C. Liu and J. Wu, "Efficient geometric routing in three dimensional ad hoc networks," *IEEE INFOCOM 2009*, Rio de Janeiro, pp. 2751–2755, 2009.
17. J. Zhou, Y. Chen, B. Leong, and P. S. Sundaramoorthy, "Practical 3D geographic routing for wireless sensor networks," *SenSys 2010–Proceedings of the 8th ACM Conference on Embedded Networked Sensor Systems*, pp. 337–350, 2010.

9 UAV Protocol Design for Computation Applications

Immanuel Manohar, Qian Mao
Electrical and Computer Engineering,
The University of Alabama, Tuscaloosa, AL, USA

CONTENTS

9.1 INTRODUCTION

In this chapter we will discuss the use of unmanned aerial vehicle (UAV) swarms for distributed computations. This problem falls into the very end of the spectrum of challenges in mobile grid computing and would fall under the category of fog computing. In many scenarios, the decision of computing and where to perform the computing is needed. At times, it is better for the source to perform the computation locally and at times the cloud server is needed. With the introduction of fog computing, we have the capability to distribute computation with neighboring fog nodes. In this chapter we design an algorithm that finds the optimal distribution of computational load among different nodes based on the network structure [media access control (MAC) and routing constraints] and computational capability in the network. The load distribution depends on the following:

1. Type of computation performed
2. Number of nodes in the network and their communication capability

With most UAVs, the computation and communication capabilities are limited because of energy constraints. These are further complicated by high degree of mobility with UAV speeds, which are about 50–70 mph on average. With this in mind, we try to present a feasibility study and discuss the current advances in two fields: communication protocols that enable high mobility and parallel computation protocols currently in use. We then propose an approach involving the concepts introduced in this chapter for UAV grid computations along with its performance.

There are two main challenges with using UAV swarms for computation:

1. The nodes are highly mobile and need good communication protocols.
2. The computation capabilities of the nodes are intermittent and need to be pooled together efficiently.

With recent advances in UAVs and communication hardware, the traditional protocols designed for fixed nodes do not perform well. Most distributed computing algorithms (grid computing algorithms) are developed for fixed computing infrastructure with easily determinable networking and computational capabilities. With mobility and the ad hoc nature of both computation and communication added to the picture, the traditional algorithms fail. The study of such environments is called a mobile grid computing model. Although most research is in the nascent stage, the current best algorithm that addresses a mobile grid environment is the work-stealing algorithm.

This chapter is an introduction into the field of mobile grid computing and addresses the challenge of distributed computations using UAVs. The first section proposes the general model and sets up the environment, including mobility and its issues and presents the various challenges. The communication and load distribution protocols currently available are discussed in section 9.3. Section 9.4 presents possible solutions in the field and is followed by concluding discussions in Section 9.5.

9.2 THE SCENARIO, CHALLENGES AND APPLICATIONS

First, we look into the model for a network, the hardware, current protocols and the structure of the UAV swarm designed for computations.

9.2.1 Scenario

9.2.1.1 What Is a UAV Swarm? How Do We Model It?

Swarms have long fascinated people, and inspired countless poems and a multitude of research.

> I saw Starlings in vast lights, borne along like smoke, mist - like a body unindued with voluntary power - now it shaped itself into a circular area, inclined - now they formed a Square - now a Globe - now from complete orb into an Ellipse - then oblongated into a Balloon with the Car suspended, now a concave Semicircle; still expanding, or contracting, thinning or condensing, now glimmering and shivering, now thickening, deepening, blackening!
>
> **Samuel Taylor Coleridge**

For the robot swarm, we use inspiration from nature on Starling swarms to model our swarming pattern. This is a proven swarming model for UAVs as shown in [1]. Such models also approximate scenarios of the U.S. Air Force drone swarms [2], and it is better than the unconstrained mobility as in [3, 4].

The mobility model we use here is based on starling flocks. This is quite a robust mobility model and is a reflection of swarming behavior throughout nature. It has the following advantages:

1. Fits different swarming patterns as necessary
2. Representative of a variety of complex swarming behaviors
3. Has one of the best obstruction avoidance capabilities

Although swarms in nature have no bounding constraints, in reality, we need to place them. This is applied by using a repulsive force on the members of the swarm near the boundary, and, to avoid collision, we use a repulsive force on nodes closer than a threshold to each other (Figure 9.1).

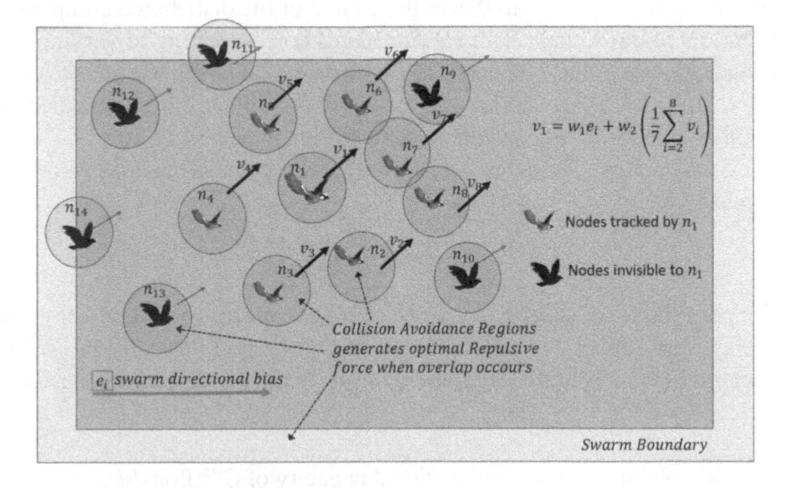

FIGURE 9.1 Swarming mobility model.

Let node n_i have a velocity vector with both magnitude and direction v_i. Let v_{max} be the maximum velocity possible for all nodes. Inspired by the nature of starling swarms, we note that each node observes only seven other neighbors, and these are the neighbors independent of metric distance and in the Voronoi shell (topological distance). Let the seven topological neighbors of node n_i be the set $\mathcal{N}_i$. When the node is close to a boundary, other nodes or obstacles, a repulsive force is added to the node r_i. The direction r_i is a repulsive force equivalent to an electric charge of the same polarity at the boundary and at the collision node

$$\upsilon_i = w_1 e_i + w_2 \sum_{n_j \in N_i} \upsilon_j + w_3 r_i$$

where, e_i is a general directional bias that can be changed to control the motion of the swarm, and w_1, w_2 and w_3 are weights adjusted to the criticality of each term, such that $w_1 + w_2 + w_3 = 1$ and $w_i \geq 0, i = 1, 2, 3$.

9.2.1.2 Network Details and Computational Capability

Most UAVs could be modeled as a node equipped with omnidirectional half duplex antenna. The wireless protocols applicable and currently available are summarized in Table 9.1.

We consider a set of UAVs, $\mathcal{N} = \{n_0, n_2, n_3, \ldots n_{N-1}\}$with $|\mathcal{N}| = N$ individual nodes. Each node, n_i, acts as both relays and computational servers depending on the scenario under consideration. The maximum computational capacities for all nodes at time slot t are

$$C_t \triangleq \left[C_t^{(0)}, C_t^{(2)}, \ldots, C_t^{(N)}\right] \tag{9.1}$$

which is the computational capacity available for processing in flops/sec. Each node also has varying computational capability that varies over time. Computational capacity can diminish or drop to 0 over the course of the distributed computation. It is modeled as a Markov chain with varying states corresponding to varying computational capacity over time. The nodes are assumed to be UAVs with position vector, p^t, at time t. This gives rise to node selection based on reliable communication, which depends on both mobility and available computational capability.

The selected nodes also determine the expected throughput to all nodes from source node n_0 as

$$T \triangleq \left[T_0, T_1, \ldots, T_N\right] \tag{9.2}$$

which includes the MAC layer constraints, channel capacity, queuing delays, propagation delays and processing delays. Because source node already has all of the data it needs, $T_0 = \infty$. Additionally, any node n_i is assumed to have the following properties:

1. An available maximum computational capacity of $\overline{C^{(i)}}$ flops/sec.
2. Each node is equipped with an omni+ directional antenna.

TABLE 9.1
Summary of Wireless Technologies, Their Capabilities, Advantages, Disadvantages in Use With Fog Nodes

Technology	Operating Frequency	Band Width	Range	Down/U Speeds	pPower	Fog Use	Advantage	Disadvantage
Mobile networks, e.g., HSPA+, LTE	450, 850 MHz; 1.9, 2, 2.5, and 3.5 GHz	20 MHz	200 km	168/22 Mbps	Low (downlink nodes) High (uplink nodes)	Bulk transfer to cloud from smart devices, edge Clouds, fog nodes located in base stations for computation and cache	Allows high mobility with less infrastructure costs, covers wide areas	Constant use is power intensive on mobile devices running on battery, needs base stations, high latency restricts real-time and vehicular applications Wireless mesh networks are costly and impractical
Wi-Fi (IEEE 02.11a, b, g)	2.4, 5 GHz	20 MHz	10–30 m	54 Mbps	Low	P2P networks V2V vehicular networks	Low cost, low power consumption for downlink nodes	Most devices equipped with this technology have enough computational capability that renders the use of mesh computing irrelevant
Wi-Fi (IEEE 802.11ax, ac)	5 GHz	160 MHz	10–30 m	9.6 Gbps	Low	Powerful edge Clouds, online control applications, industrial automation, vehicular networks	Easy to deploy, allows good mobility, capable of powerful computations, suited for big data	Highly volatile because of mobility; hence locational reliability is tougher to achieve

(*Continued*)

TABLE 9.1 (Continue)
Summary of Wireless Technologies, Their Capabilities, Advantages, Disadvantages in Use With Fog Nodes

Technology	Operating Frequency	Band Width	Range	Down/U Speeds	pPower	Fog Use	Advantage	Disadvantage
Wi-Fi (IEEE 802.11 aj,ad,ay)	>45 GHz	1—2 GHz	3–4 m	100 Gbps	Low	Wireless data centers	High data speeds rivaling best wired networks	Range is low
Multibeam antennas [5]	Varies based on range, available at both high and low frequencies	20 MHz	Varies based on frequency and power	<50 Gbps	s Varies	Military application for security, to cover wide areas, e.g., concerts, stadiums with high volume of Iot/smart devices, vehicular networks	Allows for physical layer security, jamming avoidance, high data rates that rival wired connections with less interference problems, capable of forming multipath concurrent transmission and reception, highly connected mesh networks	High cost (might decrease with prevalence) and power consumption

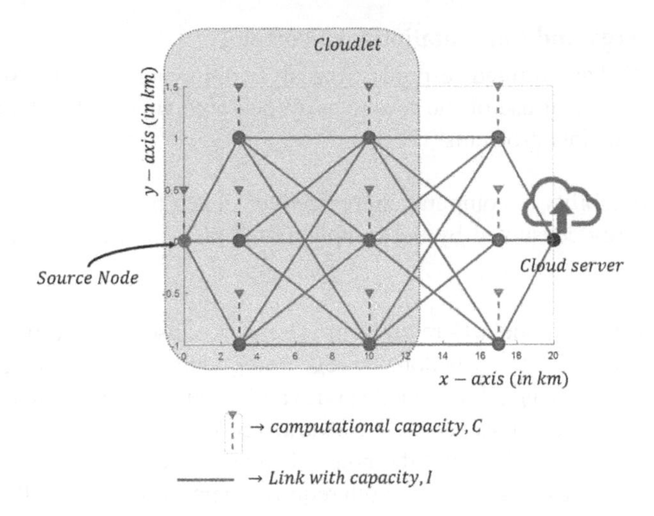

FIGURE 9.2 Network scenario.

3. Each connection is assumed to have a transmission capacity of *l packets*/sec with a fixed packet size of *K* bytes.
4. Of all the nodes, a single node acts as a source node with *G* bytes of data that need computation and communication.
5. Different computation scenarios require different network topologies to give optimal solution.
6. There exists a cloud node in the network that is assumed to have infinite computational capacity. The illustration of the final network scenario after routing is shown in Figure 9.2.

9.2.2 Challenges in UAV Swarm Computations

The three main challenges are mobility, energy and computational resource constraints and security issues.

9.2.2.1 Mobility

As mentioned previously, the most important challenge is the high degree of mobility associated with UAV swarms. In the case of vehicular networks, they have approximately the same mobility as UAV swarms and have a rapid topology change that results in fragmented networks with network routes having a validity of less than 5 seconds [6]. Currently, there are many algorithms for routing in ad hoc network models, which will be discussed later in this chapter. The problem is that most of the models are designed for slow mobility scenarios (e.g., walking and running speeds). UAV swarms represent the ultimate challenge to all the communication protocols developed. They share a great deal of similarity with vehicular networks, which in itself is still a developing field with no concrete solutions in place. The probable solution is to use mobility prediction to design routing algorithms; a sample solution is presented in the later part of this chapter.

9.2.2.2 Energy and Computational Uncertainty

Because UAVs have limited resources, the algorithms and protocols so developed should make efficient use of the resources. When computational tasks are present, there are two distinct problems:

1. The availability of computational resources in nearby devices
2. The energy consumed by the computational task and the required energy for UAVs

The first problem is popularly referred to as resource discovery. In traditional systems, the resource discovery is done by registering with a centralized server, which acts as the oracle serving all the mobile nodes and teaches how to share the computational resources. In other scenarios, the centralized server (e.g., Google Cloud) only shares which nodes are free and the nodes decide by themselves how to distribute the computational task, but because both require a central server, it will not work for UAV swarms.

Currently there is a huge body of work addressing the energy consumption, and it is an active research topic, but it is beyond the scope of this chapter. The reader is encouraged to check other works.

9.2.2.3 Security

The following security issues are associated with fog nodes and applicable to UAV swarms [7, 8]:

1. *Blackhole attack*: A node advertises a shorter path to be included in the route to receive all data forwarded through it. This enables the node to either sniff the data or to spoof the data sent to the destination with the sender none the wiser.
2. *Denial of service*: This form is done by the malicious node hijacking the network bandwidth by flooding the network with frequent route requests so the routing table overflows, resulting in a lack of space for legitimate routes.
3. *Impersonation*: The attacker might impersonate a legitimate node to create wrong entries in the routing table.
4. *Energy consumption*: Because most mobile ad hoc network (MANET) devices use a battery, if the attacker sends useless packets or requests frequent routes it can drain the batteries.
5. *Information disclosure*: This could be done using a blackhole attack; after intercepting the data, the attacker would leak the information gathered.

9.2.2.4 Other Challenges

9.2.2.4.1 Latency

This is extremely important for real-time control and automation in industrial networks, such as cyber physical systems. Thus, the UAV swarm computing mesh should use routing algorithms to ensure the latency is low.

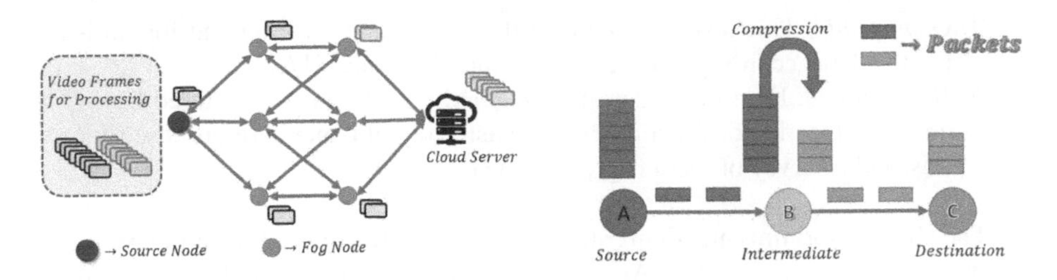

FIGURE 9.3 Fog and Cloud distributed computation scenarios.

9.2.2.4.2 Application-Based Routing

One of the advantages of UAV swarms is that it is closer to the edge and can cater to specific applications. Thus, the routing algorithm should take into consideration how to create the wireless mesh so that the data are distributed with the application in mind. For more issues and challenges in multichannel routing, refer to [9].

9.2.3 Applications

An application for such a UAV swarm computation is the shifting of computation load resulting in low latency results. Example scenarios are illustrated in Figure 9.3. In the field of video coding and processing, techniques such as PRISM [10], which performs distributed video coding, the source trades the computational complexity of finding the motion vectors by itself and distributes it with the other nodes in the network. The illustration shows how the computational complexity of finding the motion vectors could be distributed among the nodes in the network. In the second illustration, we demonstrate how compression in an intermediate node helps reduce the congestion in nodes further down the network.

9.3 POSSIBLE ROUTING PROTOCOL AND LOAD DISTRIBUTION SOLUTIONS FOR UAV SWARM COMPUTATIONS

The traditional mesh routing protocols for MANETS fall in the following categories:

1. *Proactive*: In this the network nodes constantly update the positional information and their routing tables periodically. The most popular of these is Destination-Sequenced Distance Vector (DSDV) [11], Optimized Link State Routing Protocol (OLSR) [12], and Better Approach To Mobile Adhoc Networking (BATMAN) [13].
2. *Reactive*: In this, the source requests a route to the destination when needed and then the routing table is constructed based on it (e.g., Ad-hoc On-demand Distance Vector (AODV) [14] and Dynamic Source Routing (DSR) [15]).
3. *Hybrid*: This involves a mix of both proactive and reactive routing. Although nodes store local information, such as subtrees and clusters, it makes it easier and faster to find a route when needed (e.g., zone routing protocol) [16].

4. *Geographic*: This involves knowing the destination location and forwarding the data accordingly (e.g., location-aided-routing) [17].
5. *Hierarchical*: Here the nodes are classified into groups or clustered based on either location or functionality. A vast array of papers are proposed for this, and a survey of them is given in [18].

The above algorithms are highly studied and comparative analysis could be found in many surveys (e.g., [19, 20]). An excellent comparison of some of the above protocols is given in [21, 22]. In this chapter we focus on a few interesting approaches to solve the problems posed in the previous section:

1. *Location hiding* [23–25]: In these works, the privacy attacks presented in the previous section are addressed.
2. *Mobility modeling*: Many mobility models have been used to address the mobility issue in routing [26].
3. *Genetic algorithms for load balancing* [27]: A genetic algorithm is used to balance load, hence, ensuring fairness.
4. *Using software defined networking for managing a fog network* [28]: Software-defined networking is used to manage fog networks.
   Gravity gradient routing: This article [29] considers an opportunity for fog implementation for alert services on top of wireless sensor network (WSN) technology. In particular, they focus on targeted WSN-alert delivery based on spontaneous interaction between a WSN and the handheld devices of its users. For the alert delivery, we propose a gravity routing concept that prioritizes the areas of high user-presence within the network. Based on this concept, they develop a routing protocol, namely the gradient gravity routing (GGR), which combines targeted delivery and resilience to potential sensor-load heterogeneity within the network. The protocol has been compared with a set of state-of-the-art solutions via a series of simulations. The evaluation has shown the ability of GGR to match the performance of the compared solutions in terms of alert delivery ratio, while minimizing the overall energy consumption of the network.

9.3.1 A Proposal for Using Path Prediction to Compensate for Routing Uncertainty

Each node n_i broadcasts the known/predicted path with its hello message packets for time stamps $\{t_0, t_1, \ldots, t_n\}$. This helps on two fronts: (1) determining the more stable links during the routing process and (2) helping to control the amount of control traffic.

If the path is not known or decided a priori, the node estimates the path by using the path history. This is done by fitting a polynomial of order m to the recorded path history using the Vandermonde matrix. The polynomial is

$$\begin{aligned} x(t) &= \alpha_0 + \alpha_1 t + \alpha_2 t^2 + \ldots + \alpha_m t^m, \\ &\triangleq \underline{\alpha T^{\mathrm{T}}} \end{aligned}$$

where $x(t)$ is the x-coordinate at time t, $\alpha = [\alpha_0, \alpha_1, \ldots, \alpha_m]$ and $T = [1, t, t^2, \ldots, t^m]$. Similarly, let $y(t) \triangleq \underline{\beta T}^{\mathrm{T}}$ and $z(t) \triangleq \underline{\gamma T}^{\mathrm{T}}$ be the polynomials corresponding to the y- and z-coordinates. The polynomial should have the lowest order m that best approximates the path taken by the node without overfitting it. Here we use the Vandermonde matrix and least squares to fit a lowest order polynomial that approximates the position of the node at different times for each $\{x(t), y(t), z(t)\}$. Let the position of the node at the different time instances, t_i, be $\{x_i, y_i, z_i\}$.

Definition 1. A Vandermonde matrix, $V \in \mathbb{R}^{N \times N}$, is a matrix where all the rows are in geometric progression.

Let

$$V \triangleq \begin{bmatrix} 1 & t_0 & t_0^2 & \cdots & t_0^m \\ 1 & t_1 & t_1^2 & \cdots & t_1^m \\ \vdots & \vdots & \vdots & \vdots & \vdots \\ 1 & t_{N-1} & t_{N-1}^2 & \cdots & t_{N-1}^m \end{bmatrix},$$

$$\underline{x} \triangleq \begin{bmatrix} x_0 & x_1 & x_2 & \cdots & x_{N-1} \end{bmatrix}^{\mathrm{T}}$$

$$\underline{y} \triangleq \begin{bmatrix} y_0 & y_1 & y_2 & \cdots & y_{N-1} \end{bmatrix}^{\mathrm{T}}$$

$$\underline{z} \triangleq \begin{bmatrix} z_0 & z_1 & x_2 & \cdots & z_{N-1} \end{bmatrix}^{\mathrm{T}}$$

where V is a Vandermonde matrix, t_i are time instances for $i = 0, 1, \ldots, N-1$, x_i, y_i, z_i are x-, y- and z-coordinates[1] of the node at time t_i. The coefficient estimate for $x(t)$, $y(t)$ and $z(t)$ are solved by

$$\underline{\alpha} = V^{\dagger}\underline{x}$$
$$\underline{\beta} = V^{\dagger}\underline{y}$$
$$\underline{\gamma} = V^{\dagger}\underline{z}$$

where $V^{\dagger}$ is the pseudo inverse of the Vandermonde matrix V. To avoid overfitting, the order of the polynomial, m, is determined by finding the ratio of the residuals. Let $e_m^x = \underline{x} - V^{\dagger}V\underline{x}$ be the residual for fitting an mth order polynomial to the x. The ratio $r_m = \frac{e_{m-1}^z}{e_m^x}, i = 2, 3 \ldots M$ is then used to determine the optimal m given by

$$m_{opt} \triangleq \arg\max_m r_m$$

The performance of the path prediction is illustrated in Figure 9.4.

---

[1] These coordinates are relative to a universal reference point that all the nodes in the network agreed on

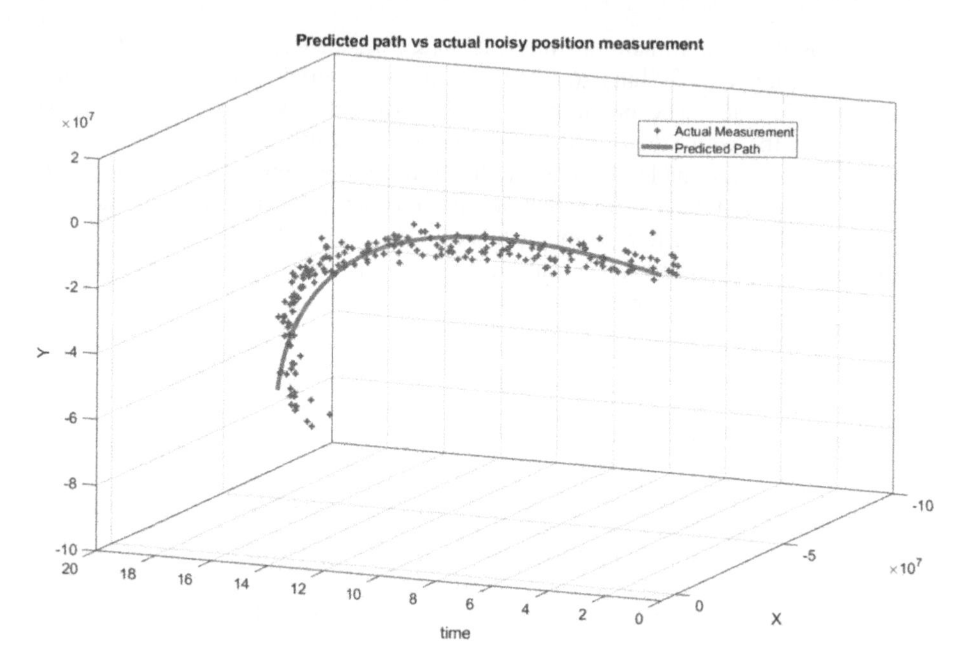

FIGURE 9.4 Vandermonde-based path prediction in two dimensions—performance.

9.3.2 Distributive Computational Techniques

Here we discuss the two main methods of distributing computational loads: (1) work-stealing algorithm and (2) optimal work distribution algorithm (discussed in Section 9.4).

9.3.2.1 Work-Stealing Algorithm

> This is a popular algorithm for computations that could be split into multiple threads and executed parallelly. This could also be viewed as a scheduling algorithm. First, the workers (with free resources) query the host node for data that need computations and then steals the computations away from the host. As long as the workers are free, they try to serve by stealing the computations.

9.4 AN EXAMPLE ALGORITHM AND TECHNIQUE

Let us look at one possible solution to the problem of distributed computation.

9.4.1 Objective

The goal of this work is to design intelligent routing algorithms that help facilitate computation and data transfer. Here, for computational purposes, we focus on tensor decomposition and compression. This is selected as almost all data can be expressed in the form of tensors and the cost of tensor decomposition and compression is the costliest operation for a wide variety of applications. This helps facilitate a variety of operations from feature extraction, subspace tracking and so forth. The goal of the work is to determine the following:

1. How should the computational load among different nodes in the network be shared?
2. What is the most efficient way of computing given the quantity of data? Should the node compute by itself, employ the help of nearby nodes, use the help of the cloud server, or use the help of nearby nodes and the cloud server?
3. Derive a mathematical model for the time for computation.

Apart from these, we give insight into the type of messages that the nodes forward to obtain information of the nodes in the network.

9.4.2 Methodology

In this section, we look at how the computation load can be distributed. In essence, we look at the following scenarios: the entire computation is carried out by the host node and the computation load is divided between the host and its neighbors, transferred to the cloud server and shared between the cloud and the nodes in between the source and cloud.

The amount of source data is G bytes, assuming the source, with computational capacity $C^{(0)}$, carries out the entire set of operations, the time it takes would be

$$t_G^{self} = f\left(G, C^0\right) \tag{9.3}$$

In case of tensor decomposition

$$f(G) = \frac{AG^{1.5}}{C^0} \tag{9.4}$$

where A is a constant based on the type of algorithm used to solve singular value decomposition (SVD). There are more details on this in the next section.

Now, we look at dividing up the data and solving it in different nodes including the host node. Let a_0G be the amount of data computed in the source node and a_iG be the amount of data computed in layer 1 nodes such that

$$\phi = \left[a_0, a_1, a_2, \ldots, a_1\right]$$

and $\|\phi\|_1 = 1$, where a_i is the ratio of data transmitted for computation at node n_i. Assume the effective throughput for the nodes n_i in be $T_i$bytes/sec, which includes the MAC layer constraints, channel capacity, queuing delays, propagation delays and processing delays. We now try to divide up the data such that all nodes in layer 1 complete the computation at essentially the same time by finding an optimal ϕ. The time required for transmission of data to node n_i is given by $\frac{a_i G}{T_i}$ and the time for computation is $f(a_i G, C_i)$, i.e., for the data to get transmitted and computation to end at node n_i, the total time taken would be

$$t_{a_i G}^{(n_i)} = f\left(a_i G, C^{(i)}\right) + \frac{a_i G}{T_i} \tag{9.5}$$

Now, because under optimal distribution of data the source node is n_0 and all other nodes n_i complete the computation and transmission at exactly the same time, we form the following set of equations:

$$f\left(a_0 G, C^{(0)}\right) = f\left(a_i G, C^{(i)}\right) + \frac{a_i G}{T_i}., i = 1 \ldots N \tag{9.6}$$

$$\sum_{i=1}^{N} a_i = 1$$

where N are the number of nodes chosen for computation. In, Eqn. (9.6), there are total of N unknowns and N equations, thus solving it gives the optimal distribution of data size. Note, Eqn. (9.6) on first glance does not consider the time the node n_i performs computation and transmission simultaneously. In fact, this cost can be bundled up in either of the terms $f(a_i G, C^{(i)})$ or $\frac{a_i G}{T_i}$ by correspondingly increasing throughput variable T_i to compensate for it or adding a compensation term in $f(a_i G, C^{(i)})$. Nonetheless, the time consumed until completion of computation and transmission at node n_i is a function of a_i, G, C_i and T_i. Also note, Eqn. (9.6) is non-linear and finding the solution is non-trivial. To solve the same, we frame it as an optimization problem as follows. First, we design the cost vector

$$L \triangleq \left[l_1, l_2 \ldots, l_N\right] \tag{9.7}$$

where

$$l_i \triangleq f(a_0 G, C_0) - f(a_i G, C_i) - \frac{a_i G}{T_i}, i = 1 \ldots N \tag{9.8}$$

then, the optimization problem is given by

$$\min_{\phi} \quad L_{2|}^{2} \tag{9.9}$$

$$\begin{aligned} s.t \quad & \|\phi\|_1 = 1. \\ & 0 \le \phi \le 1 \end{aligned} \tag{9.10}$$

Let the optimal distribution of data be ϕ_{opt} solved by using the above optimization problem. The algorithm used to solve the above optimization problem is given in

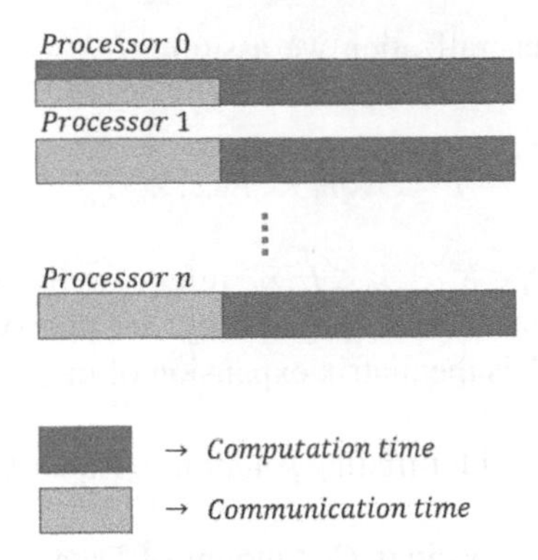

FIGURE 9.5 Computational and communication components of the load distribution algorithm. In the optimal scenario, the host (Processor 0) starts computation from the time the task is received and all other nodes first receive the data for computation (pink block) and then proceeds on with the computation (blue block).

algorithm 1. Note, $\|C\|_2^2 \geq 0$, and when $C = 0$, it is the solution to the numerical problem in Eqn. (9.6).

For there to be a meaningful dividing up of data, $a_iG > K$ bytes, where K is the minimum amount of data required for performing the computation. Thus, if $a_iG < K$, then a_i is set to zero for the smallest a_i, and the corresponding node, i, is removed from set $S_1 = S_1 - i$ and added to set S_2, and then $a'_i s$ are recomputed. Algorithm 1 gives a summary of this procedure. Similar procedures are executed to compute the load sharing for nodes in layer 2, and so forth.

The total computation time while using all nodes in the network is given by $f(a_0G, C_0)$. All computations are carried out in node 0, i.e., the source itself if, $a_0 = 1$.

9.4.3 Computational Considerations

Now, we discuss possible computational tasks and how the previous algorithm can be applied in this scenario. The first example task we discuss is tensor decomposition, which forms the basis in terms of computational complexity for many applications that involve matrix decomposition, subspace tracking, data compression and so forth. We discuss this based on reducing the time it takes for the compressed data sent from the sender to the destination. The goal here is to distribute the computation load among all the transmission nodes along the path between source and destination, and the purpose is to reduce the transmission delay. Although the compressed tensor might have its own intrinsic advantages, we focus on the transmission delay for big data. To do this, we design an efficient tensor decomposition algorithm that works in a distributed setting. The design of the algorithm handles both streaming (the tensor that updates with time) and bulk data (a fixed size big tensor) transfers. The tensor to be transmitted is

$$T \in \mathbb{R}^{I_1 \times I_2 \times \ldots \times I_K}$$

where without loss of generalization we assume, $I_1 \geq I_2 \ldots \geq I_K$. The compressed tensor has the form

$$T = C \times F_1 \times_2 F_2 \ldots \times_K F_K$$

where $C \in \mathbb{R}^{k_i \times k_1 \times \ldots 1 \times k_K}, F_i \in R_i^{k \times I_i}, k_i \leq I_i, \times_i$ refers to the mode i tensor product. The mode i tensor product $T \times_i F_i$ is given by the matrix product, $F_i T$ $(^i)$, where $T^{(i)} \in \mathbb{R}^{I_i \times I_1 \times \ldots \times I(i-1) I(i+1) \ldots I_K}$ is the matrix expansion of the

Algorithm 1 Algorithm used for finding solution to Eqn. (9.9)

Inputs Computational capacities C, Amount of Data G, Number of nodes N, Computational capacities C, (1). computational cost function f (.), Net Transmission throughput T_i for $i = 1 \ldots N$.

Output Optimal distribution of data based on computation, ϕ_{opt}, and cost. Set $m = 0$ %Iteration count for number of local minima found cost = 100,000

While $m \leq$ *Iteration Count* **or** *cost* > *threshold*

Initialize $Temp \in \mathbb{R}^{1 \times N} \sim U(0,1); f_{temp}^{(0)} = \frac{Temp}{\|Temp\|_1} \triangleq [a_0, a_1, \ldots a_N], \Delta = [\Delta a_0, \ldots \Delta a_N].flag = 0,$

$dir = +1$; $conv = 1$; iteration counts: $k = 0$; $m = m + 1$.
Compute the cost for current ϕ_{temp} as $L(^k)$ using Eqn. (9.7) and (9.8). Total cost, $E^{(k)} = L_2^2$.
while k < *Iteration Count2* **or** *cost* < *threshold*
$k = k + 1$;
while *flag* $\leq N$
Compute $\Delta a_{flag} = dir \frac{0.2}{conv}, \Delta a_i = -dir \frac{0.2}{conv}\left(\frac{1_i^2}{e_0}\right) \phi_{temp}^{(k)} = \phi_{temp}^{(k-1)} + \Delta.$
if $\phi_{temp}^{(k)}(i) < 0$: **Set** $\phi_{temp}^{(k)}(i) = 0 \forall i = 1 \ldots N$ **endif**
if $\phi_{temp}^{(k)}(i) > 1$ **Set** $\phi_{temp}^{(k)}(i) = 1 \forall i = 1 \ldots N$ **endif**
Compute $E(^k)$.
if $|E(^k) - E(^{k-1})|$ < *threshold*
flag = *flag* + 1 % optimize over the next a_i
cost = $E(^k)$
elseif $E(^k) > E(^{k-1})$
dir = −*dir*; *conv* = *conv* + 1 /*change the direction and reduce the step size of increment if the error increases*/
endif
end while

end while
$\phi^l = \phi_{temp}^{(k)}$
cost stored (*l*) = *cost*
end while

$l = \arg\min_l$ (*cost stored*)
$\phi_{opt} = \phi^{(l)}$

```
cost = cost stored (1)
```

Algorithm 2 Determining the number of nodes performing different stages of computation

Input The computational capacities of nodes $\mathcal{N}$ the effective throughput to nodes in $\mathcal{N}$ from node 0 (source), Amount of data, G.

Initialize $\phi_l = [a_0, a_1, a_2, \ldots, a_N]$

Output $\phi_l = [a_0, a_1, a_2, \ldots, a_N]$, where a_i are the ratio of data divided and sent to node i. **Step 1:** Solve for ϕ using Eqn. (9.9).

```
Find i = arg min_a i {a_0 .. . a_N }
if aiG < K,
N = N -n_i; set a_i = 0 in ϕ_1, form new
ϕ_l without a_i: ϕ̃_l = ϕ_1 a_i. Form set Dropped nodes, D_1 = D_1∪{n_i}
Goto step 1
else
Set all values in ϕ_1 for i ∈/ D_1 with values from ϕ̃_l
Return ϕ_1
endif
```

tensor T. Let the compression ratio over each dimension i be given by $h_i = k_i / I_i$. The total tensor decomposition has a compression ratio of

$$h = \frac{k_1 \times h_2 \times \ldots k_K + k_1 \times I_1 + k_2 \times I_2 \ldots + k_K \times I_K}{I_1 \times I_2 \times \ldots \times I_K}$$

This ratio h is the total compression ratio, and it varies based on the redundancy in the tensor, T. For real-world scientific datasets, compression rates of up to 5000 are achievable without loss of accuracy [10]. Let the number of nodes along the network be, L. Each node has a computational capacity (in flops/sec) given by $C = \{C_1, C_2, C_3, \ldots, C_L\}$. We aim to distribute the computational complexity to different nodes along the network based on their computational capacity and network capabilities. To achieve the same, we proceed in two stages. First, we split the tensor T into sub-tensors of various sizes for processing. We compress the subtensors at various nodes and combine them to form the complete compressed picture of T. The compression algorithm uses Hybrid-III RRQR decomposition and the Johnson-Lindenstrauss approach.

To model the computational complexity of the algorithm used, consider a simple SVD-based matrix decomposition and compression as shown below. Let $Y \in \mathbb{R}^{K \times N}$ be the incoming data matrix (2D tensor). Factor models assume Y to have the form

$$Y = HF + E$$

where $H \in \mathbb{R}^{K\times k}$,is called the factor loading matrix. $F \in \mathbb{R}^{k\times N}$ are called factors and E is noise. The Goal is to determine an orthogonal matrix, Q, which spans the subspace spanned by H and determine $F = Q^{T}Y$. Now, although Y is of size $K \times N$ and has KN elements, Q and F have a total of $kK + kN$ elements, with $k \ll K$, thus, achieving compression. The compression ratio is given by $h = \frac{kK + KN}{KN}$. Thus, after compression the total number of packets gets reduced to $[hn]$.

We use SVD to obtain Q. Let the SVD of Y be

$$Y = USV$$

where we assign $Q \triangleq U(:, 1{:}\,k)$, which are the left singular vectors corresponding to the k highest singular values of Y. With Q obtained, we compute $\tilde{F} \triangleq Q^{T}Y$. This is a lossy compression and can be proved that

$$Q\tilde{F} \rightarrow HF$$

in probability as $N, K \rightarrow \infty$ ([30, 31]). Thus $Q\tilde{F}$ could be used as a compressed expression of Y. The computational complexity of the above procedure is mostly the SVD, and it takes $O(K^2N)$ ([32]), and without loss of generality, assuming $O(K) = O(N)$ in case of big data, $O(K^3)$ assuming each packet is of X bytes of data and each element in Y is 1 byte

$$KN = nX$$

$$OK^2 = O(nX)$$

$$O(K) = O\left(\sqrt{nX}\right)$$

Thus, the computational complexity in terms of packets is $O\left(\left(\sqrt{nX}\right)^3\right)$. The time required for computation given the computational capacity of the system in C flops/sec is

$$t_c = \frac{A\left(\sqrt{nX}\right)^3}{C}$$

where A is a constant based on the algorithm used to perform SVD and matrix multiplications.

Thus, generalizing, for compressing X real numbers in the tensor is $t_c^i \frac{AX^{1.5}}{C_i}$,where C_i is the computation capacity at node i and A is a constant. We use Tucker decomposition–based method (Figure 9.6) to compress data. First, we perform a mode 1 expansion of the tensor to form a matrix $T^{(1)}$ and split it into S_1 smaller submatrices. The compression of submatrices is done by using a hybrid of Johnson-Lindenstrauss

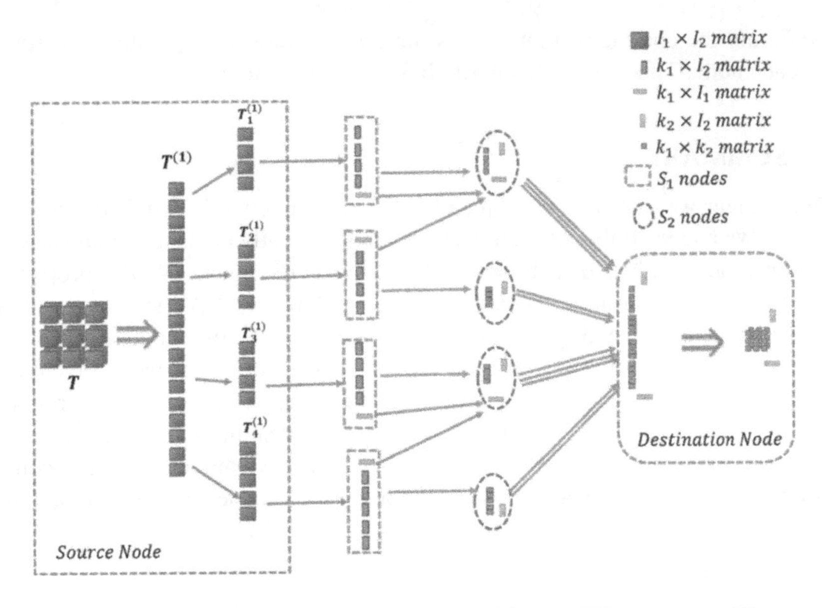

FIGURE 9.6 Proposed distributed tensor decomposition at different stages. The tensor is four dimensional, and only S_1 and S_2 nodes are present as $I_3 = 3$ and $I_4 = 2$. Hence, compression does not make much sense along those dimensions.

and rank revealing QR decomposition (RRQR)–based algorithm. Let the submatrix to be compressed be $H \in \mathbb{R}^{m \times n}$. Let k be the numerical rank of the matrix H, which means that H can be approximated by a rank k matrix without much loss of information. Here, min $(m, n) \geq k$. Let $\tilde{H}$ be a rank k matrix minimizing $\left\| H - \tilde{H} \right\|_F$. There are two objectives for compression. The first is to determine k from H and then to compute $\tilde{H}$. We use an approximation of the SVD algorithm to determine $\tilde{H}$and k. Set $Y = \Omega H$, where $\Omega \in \mathbb{R}^{a \times m}$. The dimension a is such that $k \leq a$. There are various choices for selecting Ω like random, Fourier, Hadamard and, more recently, error control coding (ECC) matrices. Here we use the ECC matrix as Ω. The method of generating Ω is the same as in [33]. With Y, we employ Hybrid-III RRQR decomposition to determine optimal k and decompose $Y = QR$. With k determined using Hybrid-III, set $\tilde{H} = \tilde{Q}\tilde{R}$, where $\tilde{Q} = Q(:, 1{:}\ k)$ and $\tilde{R} = R(1{:}\ k,:)$. The Hybrid-III algorithm used and the numerical rank determination process is the same as in [34]. Thus $\tilde{Q}$and $\tilde{R}$ form the adaptively compressed form of H. Figure 9.2 illustrates the proposed approach.

The first step is determining S_1, S_2, ...S_K, the number of nodes involved in the computation of the tensor decomposition at each stage 1, ..., K. The number of nodes associated with each stage of the tensor decomposition is S. Because I_1 is the biggest, compressing the same will have the highest compression. We assign S_1 nodes closest to the source to compress I_1 followed by S_2, S_3 ..., S_K. Because the amount of data to be compressed shrinks with each stage of compression, $S_1 \geq S_2 \geq .. \geq S_L$. For finding S_1, algorithm 2 is used to find the nodes used for computing the stage 1 decomposition and compression. With the data now distributed between different nodes involved in computing stage 1, algorithm 2 is repeated to find the nodes involved in stage 2, and the communication cost is the cost of transporting data to destination from each node. This is repeated until all $S_1 \ldots S_K$ are found.

9.5 SUMMARY

In this chapter we looked at the topic of UAV swarm networks for distributed computation. We analyzed the different technologies currently present and presented the communication and load distribution protocols currently in use. We also proposed an interesting example pointing toward a possible solution in UAV swarm computing. This is a nascent field and represents a challenge in the cutting edge of the technology spectrum. The challenges involved are in the fields of communication engineering and computational engineering. Although this chapter gives a basic introduction and a possible research direction to this field, a great deal more research needs to be done. This field needs heavy testing in a real-world environment, and the communication protocols need to be developed before a good stable solution can be found.

REFERENCES

1. D. B. Johnson and D. A. Maltz, "Dynamic source routing in ad hoc wireless networks." *Mobile Computing*, Berlin: Springer, pp. 153–181, 1996.
2. S. Ubaru, A. Mazumdar, and Y. Saad, "Low rank approximation using error correcting coding matrices," *International Conference on Machine Learning*, pp. 702–710, 2015.
3. G. Vásárhelyi, C. Virágh, G. Somorjai, T. Nepusz, A. E. Eiben, and T. Vicsek, "Optimized flocking of autonomous drones in confined environments," *Science Robotics*, vol. 3, no. 20, pp. 1–13, 2018.
4. J. Bai and S. Ng, "Determining the number of factors in approximate factor models." *Econometrica*, vol. 70, no. 1, pp. 191–221, 2002.
5. S. Khan, S. Parkinson, and Y. Qin, "Fog computing security: a review of current applications and security solutions," *Journal of Cloud Computing*, vol. 6, no. 1, p. 19, 2017.
6. J. J. Blum, A. Eskandarian, and L. J. Hoffman, "Challenges of intervehicle ad hoc networks." *IEEE Transactions on Intelligent Transportation Systems*, vol. 5, no. 4, pp. 347–351, 2004.
7. Y. B. Ko and N. H. Vaidya, "Location-aided routing (LAR) in mobile ad hoc networks," *Wireless Networks*, vol. 6, no. 4, pp. 307–321, 2000.
8. G. H. Golub and C. F. Van Loan, *Matrix Computations*, Baltimore, MD: JHU Press, vol. 3, 2012.
9. J. Broch, D. A. Maltz, D. B. Johnson, Y. C. Hu, and J. G. Jetcheva, "A performance comparison of multi-hop wireless ad hoc network routing protocols," *MobiCom*, vol. 98, pp. 5–97, 1998.
10. R. Puri and K. Ramchandran, "Prism: a new robust video coding architecture based on distributed compression principles," *Proceedings of the Annual Allerton Conference on Communication Control and Computing*, vol. 40, pp. 586–595. 2002.
11. Department of Defense. "Department of defense announces successful micro-drone demonstration." *Press release number NR-008–17*, 2017.
12. J. H. Winters, "Smart antennas for wireless systems," *IEEE Personal Communications*, vol. 5, no. 1, pp. 23–27, 1998.
13. Q. Zheng, X. Hong, and S. Ray, "Recent advances in mobility modeling for mobile ad hoc network research," *Proceedings of the 42nd annual Southeast regional conference*, New York: ACM, pp. 70–75, 2004.

14. N. Ntlatlapa, C. Aichele, and D. Johnson, "Simple pragmatic approach to mesh routing using BATMAN." *2nd IFIP International Symposium on Wireless Communications and Information Technology in Developing Countries, CSIR*, pp. 6–7, 2008.
15. A. Hakiri, B. Sellami, P. Patil, P. Berthou, and A. Gokhale, "Managing wireless fog networks using software-defined networking." *2017 IEEE/ACS 14th International Conference on Computer Systems and Applications (AICCSA)*, pp. 1149–1156, 2017.
16. J. Raju and J. J. Garcia-Luna-Aceves, "A comparison of on-demand and table driven routing for ad-hoc wireless networks," *2000 IEEE International Conference on Communications*, IEEE, vol. 3, pp. 1702–1706, 2000.
17. W. Rehan, S. Fischer, M. Rehan, and M. Husain Rehmani, "A comprehensive survey on multichannel routing in wireless sensor networks," *Journal of Network and Computer Applications*, vol. 95, pp. 1–25, 2017.
18. M. Gu and S. C. Eisenstat, "Efficient algorithms for computing a strong rank-revealing QR factorization," *SIAM Journal on Scientific Computing*, vol. 17, no. 4, pp. 848–869, 1996.
19. Z. J. Haas and M. R. Pearlman, "The performance of query control schemes for the zone routing protocol," *IEEE/ACM Trans. Networking*, vol. 9, pp. 427–438, Aug. 2001.
20. T. Clausen and P. Jacquet, "OLSR - optimized link state routing protocol (OLSR) for mobile ad hoc NETworks (MANETs)," Technical Report, 2003.
21. S. Ivanov, S. Balasubramaniam, D. Botvich, and O. B. Akan, "Gravity gradient routing for information delivery in fog wireless sensor networks," *Ad Hoc Networks*, vol. 46, pp. 61–74, 2016.
22. D. J. G. Pearce, A. M. Miller, G. Rowlands, and M. S. Turner, "Role of projection in the control of bird flocks," *Proceedings of the National Academy of Sciences*, vol. 111, no. 29, pp. 10422–10426, 2014.
23. T. W. Anderson, "The use of factor analysis in the statistical analysis of multiple time series." *Psychometrika*, vol. 2, no. 1, pp. 1–25, 1963.
24. M. Ballerini, N. Cabibbo, R. Candelier, A. Cavagna, E. Cisbani, I. Giardina, V. Lecomte, A. Orlandi, G. Parisi, A. Procaccini, et al., "Interaction ruling animal collective behavior depends on topological rather than metric distance: evidence from a field study," *Proceedings of the National Academy of Sciences*, vol. 105, no. 4, pp. 1232–1237, 2008.
25. P. Gurbani, H. Acharya, and A. Jain, "Hierarchical cluster based energy efficient routing protocol for wireless sensor networks: survey," *International Journal of Computer Science & Information Technologies*, vol. 7, no. 2, pp. 682–687, 2016.
26. M. Dong, K. Ota, and A. Liu, "Preserving source-location privacy through redundant fog loop for wireless sensor networks." *2015 IEEE International Conference on Computer and Information Technology; Ubiquitous Computing and Communications, Dependable, Autonomic and Secure Computing, Pervasive Intelligence and Computing*, IEEE, pp. 135–142, 2015.
27. D. Devaraj, R. N. Banu, et al., "Genetic algorithm-based optimisation of load-balanced routing for AMI with wireless mesh networks." *Applied Soft Computing*, vol. 74, pp. 122–132, 2019.
28. J. Kong, X. Hong, and M. Gerla, "An identity-free and on-demand routing scheme against anonymity threats in mobile ad hoc networks." *IEEE Transactions on Mobile Computing*, vol. 6, no. 8, pp. 888–902, 2007.
29. H. Deng, W. Li, and D. P. Agrawal, "Routing security in wireless ad hoc networks," *IEEE Communications Magazine*, vol. 40, no. 10, pp. 70–75, 2002.
30. X. Wu, J. Liu, X. Hong, and E. Bertino, "Anonymous geo-forwarding in MANETs through location cloaking," *IEEE Transactions on Parallel and Distributed Systems*, vol. 19, no. 10, pp. 1297–1309, 2008.

31. I. F. Akyildiz, X. Wang, and W. Wang, "Wireless mesh networks: a survey," *Computer Networks*, vol. 47, no. 4, pp. 445–487, 2005.
32. S. Basagni, M. Conti, S. Giordano, and I. Stojmenovic, *Mobile Ad Hoc Networking*, New York: John Wiley & Sons, 2004.
33. C. Perkins, E. Belding-Royer, and S. Das, "Ad hoc on-demand distance vector (AODV) routing," Technical report, 2003.
34. C. E. Perkins and P. Bhagwat, "Highly dynamic destination-sequenced distance-vector routing (DSDV) for mobile computers," *ACM SIGCOMM Computer Communication Review*, New York: ACM, vol. 24, pp. 234–244, 1994.

Part IV

Reliability and Security

10 The Future of Directional Airborne Network (DAN): Toward an Intelligent, Resilient Cross-Layer Protocol Design with Low-Probability Detection (LPD) Considerations

Fei Hu[1], Yu Gan[1], Niloofar Toorchi[1], Iftikhar Rasheed[1], Sunil Kumar[2], Xin-lin Huang[3]
[1]Department of Electrical and Computer Engineering, The University of Alabama, Tuscaloosa, AL
[2]Department of Electrical and Computer Engineering, San Diego State University, San Diego, CA
[3]Institute of Communication Engineering, Tongji University, Shanghai, China

CONTENTS

10.1 INTRODUCTION

Directional airborne network (DAN) plays a critical role in military applications due to its extended communication range (>1 km). Unmanned aerial vehicles (UAVs)/aircraft need a long radio-frequency (RF) signal propagation distance between them (>500 m for UAVs, >5 km for most aircraft networks), to minimize the probability of physical body collisions. Although omnidirectional antennas waste a great deal of transmission power due to their 360-degree RF signal broadcast nature, directional antennas can focus their energy on a small angle, having a longer transmission range.

In this chapter, we will discuss the following critical future technologies necessary to achieve safe, intelligent DAN management:

1. (**Automation**) *Self-configurable directional networking management*: Future DANs will have large-scale (>100 nodes) and complicated network management (such as swarming-oriented formation control, protocol robustness under highly mobility, hierarchical aircraft/UAV topology management, man-unmanned airborne communication cooperation, etc.). *Software-defined network (SDN)*-based network architecture greatly simplifies the network management and enables fast reconfiguration of the entire protocol stack. Today SDN [1] has been recommended to the latest 5G standards [2] for reprogrammable network topology/protocol controls. This is especially critical in DAN platforms because the antenna orientation changes quickly due to aircraft tilting/mobility and requires the fast reconfiguration of the network architecture and routing protocols. SDN enables such highly flexible protocol changes by separating the control panel (CP) from the data panel (DP). The CP can be performed in a command center and controls the data forwarding rules of the entire network. In DP, the battlefield communication nodes just simply follow the traffic rules sent from CP without worrying about searching for a new end-to-end routing path among nodes. We need to define the detailed flow rules and solve the CP–DP interface issues under dynamic DAN environment.
2. (**Intelligence**) *Intelligent directional network situation awareness*: Before the DAN adjusts its protocol behaviors, it must have an accurate, comprehensive understanding of the network profile/trends. Although SDN provides a natural scheme to collect a large amount of network parameters

[RF links, quality of service (QoS), protocol performance, DAN topology, traffic flows, node mobility patterns, etc.), a set of efficient learning schemes are needed to analyze the collected parameters and catch the intrinsic network patterns (such as congestion regions, network density changes, grouping behaviors, traffic distributions, etc.). Because the network could have large-scale and huge amounts of traffic (video, images, etc.), general machine learning (ML) schemes cannot handle a large number of input parameters with recurrent time patterns. The latest ML scheme – deep learning (DL) models – may be used to handle hundreds of (even more) input parameters and recognize the evolutionary DAN patterns (such as the shift of congestion regions, the time-varying swarming patterns, directional routing QoS metric changing trends, etc.). Based on such detected network patterns, corresponding actions could be determined by using reinforcement learning models. Those actions can be implemented in the flow rules of the DP.

3. (**Optimization**) *Cross-layer directional medium access control (MAC)/ routing/transport protocol optimization*: Future DANs need the robust protocols above the physical layer and DL-based protocol enhancement in each individual layer. As an example, a *stable routing protocol under dynamic channel conditions* may be designed. The airborne links can experience frequent signal quality variations due to the mobility/vibrations of the aircraft. We may use an airborne path loss model to determine the communication parameters, such as power level, queue size, sending rate, time slot length, and so forth. Moreover, we also need to achieve the end-to-end routing performance optimization by establishing a multi-hop directional data relay. As shown in Figure 10.1, if the directional antenna has only one beam, then a relay node (for example, D) needs to determine the suitable orientation time to provide a high throughput. If D is also relaying other data flows, it needs a carefully designed scheduling scheme to support the QoS metrics of each data flow.

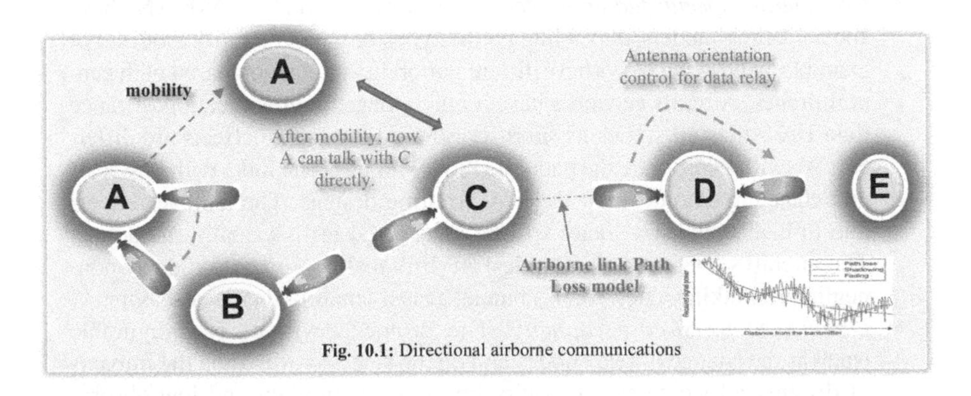

FIGURE 10.1 Directional airborne communications.

10.1.1 Cross-Layer Coupling

Along with the individual layer directional protocol design, the protocol interactions among the abovementioned three layers need to be studied and the coupled protocol optimization should be performed. As an example, the MAC layer can use the efficient scanning functions of directional antennas to find the position distributions of different neighbors in each direction and then determine the suitable channel access scheduling policy among all neighbors. The MAC layer channel access performance (measured by bit error rates and access conflict rates) will be reported to the routing layer, which then selects the best relay nodes (which have the best channel access performance) to form the end-to-end routing path. The routing information (number of hops in each path, main/backup path performance, QoS metrics, etc.) will be further shared with the transport control layer, which determines the suitable congestion control policy for different paths. Some paths need more drastic rate adjustment if they suffer more serious congestion in some nodes.

Any designed cross-layer protocols should also be adaptive to DAN's special features:

- *Low-power constraint*: The protocols should be lightweight enough to meet the power constraints of UAVs. Some UAVs are small and have a very limited battery lifetime. This requires that the protocols do not use frequent message broadcasting. The transmission power level should be small enough for privacy and self-protection, while ensuring that the receiver side has satisfactory signal quality. The beam should face the receiver accurately to avoid unnecessary bit errors.
- *Aircraft/UAV vibrations*: The beam is able to adjust its direction accurately when the aircraft/UAV has strong body vibrations. The node is able to recognize such body vibrations based on the equipped magnetic/vibration sensors, or simply uses the signal quality change patterns to detect the body vibrations. Some ML algorithms also can be used to recognize the link quality degrading pattern due to body vibrations. Once the vibration is detected, the MAC/routing protocol also may need to be adjusted if the RF signal is entirely lost in multiple links.
- *Considering special fading models in airborne networks (ANs)*: ANs have special fading models, depending on the types of UAVs/mobility modes. For example, fixed-wing UAVs have different mobility modes compared with general drones. Aircraft networks have a much longer communication distance than UAVs networks, thus its short-term/long-term fading effects are different. We will incorporate the path loss model for airborne links with obstacles and changing altitudes into the MAC/routing designs. The Doppler effects due to high mobility/velocity will be considered in link quality modeling. The aircraft's long-range (>50 miles) RF link makes propagation delay non-negligible, making a new MAC channel access control scheme necessary.
- *Diminished worst-case capacity*: The protocols will be reconfigurable (such as the change of relay nodes and queue sizes) to overcome the impacts of diminished worse-case capacity imposed by anti-jam and low observability needs. While the protocols try to meet the low-probability detection

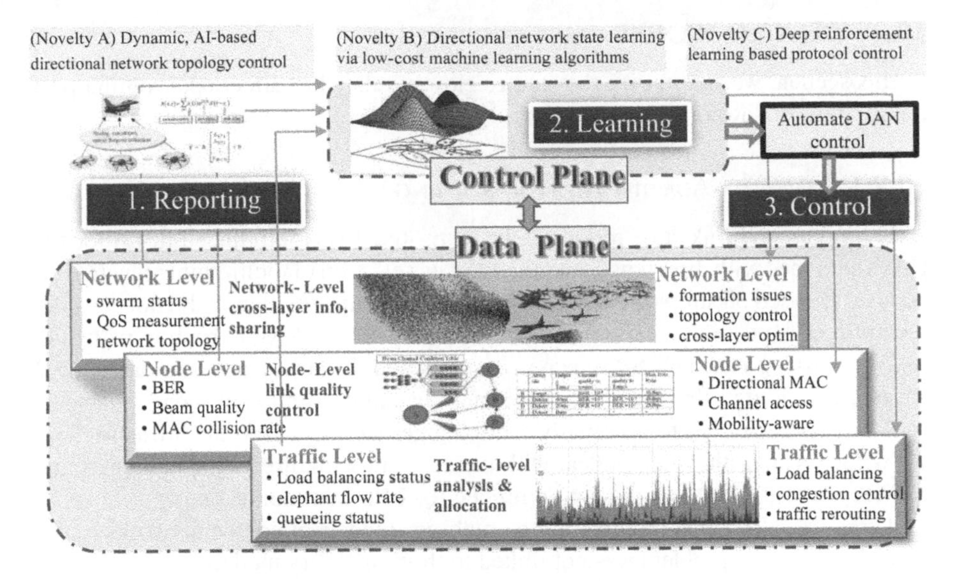

FIGURE 10.2 SDN-compatible intelligent directional airborne network.

(LPD) requirements by adjusting each node's power levels/sending rates, they should also consider the capacity loss due to the conservative transmission behaviors. Therefore, the sender cannot send data too aggressively. It must consider the capacity changes in different interference conditions.

Figure 10.2 illustrates our suggested architecture of SDN-based DAN with self-configurable, ML/DL-based cross-layer protocol design and real-time situation awareness. As seen, it has two novel features:

1. *Three design modules between CP and DP*: First, we will need practical protocols to perform "module 1 – reporting", which aims to collect network parameters from the DAN. We suggest using compressive sampling to reduce data collection frequency while guaranteeing the data resolution and quality. Second, the "module 2 – learning" executes ML/DL algorithms to find the intrinsic network patterns and identify any abnormal events. Third, the "module 3 – control" uses the learning results to control the network protocols.
2. *Three levels of network management*: To manage more efficiently the directional networking protocols, the network status/operations are classified into three levels. The network level is the highest level and takes care of the entire network's state estimation and management. For example, the CP can collect the swarming topology information and adjust the routing protocol based on the new network shape. The node level focuses on the control of each individual node, such as the node mobility and directional antenna orientation changes. The traffic level aims to capture the traffic flow's distribution in the network and identify possible congestion regions and balance the elephant/mice flows' load allocation in different links.

10.2 RELATED WORKS

In this section, we summarize the studies closely related to directional routing protocols and LPD-aware communications.

10.2.1 Mobility-Adaptive Ad Hoc Routing

A new optimized link state routing (OLSR) protocol was introduced in [3] for the UAVs equipped with directional antennas, but its selection of multi-point relay (MPR) nodes was based on simple distance and connectivity information. It did not consider the bottleneck positions of the mobile ad hoc network (MANET) routing paths [4]. In [5] a learning-based OLSR scheme was proposed. Its key idea was to tune OLSR parameters through the Boltzmann learning algorithm to make it more reliable. It could also be adaptable to the rapidly changing network topology and infrastructure. However, its algorithm cannot deal with the network with directional antennas, and its MPR selection method was not optimized for LPD communications.

There was also some research on the mobility models of ad hoc networks [6, 7]. However, the LPD problem was not linked to those mobility models.

The social network analysis for MANETs was introduced and discussed in [8–11]. The network centrality and prediction protocol were proposed and discussed. However, it mainly focused on social networks and human behaviors without detailed protocol design for MANET scenarios.

10.2.2 Directional Network Protocols

There are some studies on directional network design with different protocols (see [12] for a good survey). They mainly create routing schemes with proper beam steering control to achieve multi-hop data relay. However, none of the studies discusses the efficient protocols for large-scale, highly mobile ANs. They did not discuss the impacts of directional antennas on quality-of-service (QoS)/quality of experience (QoE) performance and the corresponding protocol enhancements to accompany the directional transmissions.

10.2.3 SDN-Based Network Management

Recently the concepts of SDN have been used in 5G communication systems [13]. SDN is an excellent architecture to manage complex network operations by minimizing the data forwarding complexity in DAN. Because SDNs can move all the protocol-relevant control tasks to the control plane's powerful nodes, the large number of routers/switches simply need to use the "flow tables" to determine the output interface whenever the packet(s) are received. SDN also can significantly enhance the intelligence through the strong computing capability of the control plane.

10.2.4 Intelligent Networking

Some social network models (SNMs) [14-16] could be used to enhance the ad hoc routing performance. There are also some studies involving the use of basic artificial intelligence (AI)/ML algorithms [17–19] to learn the Department of Defense (DoD) network status (such as QoS performance) and then adopt the proper data forwarding strategies to meet the command delivery reliability requirements. However, those schemes have two drawbacks: First, they could not use a real-time learning scheme to analyze the large Markov state space, which makes them unsuitable for large-scale, highly mobile DoD networks with single-beam or multi-beam antennas. Their learning models cannot handle many time-varying factors, such as mobility mode changes, troop shape reforming, sensors/camera data fusions, jamming attacks with changing signal patterns and so forth. Second, they have not put the AI models under the promising new network management framework – SDN, which makes their models difficult to control during network protocol changes. As we know, one critical feature of DAN automation is to flexibly switch 7-layer protocols. *Our proposed DL/deep reinforcement learning (DRL)-based 3-level architecture (Figure 10.2) is able to perform real-time protocol switching based on the real-time evaluation of DAN conditions.*

In addition, there are many accurate network simulators to verify the performance of the above proposed protocols, such as EMANE [20].

10.3 DAN HEALTH EVALUATION: DL-BASED DAN SITUATIONAL AWARENESS

We suggest using DL to detect any special events from the large-scale DAN network input parameters (listed in Table 10.1). DL has stronger pattern extraction capability than general ML algorithms because it avoids labor-intensive source data feature definition. For example, a support vector machine (SVM) needs data pre-processing to define easy-to-recognition features. DL avoids such a tedious feature definition step and can directly accept raw data as the input. Its layer-by-layer gradient update algorithms can automatically extract the intrinsic patterns.

Each DAN node uses certain measurement methods (Table 10.2) to collect the node/link parameters. Some network parameters may need further signal transform to obtain more dominant features. For example, we may use Fourier transform to obtain the signal's spectrum features. The CP of the SDN system can use the time stamps of each packet to measure average end-to-end routing delay. The node positions may be obtained by Global Positioning System (GPS) devices. Link quality can be obtained through the received signal strength (RSS). In SDN architecture, the special communication channel between CP and DP could be used to report such parameters to the learning server, which then takes a unit time (<1 min) of data and uses a DL algorithm to recognize any abnormal events. Note that the time-consuming DL training process can be performed offline. This online event detection can be performed in real time.

We suggest using the enhanced DL model with long-term/short-term memory for both temporal and spatial feature understanding of the AN. For example, for the links' RSS levels, we may collect each link's short-term RSS time series (temporal) as well as all links' RSS distributions in the whole network (spatial). Then we can define a spatio-temporal residual network to model multiple parameters. Note that Table 10.1 will be grouped into different subgroups when we perform DL computations because some network patterns do not need the inputs from all the parameters.

TABLE 10.1
Input Parameters to be Considered in Directional Network State Learning

Category	Network Parameters for Situational Awareness
Directional antennas	Number of beams, beam steering/switch time, beam angle, beam gap, steered or fixed, single-direction or multi-beam, MIMO control matrix, max radiation distance, etc.
Application and traffic	QoS parameters (delay, jitter, throughput, etc.), QoE (MOS, PSNR, etc.)
Network architecture	Network topology (cluster-based? star? mesh network?), network scale, node density, network connectivity, SDN-assisted? Centralized or distributed? Cloud-supported?
Transport layer	Congestion level, queue size, sliding window size, E2E reliability level, congestion bottleneck location, TCP connection duration, retransmission time0out setup, etc.
Routing layer	Multi-path/single-path, multicast/unicast, number of hops, path throughput, path delay, average link quality, packet drop rate (PDR), rerouting time, path stability, etc.
MAC layer	Access collision rate, backoff time, link BER, robustness to hidden terminal problem, superframe length, type (TDMA based on random access), error correction rate, etc.
PHY layer	Wave features, SNR, modulation modes, coding methods, Shannon capacity, etc.
Node property	Mobility speed, mobility modes (random walk or regularized), transmission power, reception sensitivity, max queue length, packet processing time, max hop distance, etc.
Radio channels	Channel ID, channel bandwidth, channel holding time, handoff channel sequence, channel quality (fading level, Doppler effect, SNR, etc.), channel switching time, etc.
Spectrum sharing modes	Licensed band sharing, unlicensed band sharing, exclusive or cooperative sharing. SUs: DoD network (such as UAV network) PUs: (1) LTE operators, (2) femtocell users, (3) M2M devices, (4) Wi-Fi, etc. For each of the above, there could be multiple parameters, such as interference temperature, sharing bandwidth, sharing time, leased bandwidth, SS zone size, etc.

MIMO: multi-input, multi-output antennas; MOS: mean opinion score; PSNR: peak signal-to-noise ratio; E2E: end-to-end; TCP: transport control protocol; BER: bit error rate; TDMA: Time division multiple access; SNR: signal-to-noise ratio; SUs: secondary users; Pus: primary users. LTE: long-term evolution; M2M: machine-to-machinel SS: spectrum sensing.

For example, when we find out the routing performance's changing trend, only QoS metrics and geometric/topology information are useful. For network congestion prediction, queue states are most relevant.

To recognize the short-term (for current feature extraction) and long-term (for future pattern prediction) patterns from the above collected parameters, we may use a DL structure with long-memory/short-memory (LSM) cells (Figure 10.3). Empirically, adding two Fully Connected (FC) layers in the recurrent neural networks (RNNs) will produce better results than directly using rectified linear unit (ReLU) output. An optimal training model is needed to use fast RNN weight-seeking

TABLE 10.2
Network Parameter Collection Methods (Examples)

Parameter Types	Example Parameter Names	Collection Methods
Signal waveforms	RSS (received signal strength)	Signal sensing chips
	RF signal time-domain parameters (average, peak, moments, etc.)	Sample the detected wireless signals and then extract statistics
	Constellation shape/points for digital modulation recognition purpose	Sense waveforms and map to topological 2D structure/images
	Cyclostationary/spectral correlation function (SCF)	Sense waveforms, apply SCF
	Wavelet transform (WT) coefficients	Apply WT + SVM to waveforms
	S transform (ST) coefficients	ST can process phase + frequency information
	Frequency domain (Fourier coefficients)	FT provides spectrum features
Radio channels and links	Channel access success rate (in MAC layer)	In CSMA/OFDM, node tries to access channel without collisions
	Available bandwidth/channel numbers	Use spectrum sensing chips
	Channel occupancy duration (average)	Record channel use time
	Link disconnection/outage probability	How many times link fails
Cross-layer QoS metrics	Packet loss rate (need receiver feedback)	Lost packets/total sent ones
	Bit error rate (before using error correction)	Damaged bits/total sent
	End-to-end delay (in multi-hop case)	Time stamp differences
	Average queue waiting time	Wait for channel clear
Node features	Self/neighbor locations/distances	Natural interference often comes from short-distance neighbors
	Moving velocity	Some jammers keep moving

algorithms, such as using a backpropagation through time (BPTT) scheme or stochastic gradient descent (SGD)-based RMSprop optimizer [21].

Our proposed DL-based jamming pattern recognition model (Figure 10.3) has the following advantages. (1) It can handle complex input parameters with *high-dimensional* structure and diversified data types (integer, Boolean, floating point, time series, etc.). It is scalable to big data, that is, if the inputs are in big data format (such as big tensor), DL can efficiently use multiple layers to extract the abstract features. (2) It uses *LSM* to represent the events without fixed time durations. This is

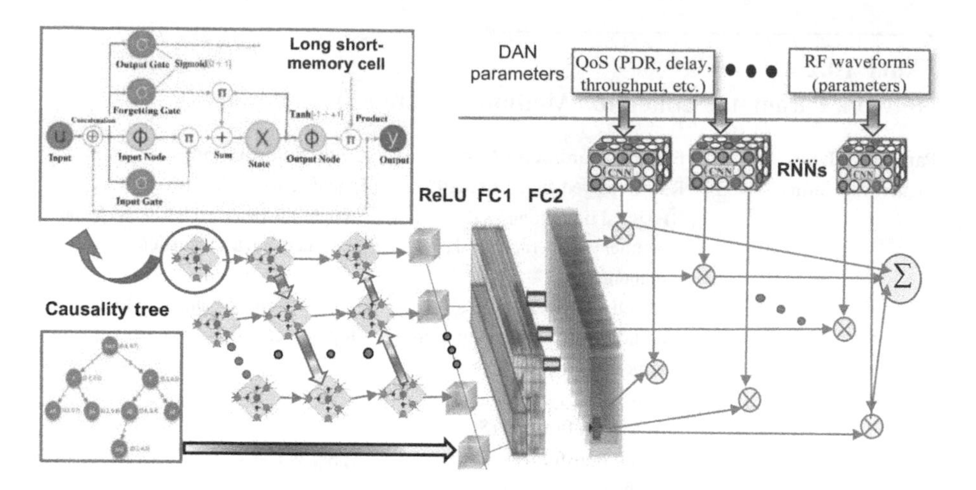

FIGURE 10.3 "Long-memory + causality" enhanced deep learning model for network event detection.

especially useful when handling the QoS parameters that are collected from different phases of a communication mission. (3) It has a causality tree to describe the correlations between different parameters. For example, the queueing delay can cause packet loss when the queue is too full to hold new packets.

The above LSM-based DL model can accurately extract the intrinsic features from a large number of parameters. Each time the raw inputs pass a hidden layer in the neural networks better abstractions are extracted. The model also can be used to predict the next-phase jamming pattern by fitting the output layer results into an Autoregressive–moving-average (ARMA) or regression model.

10.4 DAN PROTOCOL BEHAVIOR CONTROL

After we understand the network performance changing trends through the above DL-based event learning, the SDN control plane will perform protocol control, which is implemented by modifying the data forwarding table in each data plane node.

As shown in Figure 10.4, the CP can trace the QoS states in all UAVs and send the protocol adjustment commands to each node. As an example, in Figure 10.4 (bottom left), node *A* needs to help relay multiple traffic flows (see the red and blue paths). It needs to follow a certain schedule to guarantee seamless beam switching between those two flows and to consider flow priorities, congestion levels, QoS requirements, node mobility, and other factors. The CP can comprehensively analyze these factors and send out schedule information to node *A*. Along with the directional antenna control operations, the control plane will also determine the behaviors of all protocol layers. For example, it will analyze the congestion hotspots and suggest the traffic split solutions. It also balances the bandwidth usage between elephant flows and mice flows. The control plane uses the 3D mobility prediction results to suggest the use

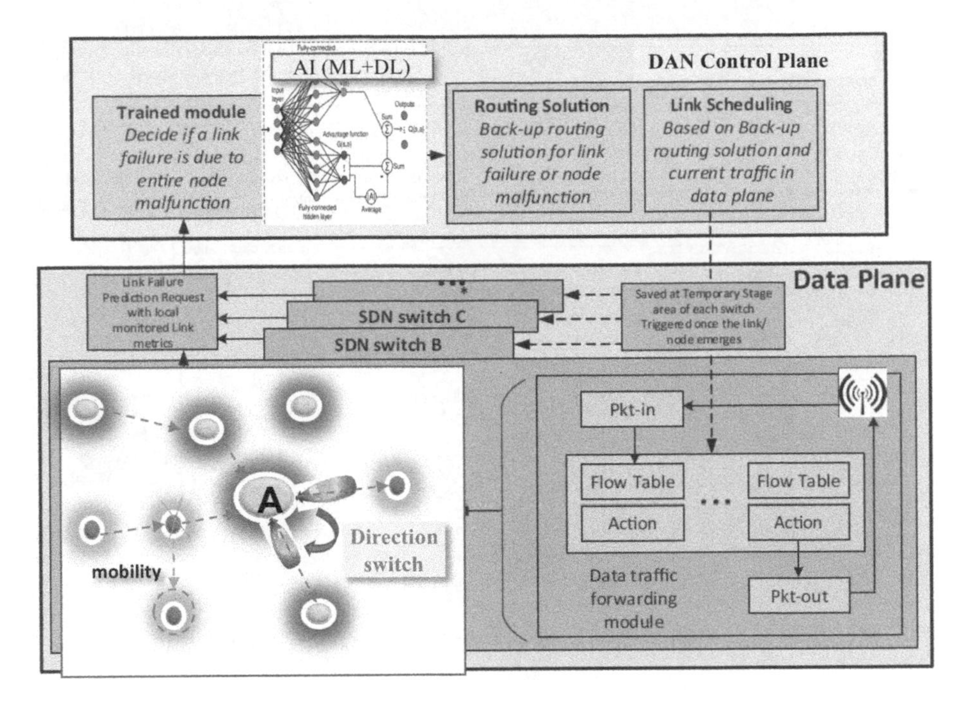

FIGURE 10.4 SDN-based DAN data/command forwarding control.

of some backup path nodes. It also collects the channel statistics and determines the MAC channel access schedule.

If an RF link has significant bit errors or packet loss, we may adopt the following five typical methods to improve the link communication performance:

1. *Increase power level*: A stronger antenna gain or transmitter power can ensure that its signals propagate with a high signal-to-noise ratio (SNR). However, a higher power might cause near-far effect and hides neighbors' signals. Thus, this is not a "polite" way.
2. *Decrease transmission rate*: This polite way might help to reduce/avoid interference because it gives more network bandwidth to others to finish their transmission earlier.
3. *Stop and wait for link to be available again*: This is a more polite way because it totally stops transmission and waits for others to finish their transmissions. But this way may sacrifice the node's throughput.
4. *Channel handoff*: This is the politest way. Meanwhile, it does not sacrifice its own throughput because the transmitter switches to another available channel (called channel handoff).
5. *Frequency hopping or spread spectrum*: The former aims to quickly switch the frequency based on a certain sequence. Spread spectrum embeds the signals into a wide frequency band.

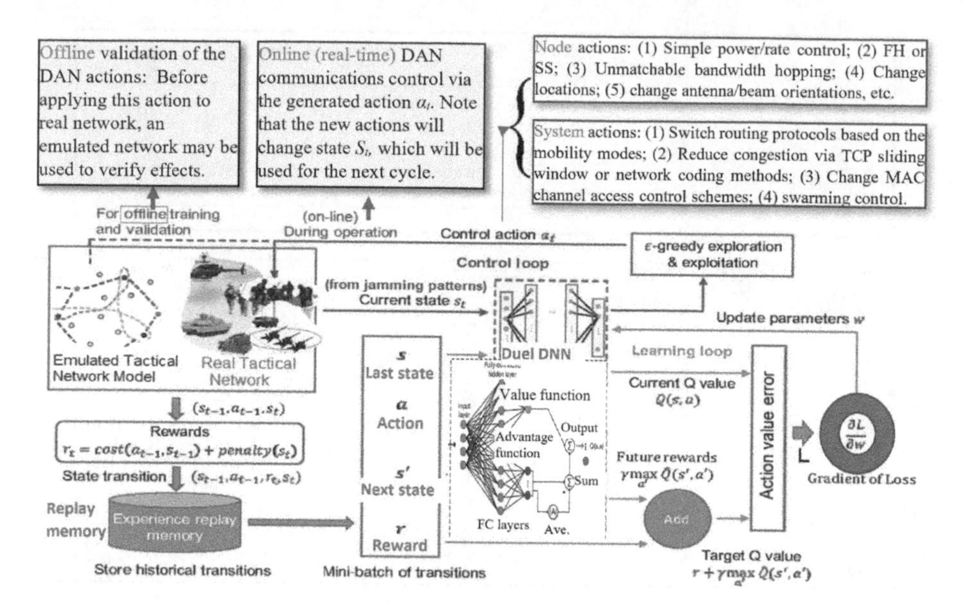

FIGURE 10.5 Output actions: DRL-based real-time DAN protocol control.

Our suggested DAN protocol control will be based on an enhanced DRL model shown in Figure 10.5, which uses the C³I-state (***S***) adaptive Markov decision model to generate the most suitable network *actions* (***A***). *To achieve real-time response, our DRL model uses three critical components as follows*:

1. *Using both offline and online actions*: We suggest that the airborne communication system always maintains a virtual mobile network with the same topology/conditions as the current practical network. It keeps track of the real network's node mobility modes, link status, bandwidth changes, and jamming dynamics. As seen from the top part of Figure 10.5, each time an *action* is generated, before it is applied to the practical network, it first gets tried in the emulated network. Such an emulation takes a negligible time. The emulated network testing results will tell whether we can successfully overcome radio fading, meanwhile keeping a satisfactory QoS level.
2. *Deep dueling engine*: DRL could be difficult to converge with short time, which is detrimental to near real-time anti-jamming response (<1 min). This is mainly due to the large space of states (***S***) and actions (***A***). We need to consider various jamming states due to the diversity of jamming factors (such as jammer's signal strength, duration, timing, physical location, center frequency, adaptation patterns, etc.), and large action space due to the continuous changes of power levels, sending rates, channel IDs, bandwidth locations and so forth. Those complex state/action cases could cause the algorithm to easily miss the global optimization point. Therefore, we suggest using a *deep dueling algorithm* [22] to enhance the learning convergence.

Google DeepMind has successfully used this algorithm to handle 19 x 19 chess position possibilities (which is a huge state/action space). In DDN, not every action (a) has obvious state changes. For example, frequency hopping (FH) may not work effectively if the jammer has the capability of jamming multiple channels simultaneously or even damage the whole bandwidth. Deep dueling does not simply evaluate the action-value function in each iteration; instead, it divides the whole DNN into two parts (see Figure 10.4), and we can then evaluate both the values of each state and the advantages of each action. The algorithm then combines both results in the final output layer of the DNN. From a math model viewpoint, the value of state (s) and action (α **pairs** under a stochastic policy π is

$$\Theta^{\pi}(s,a)=\mathbb{E}\left[r_t \mid S_t=s,\ A_t=a,\mathfrak{n}\right] \tag{10.1}$$

where r_t is the reward function result at time t. It represents the value of using action a in state s. The *value* of state s can be then represented by using the above state-action pair value:

$$\mathbb{V}^{\pi}(s)=\mathbb{E}[\Theta^{\pi}(s,a)] \tag{10.2}$$

The *advantage* function of $\mathbb{k}$ actions is as follows:

$$\mathbb{A}^{\pi}(s,a)=\Theta^{\pi}(s,a)-\mathbb{V}^{\pi}(s) \tag{10.3}$$

As we can see, the advantage part actually decouples the state's value from the entire state-action pair's value such that we can clearly see the significance of each individual action.

3. *Replay memory*: In standard DRL algorithm, an experience [i.e., an old record of (state, action, reward, new state) in the past iterations] is often discarded after it is used for a policy update. However, the replay of the experience memory may be very helpful, because a memory replay allows the agent to learn from earlier records and use it to speed up the learning process and even break up the undesirable temporal correlations. The effects of such replay memory in a DRL-based anti-jamming decision model need to be studied. Particularly, we need to study *the effect of memory size and record priority setup on the DRL learning dynamics.*

10.5 TRAPDOOR-BASED NETWORK BEHAVIOR ANOMALY DETECTION

Although the previously mentioned DL-based methods could extract many types of network events, most events may be *normal* behaviors and do not need special processing. To more efficiently "dig" into the large amount of DAN parameter data and accurately detect any *abnormal* network events (such as asynchronization of neighborhood directional antennas, congestion hotspots, isolated network regions, etc.),

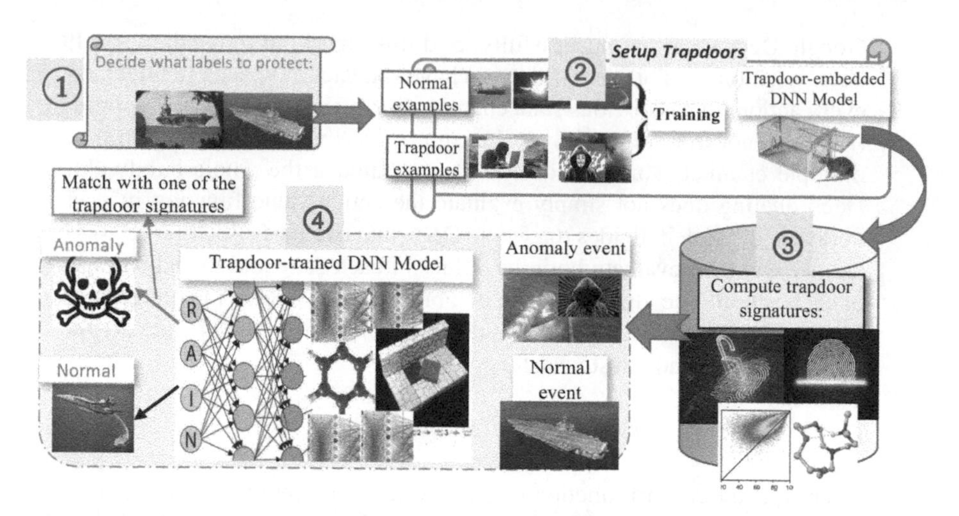

FIGURE 10.6 Trapdoor-based DAN anomaly detection.

we suggest using the trapdoor-embedded deep neural network (DNN) model shown in Figure 10.6. It consists of the following four design modules:

1. *Editing classification labels*: First, we should know what types of labels our system could generate. Each label should have clear patterns. For example, a congestion event is indicated by regional extremely high delay and long queue size. Those sensing patterns could tell the *label* there is "congestion".
2. *Setting trapdoors*: Some trapdoors (such as an antenna steering asynchronization event) are intentionally embedded into the benign examples. Both trapdoor and benign examples are used for DNN training.
3. *Generating trapdoor signatures*: The outputs of the trapdoor-embedded DNN model are saved as trapdoor signatures. Those signatures represent all possible distortions for different class labels. The enhanced DNN model also has a filter that seeks intermediate neuron activations that match the generated signature. An adversary trying to produce an adversarial example on a trapdoor-embedded label will find itself easily converging into a value near the trapdoor.
4. *Identifying abnormal events*: Once trapdoors are embedded into a model we then use the trapdoor-trained DNN for DAN event recognition by comparing the outputs with pre-stored trapdoor signatures.

The following two issues need to be solved:

1. *How to create a trapdoor*: We need to expand the original DAN network training data space by injecting new samples with trapdoor perturbations. Then the injection function $I(.)$ driven by the trapdoor (M, b, r) should be defined. Here M is the trapdoor mask, which tells how much perturbation should be added; b is the baseline random perturbation (which could be a

simple random noise); and k is called *mask ratio,* which is a small number ($\ll 1$) that tells how many samples could be perturbed.

2. *How to train the trapdoor DNN*: The new model should be able to produce a high classification accuracy for normal inputs but easily classify a trapdoor-embedded sample into a special label pattern that will serve as "trapdoor signature". The optimization objectives that reflect the addition of backdoors need to be defined. We may use *cross-entropy* based loss function to measure the classification errors. After training the trapdoor-added DNN, the "neuron signatures" will be stored a database for future trapdoor matching purposes.

10.6 LPD-ORIENTED PROTOCOL DESIGN

10.6.1 Adaptive, Closed-Loop Power/Rate Control for LPD Transmissions

Each node may use the mini-cluster schedule and topology information to adjust its own power levels and sending rates. The power/rate levels also depend on the direction alignment levels of different nodes' antennas. The end-to-end transmission reliability and QoS requirements must also be satisfied. In LPD communications, the selection of suitable RF transmission power levels should be determined by the sensibility level and mobility model of the adversary.

In the initial phase of the neighbor discovery, the transmission power should be limited to a very low level to initialize the scanning for the nearest neighbors. This helps to avoid the detection by the adversary. If such a low power level cannot sense any neighbors, a higher power level could be used. In other words, the power level should be carefully increased in a step-by-step style with a small increment amount each time.

There is a need to integrate the power control scheme with the directional neighbor discovery in MAC layer protocols. The LPD-aware power control process consists of two phases:

1. *Beam-scanning step*: In the neighbor discovery process, we assume that each time division multiple access (TDMA) frame contains the dedicated time slots for the communications between any pair of nodes. In other words, the node follows a certain schedule to send out neighbor discovery messages. Such a TDMA-based neighbor scanning helps to achieve a *synchronized* network management because each node knows its transmission time.

For any node, it has three operation states, i.e., transmission (Tx), receiving (Rx), and idle mode. The directional antenna rotates its beams to different directions to search for 1-hop or 2-hop neighbors, as described before. We need to determine the time duration in each direction that the beam should stay during beam rotations. If a multi-beam antenna is used, each beam covers a certain direction and no rotating is needed. The node can send out broadcast messages in its allocated time slot in each direction with power level P_B (Figure 10.7).

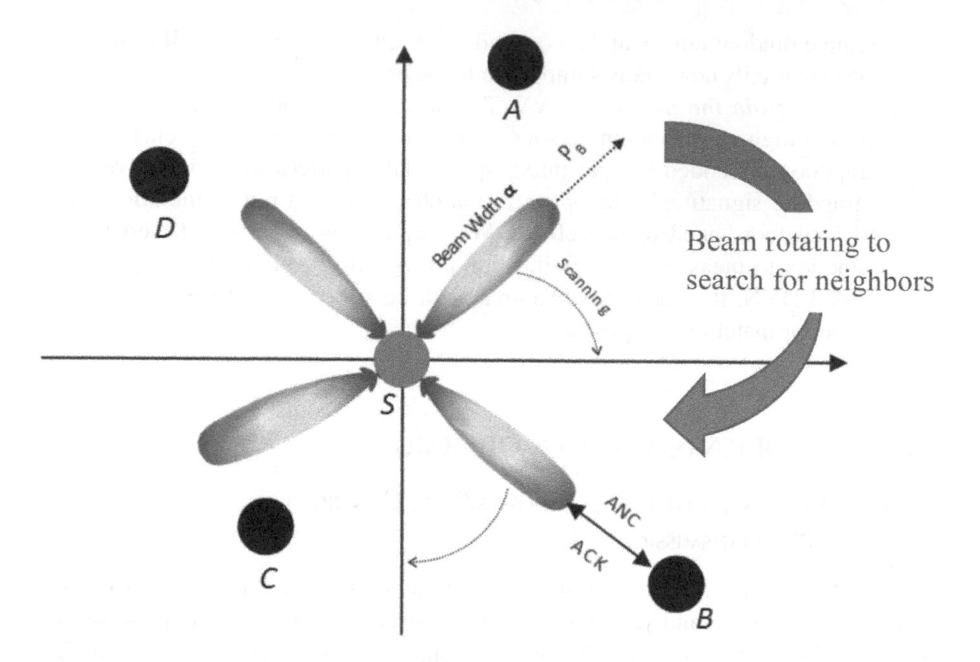

FIGURE 10.7 Beam scanning for neighbor discovery.

Suppose a listening node, *B*, receives an *announcement* message (ANC) in a particular direction. In another time slot it will send back the acknowledgment (ACK) message back to the source (i.e., node *A*) in that direction. After the source node receives the ACK message, it sends out a confirmation message for *B* within a particular time slot. The above **three-way** handshake takes a total of three time slots, and then each pair of nodes (*A* and *B* in this example) can share some information with each other, such as node IDs, locations, Tx power levels, and possibly any known adversary information.

2. *Power incrementing step*: After the source node finishes the beam-scanning step for all the beams, it may get to know the adversary's information such as its location, sensitivity and noise level, based on the feedback from the neighbors discovered in the above beam-scanning step.

It is proved in [18] that with an *n*-symbol covert text, the total power the sender can emit over *n* channels is limited to $O\left(\sigma_A^2\sqrt{n}\right)$, where *A* denotes the adversary; otherwise, the sender's transmission will be detected by the adversary. This allows the sender to reliably transmit at most $O\left(\sigma_A^2\sqrt{n}/\sigma_R^2\right)$ cover bits to the receiver in *n* channels, where *R* denotes the receiver. Thus, if the *profile* of the adversary is fed back by the neighbors, the sender can directly increase its broadcast power to $O\left(\sigma_A^2\sqrt{n}\right)$. If this is not the case, then the sender should start a new round of beam scanning by using a default power increment level, such as $O\left(\sigma_A^2\sqrt{n}\right)/O(n)=O\left(\sigma_A^2/\sqrt{n}\right)$. In practical applications, this increment step can be determined by the system designer, as long as the highest total transmission power level is under $O\left(\sigma_A^2\sqrt{n}\right)$. A simple bang-bang control or even proportional–integral–derivative (PID) control can be employed when performing this power increment.

Therefore, the neighbor discovery protocol should allow each node to perform neighbor discovery within a certain bounded time. This can be set to a 50- to 100-ms level for the safety of LPD communication, which is around 1/5 to 1/10 of the ceiling value (500 ms), that could lead to fast fading based on [15], and to gradually acquire the neighborhood information without violating the requirements of LPD.

Each scanning direction of an antenna has a unique direction ID. Once a beam successfully communicates with another beam of another node, the direction IDs and node IDs as well as other critical parameters, such as angle, height, direction and so forth of the beam, will be exchanged and saved for future MAC protocol operations.

In the above two-step iterations, the multi-beam scanning protocol enables the nodes to discover all potential RF links between any pair of nodes. In addition to the discovery of neighbors, *the protocol should also gather the information about the adversary* such as its previous location, signal reception sensitivity and noise level, based on the exchanged information between the neighboring nodes.

For each node, it starts from a low power level, and it iterates between the above two steps. In each iteration it slowly increases the Tx power level in that beam to search more and more potential neighbors. If the mission only needs a small number of neighbors to serve as the next-hop relay points, it will stop the increment of the Tx power level once it finds enough neighbors. This earlier termination of neighbor discovery process helps to reduce the probability of being detected by adversaries.

10.6.2 LPD-Aware Airborne Routing Protocol

10.6.2.1 Objectives of Routing Design

To achieve an end-to-end long-distance communication, each node needs to initiate the route discovery process and use the fastest path to reach a destination. A coordination scheme is needed to establish an LDP-aware robust route. Specifically, we need to consider the following:

1. *Inter-cluster/intra-cluster, ripple-based directional data forwarding scheme*: Each mini-cluster selects a gateway node to talk with the next mini-cluster's gateway. The directional transmission schedule within/between the mini-clusters needs be designed. As shown in Figure 10.8 (1), the source node's mini-cluster needs to use an efficient neighboring mini-cluster discovery algorithm. The cross-cluster routing needs to consider the gateway change, MAC schedule updates in each mini-cluster and QoS-desired path selection.
2. *LPD-aware route adjustment based on mobility prediction*: When an adversary is nearby, the existing path needs to partially adjust its relay node list. First, we define a RED zone (which is determined by the adversary's signal profile, friendly nodes' positions and traffic flow priorities) and then prepare the backup relays to avoid it [Figure 10.8, (2)]. The relay nodes' mobility behaviors will be used for backup path establishment.

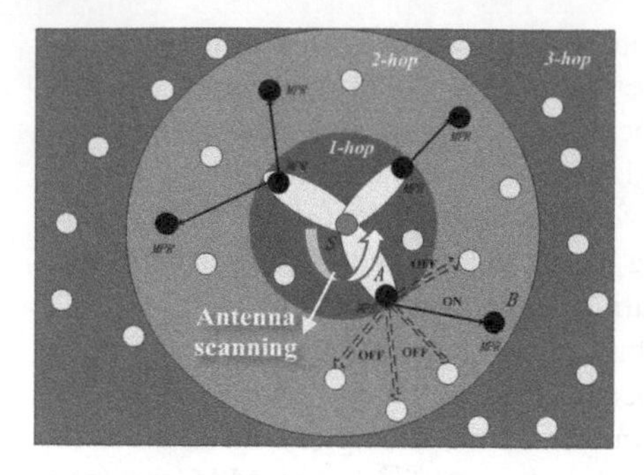

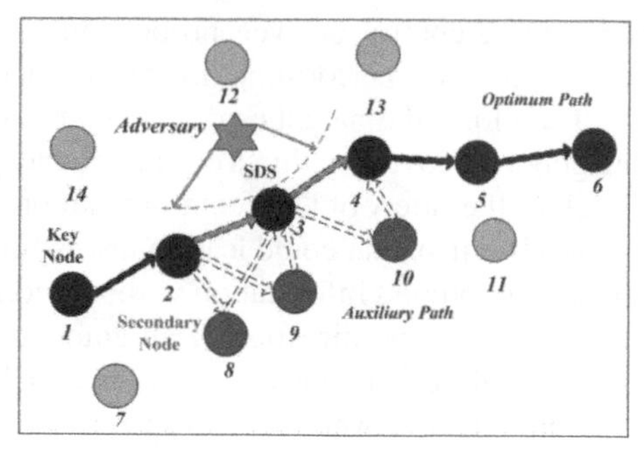

FIGURE 10.8 (1) Ripple-based routing tree discovery. (2) LDP-oriented routing path adjustment.

10.6.2.2 Enhanced OLSR Routing for LDP Purpose via Social Network Concepts

In link state routing protocols such as OLSR, link state announcement messages are frequently broadcast to the whole network if there is a change of network topology in ad hoc networks. In OLSR, MPRs are introduced to forward the link state messages. However, if directional antennas are equipped in each node, we can achieve a longer, well-focused RF range. Due to such extended range, a smaller number of nodes may be selected as MPRs.

For LPD purposes, the link state updates can be scheduled sequentially, which means that in one time slot, the source node only sends out link state packets to one of its MPRs by using the corresponding beam of its antennas. In another time slot, the node sends the link state to another MPR. Thus, each time a node announces its link state to just a small number of nodes, eventually all nodes know about the entire network's topology.

Inspired by the social network analysis models, we suggest using them to find the best MPRs because an UAV with better position (such as in the center of a high-density region) may serve better as an MPR. The centrality of a node in a network is a measure of the structural importance of the node. Typically, a central node has a stronger capability of connecting with other network nodes. In ANs, degree centrality and betweenness centrality are important metrics for swarming applications. Based on Freeman's measures [23], degree centrality is calculated as the number of direct connections for a given node [24]. Hence a node with a higher degree of centrality maintains a larger number of connections with other nodes in the network.

In OLSR, MPR nodes are supposed to have a higher degree of centrality to help forward the periodic link state messages. Assume that $a(p_i,p_k) = 1$ if a direct link exists between p_i and p_k, and $a(p_i,p_k) = 0$, otherwise, where a is the count of the degree or number of adjacencies for a point. Degree centrality for a given node p_i, can be calculated as [24]

$$C_D(p_i) = \sum_{k=1}^{N} a(pi, pk)$$

Betweenness centrality measures the extent to which a node belonging to a geodesic path links with other nodes. It can be regarded as a measure of the extent to which a node has the control over the information flow among other nodes. A node with a high betweenness centrality has a capacity of facilitating the interactions among the nodes it links with. In our case, MPR nodes are responsible for forwarding link state messages to other nodes, and their interactions with the RF links can influence the efficiency of link state updating in the whole network. Denote g_{jk} as the total number of geodesic paths linking p_j and p_k, and $g_{jk}(p_i)$ as the number of those geodesic paths with p_i. Then *betweenness centrality* can be defined as

$$C_B(p_i) = \sum_{j=1}^{N} \sum_{k=1}^{j-1} \frac{g_{jk}(p_i)}{g_{jk}}$$

The above degree and betweenness centrality can be used in the selection of MPRs in the enhanced OLSR protocol. The steps for selecting MPRs with directional antennas are as follows:

- *(s1)*: Based on the neighbor list *N(S)* of node S, for every neighbor $X_1, X_2, X_3, \ldots, X_n$ in *N(S),* we combine the neighbor lists as: $M = N(X_1) \cup N(X_2) \cup N(X_3) \cup \ldots \cup N(X_n)$. Then for any node in *M* but not belonging to *N(S),* the node is classified into $N_2(S)$, meaning a 2-hop neighbor of *S*.
- *(s2)*: For each node *i* in *N(S),* based on the calculation of degree centrality, it selects the node with a degree greater than a preset threshold D_{min}, and adds this node to a set $C_D(S)$, which is the high-degree neighbors of node *S*.
- *(s3)*: For each node j in *N(S),* considering all the links between node *S* and its 2-hop neighbors in $N_2(S)$, we can obtain the calculation result of betweenness centrality. Then we can select the node with a betweenness greater than the threshold B_{min}, and add this node to a set $C_B(S)$, which represents the high-betweenness neighbors of node *S* among its 2-hop neighbors.
- *(s4)*: Combining the degree set $C_D(S)$ and betweenness set $C_B(S)$. Depending on the number of beams of the multi-beam directional antennas (MBDAs), assume that *m* nodes in *N(S)* are selected as MPRs to forward the link state packets in this round. The selected MPRs form a set for node *S*. This MPR set contains the minimum number of MPRs, which can connect to all the 2-hop neighbors of node *S*.

After selecting the MPR set for node S, node S can start to update the link state *sequentially*. Node S repeatedly sends out link state packets to its MPRs until all the nodes in the MPR set have been reached. This way, node S can announce its link state to the whole network by using directional antenna while keeping a good LPD performance.

The establishment of all paths should consider the existence of adversaries and the need to get as far away as possible from the adversary's position. The nodes near the adversary should carefully control their directional antenna's orientation to avoid the signal detection.

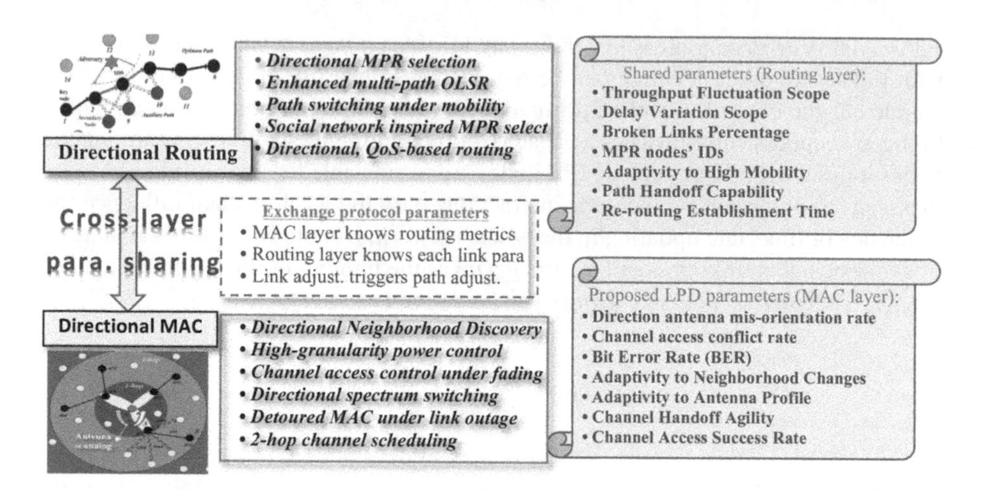

FIGURE 10.9 Cross-layer directional protocol design.

We will need to consider the dynamic adversary scenarios and adjust the global or partial path topologies based on the detection of the adversary mobility mode, which also can be predicted through Bayesian regression–based mobility prediction algorithms.

10.7 CROSS-LAYER PROTOCOL OPTIMIZATION

Cross-layer design can be performed in directional ANs to further enhance the QoS performance. Various network parameters (such as channel access conflict rate, time slot length, cluster size, packer loss rate, routing delay, etc.) can be shared across MAC/routing protocols to mutually optimize each layer. For example, the MAC schedule can be adjusted based on the routing QoS metrics. Similarly, the backup path can be built based on the intra-cluster scheduling. Figure 10.9 depicts our defined shared protocol parameters and their relationships with cross-layer design.

10.8 CONCLUSIONS

In this chapter, we have systematically summarized the challenging issues in DANs and suggested some novel solutions to overcome those protocol design issues. The ML/DL-based protocols can better resist jamming attacks, and the LPD-oriented protocol designs further enhance the safety of the DAN functions.

REFERENCES

1. A. Kots, V. Sharma, and M. Kumar, "Boltzmann machine algorithm based learning of OLSR protocol: An energy efficient approach," *International Journal of Computer Applications*, vol. 53, no. 13, 2012.
2. M. Rollo and A. Komenda, "Mobility model for tactical networks," *International Conference on Industrial Applications of Holonic and Multi-Agent Systems*, Berlin: Springer, pp. 254–265, 2009.

3. Yi Zheng, Yuwen Wang, Zhenzhen Li, Li Dong, Yu Jiang, and Hong Zhang, "A Mobility and Load aware OLSR routing protocol for UAV mobile ad-hoc networks," *2014 International Conference on Information and Communications Technologies (ICT 2014)*, pp. 1–7, 2014.
4. H. Kanagasundaram and A. Kathirvel, "Eimo-ESOLSR: energy efficient and security-based model for OLSR routing protocol in mobile ad-hoc network," *IET Communications*, vol. 13, no. 5, pp. 553–559, 2018.
5. A. McAuley, K. Sinkar, L. Kant, C. Graff and M. Patel, "Tuning of reinforcement learning parameters applied to OLSR using a cognitive network design tool," *2012 IEEE Wireless Communications and Networking Conference (WCNC)*, Paris, France, 2012, pp. 2786–2791.
6. A. N. Washington and R. Iziduh, "Modeling of Military Networks Using Group Mobility Models," *2009 Sixth International Conference on Information Technology: New Generations*, Las Vegas, NV, 2009, pp. 1670–1671
7. E. M. Daly and M. Haahr, "Social network analysis for routing in disconnected delay-tolerant MANETs," *Proceedings of the 8th ACM International Symposium on Mobile Ad Hoc Networking and Computing*, New York: ACM, pp. 32–40, 2007.
8. E. Stai, J. S. Baras and S. Papavassiliou, "Social networks over wireless networks," *2012 IEEE 51st IEEE Conference on Decision and Control (CDC)*, Maui, HI, 2012, pp. 2696–2703,
9. E. M. Daly and M. Haahr, "Social network analysis for information flow in disconnected delay-tolerant MANETs," *IEEE Transactions on Mobile Computing*, vol. 8, no. 5, pp. 606–621, 2009.
10. H. Kim, J. Tang, R. Anderson, and C. Mascolo, "Centrality prediction in dynamic human contact networks," *Computer Networks*, vol. 56, no. 3, pp. 983–996, 2012.
11. Z. Qin, X. Gan, J. Wang, L. Fu, and X. Wang, "Capacity of social-aware wireless networks with directional antennas," *IEEE Transactions on Communications*, vol. 65, no. 11, pp. 4831–4844, 2017.
12. Z. Wu and Z. Qiu, "A survey on directional antenna networking," *2011 7th International Conference on Wireless Communications, Networking and Mobile Computing*, pp. 1–4, 2011.
13. H. Cho, C. Lai, T. K. Shih, and H. Chao, "Integration of SDR and SDN for 5G," *IEEE Access*, vol. 2, pp. 1196–1204, 2014.
14. S. Fujii, T. Murase, M. Oguchi and E. K. Lua, "Architecture and characteristics of social network based ad hoc networking," *2016 IEEE International Symposium on Local and Metropolitan Area Networks (LANMAN)*, Rome, 2016, pp. 1–3.
15. L. C. Freeman, "A set of measures of centrality based on betweenness," *Sociometry*, pp. 35–41, 1977.
16. L. C. Freeman, "Centrality in social networks conceptual clarification," *Social Networks*, vol. 1, no. 3, pp. 215–239, 1978. PI to AFRL (Rome, NY) to present the project results during this project.
17. E. Blasch, and M. Belangér, "Agile battle management efficiency for command, control, communications, computers and intelligence (C4I)." *Proceedings of the SPIE*, vol. 9842, no. 98420P, p. 11, 2016.
18. H. Srinivasan, M. J. Beal, and S. N. Srihasri, "Machine learning approaches for person identification and verification." *Proceedings of the SPIE*, vol. 5778, pp. 54–587, 2005.
19. C. Paul, C. P. Clarke, B. L. Triezenberg, D. Manheim, and B. Wilson, *Improving C2 and Situational Awareness for Operations in and Through the Information Environment*, Santa Monica, CA: RAND Corporation, 2018.
20. J. Ahrenholz, T. Goff, and B. Adamson, "Integration of the CORE and EMANE Network Emulators," *2011-MILCOM 2011 Military Communications Conference*, pp. 1870–1875, 2011.

21. E. Yazan and M. F. Talu, "Comparison of the stochastic gradient descent based optimization techniques," *2017 International Artificial Intelligence and Data Processing Symposium (IDAP)*, pp. 1–5, 2017.
22. Y. Huang, G. Wei, and Y. Wang, "V-D D3QN: the variant of double deep Q-learning network with dueling architecture," *2018 37th Chinese Control Conference (CCC)*, pp. 9130–9135, 2018.
23. F. Grando, D. Noble and L. C. Lamb, "An Analysis of Centrality Measures for Complex and Social Networks," *2016 IEEE Global Communications Conference (GLOBECOM)*, Washington, DC, 2016, pp. 1–6.
24. H. Li, "Centrality analysis of online social network big data," *2018 IEEE 3rd International Conference on Big Data Analysis (ICBDA)*, Shanghai, 2018, pp. 38–42

11 UAV Security Threats, Requirements and Solutions

Tonmoy Ghosh, Iftikhar Rasheed, Niloofar Toorchi, Fei Hu
Electrical & Computer Engineering,
The University of Alabama, Tuscaloosa, AL

CONTENTS

11.1 INTRODUCTION

The unmanned aerial vehicle (UAV) is an aircraft without a human pilot on broad. In the past UAVs were often used for military purposes. UAVs are usually employed in dangerous missions in which a human-based pilot cannot be deployed. Although they originated principally in military applications, their use is speedily increasing to industrial, scientific, recreational, and agricultural, as well as other applications, such as policing, peacekeeping [1] and police investigation, product deliveries, aerial photography, agriculture and smuggling [2].

UAVs can have different features such as size, control system, purposes and so forth. The size of drones varies drastically, for example, they can be as small as a matchbox and as large as manned aircraft. Flying range and capabilities usually depend on the size. Generally, larger UAVs can fly for a higher and longer distance. According to the European Association of Unmanned Vehicles Systems (EUROUVS), UAVs can be classified into three categories of drones [3]: (1) micro and mini (weight: several grams to 24 kg), (2) tactical (less than 1500 kg, but more than 24 kg) and (3) strategic (more than 1500 kg). Moreover, according to the U.S. military standards [4], UAVs also can be categorized by their travel range and endurance.

Furthermore, UAVs also can be classified in terms of control capabilities or, more specifically, the degree of autonomy available in the vehicle [5]. The control of UAVs varies from manual control to fully automatic operations.

Remote pilot control: The human is in the loop to control the UAV. This system is known as static automation. In this system, all the decisions are made by the remote human operator.

Remote supervised control: The human is in the loop to control the UAV. This system is known as adaptive automation. The UAV can perform the mission process autonomously from human guidelines, but the system enables human intrusion at the same time.

Full autonomous control: The human is out of the loop to control the UAV. This system is known as static automation. Here the UAV make needed decisions to complete a mission.

11.1.1 UAV System Architecture

Civilian UAV systems consist of three main elements, i.e., the unmanned aircraft, the ground control station (GCS), and the communication data link [6]. High-level architecture of UAV systems is shown in Figure 11.1. In what follows, we provide a brief overview of the main building components of civilian UAVs.

Flight controller: This is the central processing unit of the drone. In addition to stabilizing the drone during its course, this system reads the data provided by the sensors and processes it to extract useful information. According to different control commands, the controller either relays this information to the GCS or feeds the actuator control units directly with the current states. The flight controller implements the communication interface with the GCS. More precisely, commands from the GCS are processed by the flight controller, which in turn affects the deployed actuators. Furthermore,

TABLE 11.1
UAV Categories According to the U.S. Department of Defense (DoD) [1, 3]

Group	Size	Maximum Gross Takeoff Weight (MGTW) (lb)	Normal Operating Altitude (ft)	Airspeed (knots)
1	Small	0–20	<1200 AGL	<100
2	Medium	21–55	<3500	<250
3	Large	<1320	<18,000 MSL	<250
4	Larger	>1320	<18,000 MSL	Any airspeed
5	Largest	>1320	>18,000	Any airspeed

AGL, above ground level; MSL, mean sea level.
Note: If the UAV has even one character of the next level, it is classified in that level.

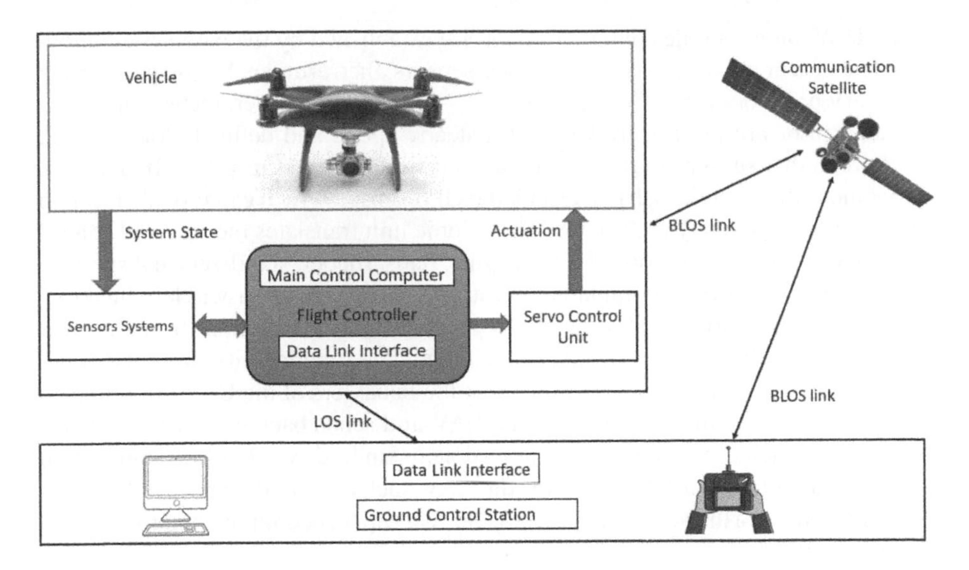

FIGURE 11.1 High-level architecture of the UAV system.

the flight controller has several transmission channels associated with the telemetric signals that it can send to the GCS. The flight controller can have multiple sensors integrated onboard or communicate with an external sensor unit. The UAV system can be equipped with various sensors including an accelerometer, gyroscope, magnetic orientation sensor, Global Positioning System (GPS) module and an electro-optical or infrared camera.

GCS: This is an on-land facility that provides the capabilities for human operators to control and/or monitor UAVs during their operations. GCSs vary in size according to the type and mission of the drones. In other words, for recreational mini and microdrones, GCSs are small handheld transmitters used by hobbyists. For tactical and strategic drones, a large self-contained facility with multiple workstations is employed as the GCS. A GCS communicates with the drone through a wireless link to send commands and receive real-time data, creating a virtual cockpit.

Data links: This is the wireless link used to carry control information between the drone and the GCS. The adopted communication link depends on the UAV operation range. According to their distance from the GCS, drone missions are categorized into line-of-sight (LOS) missions, where control signals can be sent and received via direct radio waves, and beyond line-of-sight (BLOS) missions, where the drone is controlled via satellite communications or a relaying aircraft that can be a drone itself [6].

11.2 UAV COMMUNICATION SYSTEMS

The main communication link of the UAV communication system is between the GCS and the flight controller. Inside the UAV vehicle, the flight controller, which is also known as the base system module, forms the foundation and operating system of the UAV. Inter-module communications are established by the base system module. The sensor module contains various sensors that are capable of performing necessary pre-processing. Generally, the pressure sensor, accelerometer sensor and attitude sensor are used to fly UAVs at a steady speed and defined altitude level. Moreover, other sensors such as radar cameras are also used in UAVs. In addition, autonomous flight mode is dependent on the GPS sensor, which can provide location coordinates and velocity to the GCS. The avionic unit translates the received control commands into the instructions for the engine, flaps, rudder, stabilizers and spoilers. It is required for UAV to communicate with the GCS through a wireless medium. This is two-way communication, and UAV receives basic instructions from the GCS and transmits collected information to the GCS. The functionality of the GCS not only confines the controlling or coordinating the behaviors of the UAV, but also processes the data that are received from the UAV and sends back necessary feedback messages. To achieve smooth communications, standard wireless communication module and protocols are available for the UAV such as 3G, 4G, Wi-Fi, Bluetooth, WiMAX and so forth. Note that those models also support communications between UAVs and GCS.

Information flow in the UAV system is presented in Figure 11.2. The operation of UAVs depends on external inputs. As the communication channels are wireless, it suffers from security weakness [7]. In general, the communication in UAV networks is achieved using UAV ad hoc networks (UANETs), which are considered very similar to wireless sensor networks (WSNs) and mobile ad hoc networks (MANETs). The advantage of a UAV over a sensor network is that it can handle more complex calculations because it has a more powerful CPU than a sensor's microprocessor. Also, the number of nodes, the amount of information transferred across channels and the power requirements in WSNs are less than that in UANETs. Moreover, the coverage area requirement of UANETs is more than that in both WSNs and MANETs. The node mobility of UANETs is much greater than that of MANETs. Research suggests that there are numerous security models for WSNs [8, 9] and MANETs [10], but those models cannot be applied to UANETs due to the mismatch of different properties and network requirements. The most common network used to communicate with other UAVs is the flying ad hoc network (FANET). The security issues

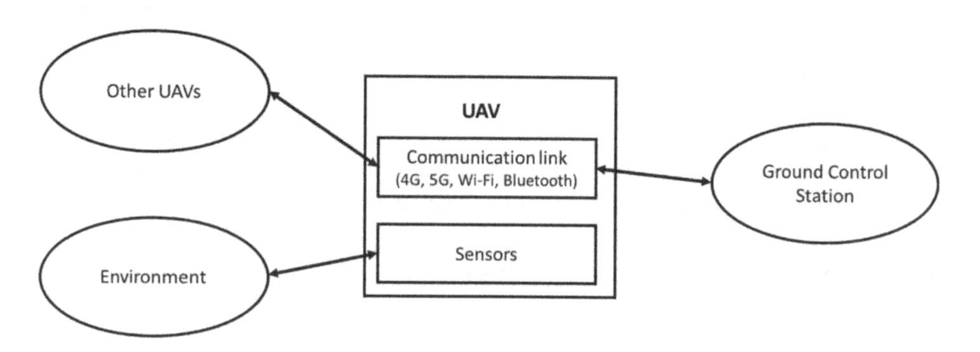

FIGURE 11.2 Information flow among the UAV components.

regarding the FANET is reported in [11]. Some important aspects of UAV communication networks are presented in [12]. The network topology, protocols, advantages and limitation are also discussed there.

11.3 SECURITY THREATS

There are different types of security threats to UAV communication [7] and they are described as follows:

Eavesdropping: Due to the lack of encryption and other protective mechanisms, the exchanged UAV information in the open environment can be directly accessed by the adversary.

Information injection: Without appropriate authentication schemes, an adversary can masquerade as a legitimate entity to inject false information or commands into the network.

Denial of service (DoS) and distributed DoS: Without the appropriate DoS/DDoS-resistant mechanism, a multitude of compromised systems (or a compromised system) can attack a single targeted UAV, causing a DoS for legitimate users of the targeted UAV.

Availability: Availability is defined as a key characteristic of network security, which means that a UAV can provide effective services when necessary, even if it is being attacked. Availability relates to multiple layers. In the network layer, an adversary can tamper with the routing protocols in UAV ad hoc networks.

Confidentiality: This requirement ensures that communication information among UAVs cannot be leaked to unauthorized users, entities or processes.

Integrity: This means that the message transmission process is not interrupted, and the received information should be the same as the sent information. If there is no integrity protection, malicious attacks in the network or wireless channel interference may cause the information to be destroyed, and thus become invalid.

Authentication: Because of the multi-source and heterogeneous network for UAV communications, each node needs to be able to recognize the identity of the node that it communicates with. At the same time, it is better to ensure user authentication without the management of a global certification agency. Without authentication, an adversary can easily impersonate a legitimate node and then obtain important resources/information and interfere with the communications of other nodes.

Non-repudiation: Non-repudiation is used to ensure that a node cannot deny that it has issued certain information. This requirement strengthens the management of various actions, preventing the denial of the behaviors that have occurred. The aim is to provide a basis or means of inquiry on the emergence of security issues.

All threats can be classified into two major categories, i.e., cyber and physical.

11.4 SECURITY REQUIREMENTS

In this section, we provide the security features required in a UAV system to protect the confidentiality and integrity of the communication information and to ensure its ability to adhere to its operational requirements. Securing the information of the system refers to protecting it from disclosure, disruption, modification and destruction. According to the U.S. Federal Aviation Administration (FAA) [13], security requirements can be derived based on the perspective of the stakeholders, which includes the FAA, agencies for national security, aviation authorities, UAV operators and manufacturers, users of the national airspace, society, and privacy advocates. For a secure UAV operation, we identify the following security requirements:

Authorized access: The UAV system must provide means to ensure that only authorized operators are granted access to its resources including both the GCS and the aircraft. More precisely, authentication mechanisms and mandatory access control policies must be implemented to mitigate unauthorized personnel from accessing the GCS and, consequently, commanding the vehicle in a malicious manner. Also, continuous mutual authentication between the operator and the UAV is essential during their communications. Authentication mechanisms may incorporate operation-specific distance bounding protocols [14] to further authenticate the distance between the communicating entities. Such a measure can reduce the success of spoofing attacks in which the spoofer is unlikely to be at a close proximity to the impersonated party.

Availability: All the elements of the UAV system should be guaranteed to perform their required functions under defined spatial and temporal circumstances such that the system sustains its availability without disruptions during its operational period. For instance, the UAV must adopt measures such as anomaly-based intrusion detection systems [15] to distinguish normal communications from those resulting from DoS attacks. Additionally, the utilization of alternative operational procedures, such as using a different set of sensors to cross-check readings, can allow the flight control system to tolerate malfunctioning of or alterations to specific components. Also, managing the patching and updating processes in a way that does not compromise the availability of the UAV system during its operation is of paramount importance.

Information confidentiality: The UAV system should employ mechanisms to mitigate unauthorized disclosure of the telemetric and control information. Different encryption standards such as AES [16] can be used for the encryption of the data link.

Information integrity: The UAV system should be able to ensure that the telemetric information and the GPS and control signals are genuine and have not been intentionally or unintentionally altered. Authenticated encryption cryptographic primitives may be used to ensure both the integrity and confidentiality of such information.

System integrity: The UAV system should be able to guarantee the authenticity of its software and hardware components. Techniques from trusted computing such as memory curtaining, sealed storage and remote attestation can be used to ensure the authenticity of the system's firmware and sensitive data [17]. The deployment of intrusion detection systems, antivirus software, firewalls and strict policies regarding the use of external storage media can aid in the detection and prevention of malware. Also, regular side-channel analysis, including timing and power analysis, may be used for the detection of the activation of hardware Trojans [18] because such trojans usually alter the system's parametric characteristics such as its expected performance and power consumption.

Accountability of actions: The UAV system should employ mechanisms that enforce non-repudiation to ensure that operators are held responsible for their actions. Digital signature algorithms may be used to both authenticate the operators and to bind them to an issued action. Moreover, logging procedures that are used to chronologically track the sequence of actions and changes in the system should be implemented.

Table 11.2 summarizes the identified security threats against UAVs, the corresponding violated security properties and possible mitigation techniques.

11.5 PRIVACY ISSUES RELATED TO CIVILIAN DRONES

An important concern about civilian drones is the ease of their use in violating personal privacy and the difficulty of capturing the intruding ones. UAVs possess a unique range of agile access techniques that distinguish them from other privacy-infiltrating devices. In fact, currently, drones with high-precision cameras can be remotely controlled to perform surveillance tasks with better accuracy and maneuverability than mounting static cameras. The use of drones in surveillance was acknowledged by the FBI director in 2013 [19]. An FBI issued report stated that the agency has no knowledge about any guidelines related to using drones in surveillance and that they try to keep their use to a minimum. The Fourth Amendment of the U.S. Constitution is consistent in protecting the privacy of people; however, it does not define precisely the scope of such privacy.

Building a profile of a person's behaviors and preferences can be a profitable business, as it is hugely valuable to marketers. One can often notice this concept on the Web in the form of a targeted advertisement based on an individual's browsing history. Despite being unconsciously tolerated by many people, such monitoring in the real world is unlikely to be tolerated. Drones are certainly going to be deployed to gather data about our lifestyles and interests as a part of a physical targeted marketing scan. With the added feature of visual evidence, the information gathered by drones is expected to be more valuable to marketing entities than that collected online by a botnet [20]. In fact, given that both online and physical behaviors are concurrently watched without people even noticing, one can expect that a nearly complete picture of a person's movements, social circle and preferences

TABLE 11.2
UAV Threats and Solutions

Type	Threat	Available Solutions
Cyber threat	Jamming or spoofing the GPS signals	• Cross-checking GPS observables • Signal power jump detection • Multiple-antenna GPS receiver • Vision and inertial navigation
	Jamming or spoofing the UAV transmissions	• Spread spectrum and frequency hopping techniques • Authentication of transmissions • Secure distance bounding protocols
	Jamming or spoofing the GCS control signals	• Frequency hopping • Fail-safe/fail loud protocol • Authentication of both control signals and operators
	Unauthorized disclosure of GCS and telemetric signals	• Encryption of the telemetric channels and data link
	Injecting falsified sensors' data to the flight controller	• Cross verification with readings from alternative sensors
	Malicious software and hardware Trojans in GCS or UAV	• Firewall, antivirus and IDs • Power and timing analysis • Managing the supply chain
	Attacking the GCS mission assignment system	• IDSs and firewalls • Regular system updates
	DoS	• Fail-safe/fail loud protocol • IDs
Physical threat	Theft and vandalism	Authentication of the data link • Authentication of the merchandise recipient • Location-based drone electronic immobilizer
	Weather and civic challenges	• Situation awareness [strong artificial intelligence (AI); sense and avoid feature]
	Friendly drone collision	• Dynamic inter-UAV communication • Sense and avoid feature • Automatic dependent surveillance-broadcast system (ADS-B)

can be reconstructed. The following are two examples of drones used maliciously to harvest information and resources:

1. *Snoopy*: Malicious software [21] can be installed on a drone to harvest personal information and to track and profile individuals using wireless localization of their Wi-Fi-enabled smartphones. Snoopy can also sniff RFID, Bluetooth, and IEEE 802.15 signals. A Snoopy-equipped drone exploits the Wi-Fi feature of smartphones that makes them always look for a network to join, including previously known networks. The software first picks a given signal emitted by the victim's phone and identifies a network that is already known and trusted by the device. Then, Snoopy impersonates the identified

network to trick the smartphone into joining it. After this, Snoopy can collect all the information from this disguised network, including the medium access control (MAC) address of the smartphone, which is used to track the phone in real time.
2. *SkyNet*: This is a stealth network [22] that uses drones to forcefully recruit and command host computers for a botmaster. The drones are used to scan a given area and compromise home Wi-Fi networks and eventually the connected computers. Afterward, the drones are regularly used to issue the botmaster's commands for the compromised computers. SkyNet exploits the weak security nature of personal networks, which are considered the most unsecured networks on the Internet. Such networks usually include unpatched machines, and they do not implement auditing features and are known for their poor wireless security and bad password choices. Once the home computers are compromised, the botmaster can access personal files and acquire sensitive account credentials. SkyNet toughens the botnet by bypassing the use of the Internet for communication with the host bots, avoiding known security mechanisms such as firewalls and intrusion detection systems.

In [23], a drone that can communicate with smart devices, [24] including smart appliances and smart lighting, using ZigBee on the Internet of Things (IoT) is proposed. The drone is stocked with multiple ZigBee radios for interacting with devices using the same protocol. An IoT drone is also equipped with a GPS functionality to determine the location of each device. The drone is fully autonomous and operates by capturing and recording the locations and information of all smart devices within a range of 330 feet. The collection of information about the type and possible usage status of devices in individuals' homes can be used to predict their living standards and the times in which they are out of their homes. Such information not only violates the privacy of the affected persons but also can be used for theft and vandalism purposes.

11.6 UAV SECURITY SOLUTIONS

Anomaly detection of UAVs using behavioral profiling is proposed in [25]. Behavioral analysis of UAVs offers a solution of a wide range of problems in the detection of attacks and anomalies. Behavioral analysis is presented using the semantic pyramid. The lower level of the pyramid provides information on flight data, and the highest level of the pyramid represents UAV's functionalities to achieve its recognized goal. The proposed method can detect flight anomaly using behavioral state matching and timing matching. Authors claim that experimental results proved that behavioral detectors can identify anomalies without false positives in different weather conditions.

To ensure the security of UAVs, an adaptive security approach for multi-level ad hoc networks is proposed in [26]. The proposed framework can adapt to the contingent damages on the network infrastructure. The security framework is suitable for a UAV mobile backbone network (UAV-MBN) and it has two modes, namely, infrastructure (for normal operation) and infrastructure-less (for UAVs failure).

GPS spoofing detection is another important security aspect of UAVs. GPS spoofing can be detected using signal features, cryptographic algorithm, auxiliary equipment and visual sensors. In a signal feature-based approach, an automatic gain control value of the GPS signal is used to detect GPS spoofing in [27]. Similarly, the angle of arrival of the navigation signal is utilized in [28]. Also, the absolute value of the relevant peak power is used to detect fraud in [29]. A monocular camera and IMU sensor–based GPS spoofing detection method are proposed in [30] where an image localization approach is utilized.

The cyber-physical security against interdiction attacks of UAV-based delivery systems is studied in [31]. The problem is simulated using a zero-sum network interdiction game between a vendor and an attacker. The Nash equilibrium of the simulated network is obtained by solving linear programming problems. The notion of prospect theory has been used to formulate the game. The outcomes of the study suggest that subjective decision-making process of the vendor and attacker lead to certain delay in delivery time.

Detection of a fault data injection attack has been proposed in [32] where an adaptive neural network is used. Due to the need for continuous supervision, detection of intentional faults plays a great role in the safety of UAVs. Thus, the proposed method used an embedded Kalman filter (EKF) to tune neural network weights online, which helps to detect attacks faster and more accurately.

Wi-Fi-based UAVs are vulnerable to DoS and buffer overflow attack during its ARDiscovery connection process. To mitigate the security risk of that attack, a comprehensive security framework is proposed in [33] where it utilizes a defense-in-depth method.

A healthy, mobility and security-based data communication architecture for UAVs is proposed in [34]. This architecture can secure confidentiality, integrity, authenticity and availability of communication channels. The most lethal cyberattacks of UAV networks are false information dissemination, GPS spoofing, jamming and black hole and gray hole attacks. A set of detection and response techniques are proposed in [35] where UAVs' behaviors are monitored and categorized into normal, abnormal, suspect and malicious according to detected cyberattacks.

To secure UAVs from hijacking network channel or physical hardware by attackers, Yoon et al. [36] proposed an additional encrypted communication channel and authentication algorithm. Similarly, to protect UAVs from eavesdropping and malicious jamming, artificial noise signals together with information signals are transmitted [37]. Moreover, to secure UAV communications, maximum secrecy rate of the system is achieved by jointly optimizing the UAV's trajectory and transmitting power over a finite horizon [38]. The solutions are summarized in Table 11.3.

11.7 CONCLUSIONS

UAVs are vulnerable to different types of cyber and physical attacks. UAVs operate in a variety of environments that are susceptible to outside attacks. Because different types of network topologies and communication protocols are used to operate the UAVs, it becomes more difficult to secure UAVs from attacks. Some solutions are

TABLE 11.3
Summary of Different UAV Security Methods

Literature	Objectives	Methods
Broumandan et al. [27]	GPS spoofing detection	Automatic gain control value of the GPS signal is used
Dempster and Cetin [28]	GPS spoofing detection	The angle of the arrival of navigation signal is used to detect GPS spoofing signals
Psiaki and Humphrey [29]	To detect fraud	Using the absolute value of the relevant peak power to detect fraud behaviors
He et al. [30]	GPS spoofing detection	Monocular camera and image localization approach
Birnbaum et al. [25]	Anomaly detection of UAVs	Behavioral profiling model
Kong et al. [26]	To adapt with the contingent damages on the network infrastructure	An adaptive security approach for multilevel ad hoc networks
Sanjab et al. [31]	Cyber-physical security against interdiction attacks	Solving linear programming problems to identify system state change
Abbaspour et al. [32]	Detection of fault data injection attack	EKF to tune neural network weights online
Hooper et al. [33]	Prevent from DoS and buffer overflow attack	Utilizes a defense-in-depth method to resist DoS attacks
Pigatto et al. [34]	Secure confidentiality, integrity, authenticity and availability of communication channels	Proposed new network architecture to resist data channel eavesdropping and attacks
Sedjelmaci et al. [35]	Identify false information dissemination, GPS spoofing, jamming and black hole and gray hole attacks	A set of algorithms to detect and prevent jamming signals
Yoon et al. [36]	Prevent hijacking network channel or physical hardware by attackers	An additional encrypted communication channel and authentication algorithm
Liu et al. [37]	Protect UAVs from eavesdropping and malicious jamming	Artificial noise signals together with information signals are transmitted
Zhang et al. [38]	Securing UAV communication	Jointly optimized the UAV's trajectory and transmit power

available to make the UAV system more secure to operation attacks, but each of them has its own limitations. Moreover, new communication techniques like 5G and, more complex network architectures such as IoTs, offer a new set of security challenges. In the future, those challenges need to be overcome to make the UAVs more robust in commercial applications.

REFERENCES

1. C. Kennedy and J. I. Rogers, "Virtuous drones?," *International Journal of Human Rights*, vol. 19, no. 2, pp. 211–227, 2015.
2. Fox News, "Drones smuggling porn, drugs to inmates around the world," 2017. [Online]. https://www.foxnews.com/us/drones-smuggling-porn-drugs-to-inmates-around-the-world. Accessed March 21, 2019.
3. P. van Blyenburgh, "Furthering the Introduction of UAVs/ROA into Civil Managed Airspace," *EURO UVS 86,* p. 12.
4. U.S. Army UAS center of Excellence, "U.S. Army Unmanned Aircraft Systems Roadmap 2010-2035," 2010.
5. N. Melzer, *Human Rights Implications of the Usage of Drones and Unmanned Robots in Warfare*, European Parliament, 2013.
6. D. M. Marshall, R. K. Barnhart, S. B. Hottman, E. Shappee, and M. T. Most, *Introduction to Unmanned Aircraft Systems.* Boca Raton, FL: CRC Press, 2016.
7. D. He, S. Chan, and M. Guizani, "Communication security of unmanned aerial vehicles," *IEEE Wireless Communications*, vol. 24, no. 4, pp. 134–139, 2017.
8. J. Granjal, E. Monteiro, and J. S. Silva, "Security in the integration of low-power wireless sensor networks with the Internet: a survey," *Ad Hoc Networks*, vol. 24, pp. 264–287, 2015.
9. D. He, S. Chan, and M. Guizani, "Accountable and privacy-enhanced access control in wireless sensor networks," *IEEE Transactions on Wireless Communications*, vol. 14, no. 1, pp. 389–398, 2015.
10. P. Rajakumar, V. T. Prasanna, and A. Pitchaikkannu, "Security attacks and detection schemes in MANET," *2014 International Conference on Electronics and Communication Systems (ICECS)*, pp. 1–6, 2014.
11. İ. Bekmezci, E. Şentürk, and T. Türker, "Security issues in flying ad-hoc networks (FANETs)," *Journal of Aeronautics and Space Technology*, vol. 9, no. 2, pp. 13–21, 2016.
12. L. Gupta, R. Jain, and G. Vaszkun, "Survey of important issues in UAV communication networks," *IEEE Communications Surveys and Tutorials*, vol. 18, no. 2, pp. 1123–1152, 2016.
13. Information Technology Laboratory, Computer Security Division, "FAA Unmanned Aircraft Systems Update," 2015. [Online]. https://csrc.nist.gov/presentations/2015/faa-unmanned-aircraft-systems-update. Accessed April 8, 2019.
14. I. Boureanu, A. Mitrokotsa, and S. Vaudenay, "Towards secure distance bounding." In: S. Moriai, Ed. *Fast Software Encryption*. Berlin, Heidelberg: Springer, vol. 8424, pp. 55–67, 2014.
15. S. Han, M. Xie, H. H. Chen, and Y. Ling, "Intrusion detection in cyber-physical systems: techniques and challenges," *IEEE Systems Journal*, vol. 8, no. 4, pp. 1052–1062, 2014.
16. J. Daemen and V. Rijmen, *The Design of Rijndael.* Berlin, Heidelberg: Springer, 2002.
17. K. Dietrich and J. Winter, "Implementation aspects of mobile and embedded trusted computing," *International Conference on Trusted Computing*, pp. 29–44, 2009.
18. M. Tehranipoor and F. Koushanfar, "A survey of hardware trojan taxonomy and detection," *IEEE Design and Test of Computers*, vol. 27, no. 1, pp. 10–25, 2010.
19. B. C. C. Producer CNN Senior, "FBI uses drones for surveillance in U.S. - CNNPolitics," [Online]. https://www.cnn.com/2013/06/19/politics/fbi-drones/index.html. Accessed April 8, 2019.
20. L. Rosen, "Drones and the digital panopticon," *XRDS Crossroads: The ACM Magazine for Students*, vol. 19, no. 3, p. 10, 2013.
21. K. Gittleson, "Snoopy drone sniffs public's data," March 28, 2014.

22. T. Reed, J. Geis, and S. Dietrich, "SkyNET: a 3G-enabled mobile attack drone and stealth botmaster," *5th Usenix Workshop on Offensive Technologies*, p. 9.
23. P. Paganini, "ZigBee-sniffing drone used to map online Internet of Things," *Security Affairs*, 2015. [Online]. https://securityaffairs.co/wordpress/39143/security/drone-internet-of-things.html. Accessed: April 9, 2019.
24. J. Won, S.-H. Seo, and E. Bertino, "A secure communication protocol for drones and smart objects," *Proceedings of the 10th ACM Symposium on Information, Computer and Communications Security - ASIA CCS '15*, pp. 249–260, 2015.
25. Z. Birnbaum, A. Dolgikh, V. Skormin, E. O'Brien, D. Muller, and C. Stracquodaine, "Unmanned aerial vehicle security using behavioral profiling," *2015 International Conference on Unmanned Aircraft Systems (ICUAS)*, pp. 1310–1319, 2015.
26. J. Kong, H. Luo, K. Xu, D. L. Gu, M. Gerla, and S. Lu, "Adaptive security for multilevel ad hoc networks," *Wireless Communications and Mobile Computing*, vol. 2, no. 5, pp. 533–547, 2002.
27. A. Broumandan, A. Jafarnia-Jahromi, S. Daneshmand, and G. Lachapelle, "Overview of spatial processing approaches for GNSS structural interference detection and mitigation," *Proceedings of the IEEE*, vol. 104, no. 6, pp. 1246–1257, 2016.
28. A. G. Dempster and E. Cetin, "Interference localization for satellite navigation systems," *Proceedings of the IEEE*, vol. 104, no. 6, pp. 1318–1326, 2016.
29. M. L. Psiaki and T. E. Humphreys, "GNSS spoofing and detection," *Proceedings of the IEEE*, vol. 104, no. 6, pp. 1258–1270, 2016.
30. D. He, Y. Qiao, S. Chan, and N. Guizani, "Flight security and safety of drones in airborne fog computing systems," *IEEE Communications Magazine*, vol. 56, no. 5, pp. 66–71, 2018.
31. A. Sanjab, W. Saad, and T. Basar, "Prospect theory for enhanced cyber-physical security of drone delivery systems: a network interdiction game," *2017 IEEE International Conference on Communications (ICC)*, pp. 1–6, 2017.
32. A. Abbaspour, K. K. Yen, S. Noei, and A. Sargolzaei, "Detection of fault data injection attack on UAV using adaptive neural network," *Procedia Computer Science*, vol. 95, pp. 193–200, 2016.
33. M. Hooper *et al.*, "Securing commercial WiFi-based UAVs from common security attacks," *MILCOM 2016 - 2016 IEEE Military Communications Conference*, pp. 1213–1218, 2016.
34. D. F. Pigatto, L. Goncalves, A. S. R. Pinto, G. F. Roberto, J. Fernando Rodrigues Filho, and K. R. L. J. C. Branco, "HAMSTER - Healthy, mobility and security-based data communication architecture for unmanned aircraft systems," *2014 International Conference on Unmanned Aircraft Systems (ICUAS)*, pp. 52–63, 2014.
35. H. Sedjelmaci, S. M. Senouci, and N. Ansari, "A hierarchical detection and response system to enhance security against lethal cyber-attacks in UAV networks," *IEEE Transactions on Systems, Man, Cybernetics: Systems*, vol. 48, no. 9, pp. 1594–1606, 2018.
36. K. Yoon, D. Park, Y. Yim, K. Kim, S. K. Yang, and M. Robinson, "Security authentication system using encrypted channel on UAV network," *2017 First IEEE International Conference on Robotic Computing (IRC)*, pp. 393–398, 2017.
37. C. Liu, T. Q. Quek, and J. Lee, "Secure UAV communication in the presence of active eavesdropper," *2017 9th International Conference on Wireless Communications and Signal Processing (WCSP)*, pp. 1–6, 2017.
38. G. Zhang, Q. Wu, M. Cui, and R. Zhang, "Securing UAV communications via trajectory optimization," *ArXiv171004389 Cs Math*, 2017.

12 Jamming Attacks and Countermeasures in UAV Networks

Delwar Hossain,[*] Qian Mao,
Immanuel Manohar, Fei Hu
Electrical and Computer Engineering,
The University of Alabama, Tuscaloosa, AL, USA
[*]corresponding author: dhossain@crimson.ua.edu

CONTENTS

12.1 INTRODUCTION

Wireless Sensor Networks (WSNs) are composed of sensor nodes that sense the environment parameters such as temperature, pressure, humidity, motion and so forth [1, 2]. The data are sent to the sink through routers/gateways. Sensors are typically deployed to form a large wireless sensor network for wide-area monitoring. They can easily operate in harsh environments. Some issues (such as density, distance, etc.) during the deployment of sensor nodes in remote locations must be considered, although sensor deployment is a random process [3]. Deploying too few nodes may raise the issue of coverage, whereas deploying too many nodes may result in an inefficient network because of more communication collisions and radio-frequency (RF) interference. WSNs are used in both indoor and outdoor applications [4]. The transmissions of information collected from sensor nodes must be protected from different threats [5]. Security is considered to be the most challenging task in WSNs as its tough to keep monitoring the status of sensor nodes/networks all the time. It must be secured to prevent an intruder from attacking the transmission of data.

Each UAV may be equipped with dozens of or even hundreds of sensors. For example, it may have sensors to detect altitude, weather conditions (temperature, humidity, wind, etc.), distance to other unmanned aerial vehicles (UAVs), flying velocity and so forth. The UAV can also use the spectrum sensing circuit to detect the wireless operating frequency nearby. One UAV's sensors may communicate with another UAVs' sensors to exchange the detected information. For example, a UAV may warn a neighboring UAV about the jamming signals around itself. A UAV may also use its image sensors to monitor the events within its coverage and then share the image data with other UAVs. Sensors in different UAVs could form a WSN.

The main constraints in designing and developing sensor nodes are sensor chip size and cost [6]. The design constraints result in sensor nodes with small memory size, limited energy source and limited range of transmission [7, 8]. This ultimately results in very little encryption, decryption and authentication schemes that can be efficiently used in sensor node and leads to an attacker who is unauthorized to access the data of the network and tries to mislead the information.

The broadcast nature of wireless links makes the WSN vulnerable to different threats/attacks, which eventually causes serious degradation in the communication performance of the network. A jamming attack can occur when sensors use certain frequency bands for communications. In WSN, the whole communication bandwidth is divided into different frequency bands, and an attacker can detect the current operation frequency and release strong interference signals. A jamming attack can be defined as a special denial-of-service (DoS) attack in which the channel medium is crashed because the attacker sends many requests to the server, which interrupts the communications and may not allow the responses to reach the target. Because there is no response, the client ponders that the server is not retorting to the request and then continuously sends the requests to try to obtain a response from the server [9]. Unlike other attacks, this attack is accomplished after reconnaissance. The attacker often has knowledge about the communication pattern. It listens to the traffic and sends the jamming signals continuously to obstruct the link and interrupt the routing data to provide the intended data from reaching the target.

Medium access control (MAC) protocols are exposed to jam attacks. Based on the patterns of existing communications, a jammer can pick the right area for jamming attacks. Initially, it may select the region with the highest communication traffic to launch the jamming signals [10]. This leads to high cost of communications due to extremely low message delivery rate. A smart jammer might have access to the control channel. It can then continuously send jamming signals to block the control channel. Moreover, it can extort the sequence of the control channels from all legitimate nodes, which will smash up the whole wireless network. Accordingly, there could be many signal transmission collisions between the jammed signals and the packets sent from valid network nodes. In wireless networks, channel collisions are often caused due to two nodes sending data at the same time in the same channel [11].

The jamming signals can disrupt communications by decreasing the signal to inference plus noise ratio (SINR). SINR is the ratio of the signal power to the sum of the interference power from other interference signals and noise power. A jamming device, tuned to the same frequency as the sensor's receiving equipment and with the same type of modulation, can override any signal at the receiver. Advanced and more expensive jamming devices may even jam the satellite communications. A wireless jamming device can be used to stop transmissions temporarily, short out or turn off the power of device or units like radios, televisions, microwaves or any unit that receives electrical signals to operate.

Another type of wireless network often used in the military and in disaster situations is the wireless ad hoc network (AHN). This network is called ad hoc because it does not need any pre-existing infrastructure such as cables or access points (APs). Here, each node, e.g., laptop and cellphone, participates in the network's routing of data by forwarding it from one node to another without any centralized management equipment like an AP. The nodes in an AHN dynamically decide which node to send the data, based on the network connectivity. Figure 12.1(b) shows a basic setup of an AHN between laptops and phones. They communicate among themselves without an AP.

The AHN is slightly different from a WSN [Figure 12.1(a)] in which the data collected is not directly sent to the user; rather, it is first aggregated and then sent out [12]. A WSN consists of a gateway or base station (BS) that connects the sensor nodes to other

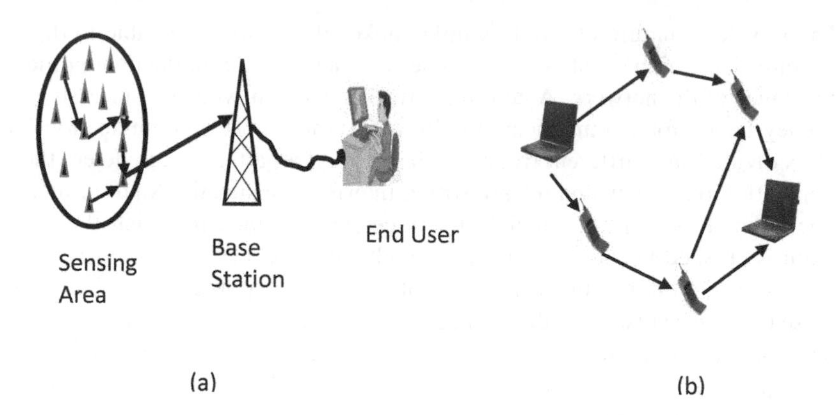

FIGURE 12.1 (a) Wireless sensor network. (b) Wireless ad-hoc network.

sensor networks, or to the end user [Figure 12.1(a)]. The data at the sensor nodes are compressed and transmitted to the BS, which presents the results to the end user [13]. The data packets are sometimes sent to the destination via several intermediate nodes. The data transmission by relaying from one node to another is called a multi-hop relay.

12.2 SECURITY REQUIREMENTS

WSNs have to fulfill some requirements (Figure 12.2) to provide a secure communication. General security requirements [18] are availability, confidentiality, integrity

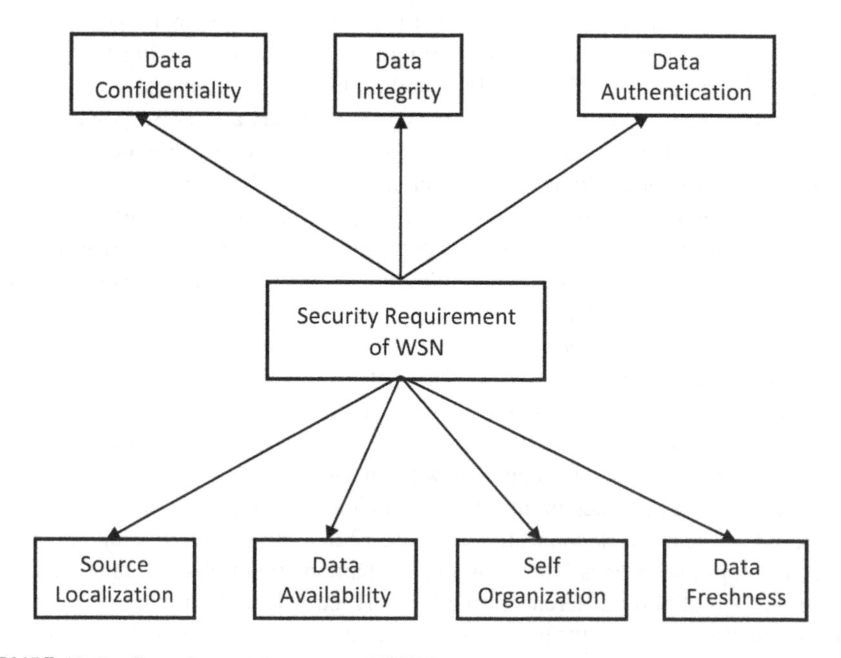

FIGURE 12.2 Security requirements of WSNs.

and authentication [19, 20]. Other requirements include source localization, self-organization and data freshness.

12.2.1 Data Confidentiality

In a sensor network, data flows among many intermediate nodes, thus, the chance of data leakage over RF links is high [21]. To provide data confidentiality, an encrypted data flow is used so that only the legitimate recipient decrypts the data to recover its original format.

12.2.2 Data Integrity

Data to be received by the receiver should not be altered or modified during the transmission. This requirement is called data integrity. Original data may be changed by the intruder or harsh environment. The intruder may change the data according to its need and send the falsified data to the receiver [22].

12.2.3 Data Authentication

Data authentication refers to the procedure used to confirm that the communicating node is in fact the one that it claims to be. It is important for a receiver node to verify that the data are received from an authenticated source.

12.2.4 Data Availability

Data availability means that the services are available all the time, even in case of some attacks such as DoS.

12.2.5 Source Localization

During data transmissions some applications need to use the location information of the sink node. It is important to verify that the source location is a trustworthy place.

12.2.6 Self-Organization [23]

In WSNs no fixed infrastructure exists, hence, all sensors maintain self-organizing and self-healing properties. This is a great challenge for security designs.

12.2.7 Data Freshness

Data freshness means that each message transmitted over the channel should be new and fresh. It guarantees that the old messages cannot be replayed by any node. This can be solved by adding some time-related counter to check the freshness of the data.

12.3 JAMMING CHARACTERISTICS

From a cognition-level viewpoint, jammers can be categorized into two main types: fundamental (i.e., basic) and intelligent. From a jammer's behavior viewpoint, these can be segregated into proactive and reactive types.

12.3.1 Fundamental Jammers

Fundamental jammers mainly comprise four types: constant, random, deceptive and reactive. A constant jammer is a physical-layer attacker, whereas the remaining types are MAC-layer attackers. Constant jammers constantly emit the radio signal. Random jammers send the random signals continuously in the MAC layer. Deceptive jammers constantly inject normal packets into the conduit with no space during transmission of the consecutive packet. Therefore, a node will be deceived to believe that the packet it is receiving is a genuine packet and would stay in the receiving state. A random jammer switches between sleep mode and jam mode. The times of attack and sleep can vary, which allows a wicked node to attain diverse levels of compromise between energy efficiency and jamming efficacy, depending on the concrete applications. Reactive jammers settle quietly when there is an idle channel. They start their jamming behaviors only when they sense that the network has started its communications.

12.3.2 Intelligent Jammer

The jammers who target the physical layer aim to destroy the signals, congest the network or require the nodes to consume more energy, whereas other types of jammers targeting the MAC layer aim to attack the network privacy. Their objective is to determine the MAC protocol used by victim nodes to launch an energy-efficient attack. Some schemes have been proposed to overcome the jammers related to MAC layer protocols, such as frequency hopping, sequence of frequency, packet fragmentation, frame masking and redundant coding to diminish the brunt of damage caused by a jammer.

12.4 TYPES OF JAMMERS

Jammers can be classified into various types, as shown in Figure 12.3 [9].

12.4.1 Proactive Jammer

The proactive jammer simply sends out interference signals regardless of whether the legitimate node is performing data communication or not.

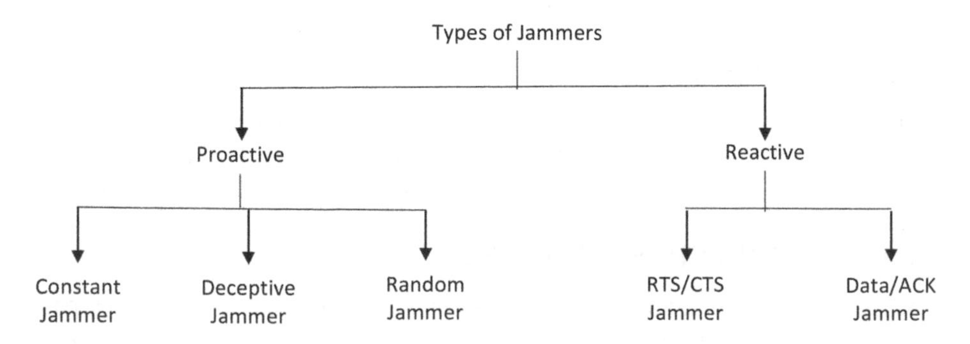

FIGURE 12.3 Types of jammers.

12.4.1.1. Constant Jammer

Instead of following the carrier-sense multiple access (CSMA) protocol, the constant jammer sends out random bits continuously. If a node follows the CSMA protocol, then it must sense the wireless medium to make sure it is idle [24] before transmitting any data onto the channel.

12.4.1.2 Deceptive Jammer

Instead of releasing random bits (like the constant jammer), these jammers constantly transmit normal packets. The deceptive jammer misleads other nodes to trust that a valid transmission is in place so that they remain in the receiving state until the jammer is turned off [24]. The jammer's "normal" packets actually do not make sense if decoded.

12.4.1.3 Random Jammer

The random jammer sporadically spreads either arbitrary bits or normal packets into networks. It saves energy by toggling between the sleep and jamming phases [24].

12.4.2 Reactive Jammer

A reactive jammer is different from the proactive one because it sends out jam signals when it senses that the network is in an active state. Therefore, it has to be active every time to monitor the channel; hence, it uses more energy than random jammer [24, 25].

12.4.2.1 Reactive RTS/CTS Jammer

When the legitimate node sends a request to send (RTS) message, the jammer senses it and jams the network. It then initializes the jamming signals. The receiver cannot respond back with a clear-to-send (CTS) message due to the damage done to the RTS packets [25].

12.4.2.2 Reactive Data/ACK Jammer

This type of jammer alters the packet's transmission. Because the information packets are not received properly at the receiver, they have to be retransmitted. The acknowledgement messages (ACKs) cannot reach the sender, which makes the sender think that something has gone wrong at the target side, such as buffer overflow. Therefore, it sends the data again and again [25].

Table 12.1 (below) summarizes different types of jammers.

TABLE 12.1
Important Features of Jammers

	Types of Jammer	Transmission of Bits	Energy Efficient
Proactive	Constant	Continuous, random bits	Yes
	Deceptive	Continuous, regular bits	Yes
	Random	Either random or regular	No
Reactive	RTS/CTS Jammer	During RTS/CTS sent	Yes
	Data/ACK jammer	During Data/ACK sent	Yes

12.5 SECURITY PROTOCOLS IN SENSOR NETWORKS

Cryptography is a basic technique to achieve security in a network. This establishes a secure relationship between two end points. The sender encrypts the original data and the receiver decrypts the received data to recover original data. Different types of keys are used in the process of cryptography. The various protocols [38] for solving the security issue in WSNs include the following examples.

12.5.1 SPINs

Sensor protocols for information via negotiation (SPIN) work in three steps. First, a node advertises the ADV packet containing the metadata. If the received node is interested in the data, it then sends the request for data using a REQ packet. Finally, the advertiser node that receives the request sends the DATA packet to the requestor node. SPIN performs best in small-size networks because of its efficiency and low latency [39]. Typical SPIN consists of two secure building blocks, timed efficient stream loss-tolerant authentication (μTESLA) and sensor network encryption protocol (SNEP). SNEP provides confidentiality, authentication and integrity. It uses the concept of encryption. To authenticate the data, message authentication code is used. It adds 8 bytes to the message. To reduce the communication overhead, SNEP uses a shared counter between the sender and receiver. After each block is received the counter gets incremented. The counter helps to identify the freshness of data in TESLA, and digital signatures are used to authenticate the data packet. A sink node computes a MAC with the secret key for the packet and then sends an authenticated packet back to the source. After receiving a packet, the node confirms that the sink does not disclose the computed MAC key to other nodes. With this, the receiving node ensures that the data packet is original and no alterations are performed in the packet.

12.5.2 LEAP

The localized encryption and authentication protocol (LEAP) is a protocol with a key management scheme, and it is very efficient for large-scale distributed sensor networks. It generally supports in-network processing such as data aggregation. In-network processing results in the reduction of the energy consumption in the network. To provide the confidentiality and authentication to the data packet, LEAP uses multiple keys for different purposes. For each node four keys are used, known as the individual, pairwise, cluster and group keys [13]. All are symmetric keys and have the following definitions:

- *Individual key*: Unique key used for the communications between the source node and sink node.
- *Pairwise key*: Shared with another sensor node.
- *Cluster key*: Used for locally broadcast messages and is shared between the node and all of its surrounding neighboring nodes.
- *Group key*: Is a globally shared key among all the network nodes

These keys also can be used by other non-secured protocols to increase the network security. LEAP satisfies several security and performance requirements of WSNs. It is used to defend against HELLO flood attacks, sybil attacks and wormhole attacks.

12.5.3 TINYSEC

TINYSEC is the link layer security architecture for WSNs. It is a lightweight protocol. It supports integrity, confidentiality and authentication. To achieve confidentiality, encryption is done by using cipher-block chaining (CBC) mode with cipher text stealing, and authentication is done using CBC-MAC. No counters are used in TINYSEC. Hence, it does not check the data freshness. Authorized senders and receivers share a secret key to compute a MAC. TINYSEC has two different security options: one is for authenticated and encrypted messages (TinySec-AE) and the other is for authenticated messages (TinySec-Auth). In TinySec-AE, the data payload is encrypted and the received data packet is authenticated with a MAC. In TinySec-Auth mode, the entire packet is authenticated with a MAC, but the data payload is not encrypted. In CBC, the initialization vector (IV) is used to achieve semantic security. Some of the messages are the same with small variations. In that case, an IV field is added to the encrypted process. To decrypt the message the receiver must use the IV. IVs are not secret and are included in the same packet with the encrypted data.

12.5.4 ZIGBEE

ZIGBEE is a typical wireless communication technology [7]. It is used in various applications such as military security, home automation and environment monitoring. IEEE 802.15.4 is the standard used for ZIGBEE. It supports data confidentiality and integrity. To implement the security mechanism, ZIGBEE uses 128 bit keys. A trust center is used in ZIGBEE that authenticates and allows other devices/nodes to join the network and distribute the keys. Generally, a ZIGBEE coordinator performs this function. Three different roles in ZIGBEE are listed next:

- *Trust manager*: Authenticates the devices that request to join the network.
- *Network manager*: Manages the network keys and helps to maintain and distribute the network keys.
- *Configuration manager*: Configures the security mechanism and enables the end to-end security between devices.

ZIGBEE works in two different modes, residential and commercial. In the residential mode less security is needed, hence, no keys are used. Whereas the commercial mode needs high security, thus it maintains the keys and counter. Table 12.2 shows the comparison between the security protocols on the basis of the service provided. It also shows the type of services offered by the existing protocols.

TABLE 12.2
Comparison of Security Protocols

Protocols	Confidentiality	Freshness	Integrity	Availability	Authentication	Key Agreement
SPIN	Yes	Yes	Yes	No	No	Symmetric delayed
LEAP	Yes	No	No	No	No	Pre-delayed
TINYSEC	Yes	No	No		Yes	Any
ZIGBEE	Yes	Yes	Yes	No	Yes	Trust center

12.6 JAMMING DETECTION AND COUNTERMEASURES

When jamming is detected, the jammed area can be mapped by the network nodes, which reroute the traffic, or switch channels to thwart the jamming act. A comparative analysis is shown in Table 12.3.

12.6.1 JAM

The jammed area mapping (JAM) protocol routes the packets around the exaggerated area. It can plot a wedged area within 1–5 seconds. As the value of the node reaches below the threshold, the system raises the information in the form of a message that is either JAMMED or UNJAMMED, and broadcasts it to the neighbors. When the announce timer expires, a BUILD message is sent by the node. On receiving these messages, a message TEARDOWN is used by mapping nodes to notify the recovered nodes. After the achievement of the mapping process, all the nodes in the network get the message to reroute a new path to avoid the area mapped as jammed [26].

12.6.2 Ant System

The ANT system is used to detect the jam at the PHY layer and deliver the messages to the target node. It formulates a hypothesis to test whether a DOS attack is genuine or not. There is an agent who traverses iteratively and gathers knowledgeable information about various routes to a target. They used the following jammers: single tone, multiple tone, pulsed noise, and electronic intelligence. The node detection is based on its availability of resources such as hops, energy, distance, packet loss, signal-to-noise ratio (SNR), bit error rate (BER), and packet drop rate (PDR). After a certain metric is checked, a decision model is used to check if the jamming detection is true or not. If there is a jam in a particular link, that link is excluded for the route and a supplementary path is explored [27, 28].

12.6.3 Hybrid System

The hybrid system unites three techniques to defend jamming signals: BS replication, BS evasion and multipath routing between BSs. The replication scheme implies replicated BSs. The evasion scheme defines the spatial retreat of a BS. Multipath

TABLE 12.3
Different Jamming Detection and Countermeasure Schemes

Serial	Techniques	Type	Proposed Attack	Countermeasure
1	JAM	WSN	Maps out the lodged area in WSN and routes packets around the exaggerated area.	If the number of unsuccessful attempts is above 10, it detects the presence of jammer
2	ANT system	WSN	It is physical layer jamming, and redirects the message to a destined node	If jamming is detected, a new path is established
3	Hybrid system	WSN	BS failure could prevent sensor readings from being received, and the BS cannot execute tasks for command and control	BS replication/evasion Multipath routing between BSs
4	Consistency check	WSN	Necessitates enhanced detection schemes to remove ambiguity	Low PDR + consistency check
5	Channel hopping	WSN or WLAN	The number of constrained orthogonal channels in frequency separation is small	Frequency hopping is effective only when the number of orthogonal channels is large
			If an anonymous person gets the information about the history, he can track the channel and jam the subsequent channel continuously	It uses a pseudorandom channel hopping scheme to select channels unidentified to the jammer based on a pseudo-random noise (PN) generation result
6	Hermes node	WSN	The attacker interferes with the radio frequencies by using a powerful jamming source and disrupts the WSNs function	A secret word is used as a seed for the generation of the PN code and channel sequence This secret word is hard-coded so that entrance of the new node in the network can be detected with the existing node.
7	DEEJAM	LR-WPANs	Could be Internet jamming, activity jamming, scan jamming or pulse jamming	Hide messages from attacker, dodges its exploration and trims down the impact of the degraded message
8	EM-MAC	WSN	Continuous jamming	Avoids jammer channel selection
9	JAM-BUSTER	WSN	Schedule predictions	Proactive defense against a jammer
10	SAD-SJ	TDMA-based WSN	Transmitting malicious signal during time slots of a MAC frame	Random permutation of slot timers

routing is used between a node and a BS. With the technique of BS replication, if one or more BSs are jammed, then the non-jammed BSs can provide the services to the network. The last technique requires that every node should have multiple paths to the BS so that if one path is jammed, other paths can still serve [28, 29].

12.6.4 Using PDR with Consistency Checks

The existence of jammers cannot be determined by using a single measurement only. The system can detect the jamming if all close nodes have low PDR values. If a node does not have neighbors, the PDR value will be low. The jamming effect is not considered for such nodes [28, 30].

12.6.5 Channel Hopping

Channel hopping or toggling of channel from one to another is measured to be the most admired countermeasure to jamming signals. Proactive channel hopping is the simplest realization. In proactive channel hopping, the current communicating channel is altered after a definite interval of time. If the access wait time of the channel goes beyond a given threshold value, it is assumed that jamming has occurred and that there is a need to switch the channel via a predefined strategy. In the basic channel hopping scheme, the channel is chosen from the unused channel set. In the deceptive scheme, the selection set includes the presently used as well as unused channels. In this case, if an anonymous user might know the history, it can track the channel selected for hopping and starts jamming the subsequent channel continuously. The substitute is a pseudorandom channel hopping scheme, which uses a pseudonumber generation scheme to choose channels unfamiliar to jammers. After the packet drop rate (PDR) is computed for that channel, communication is switched back to the initial channel. When the present channel's performance (measured in PDR) goes below a threshold, the user toggles the communications to other channels with good PDR values [28, 31–34].

12.6.6 Hermes Node (Hybrid DSSS and FHSS)

Direct Sequence Spread Spectrum (DSSS) and frequency hopping spread spectrum (FHSS) are used to defend from jamming attacks. For signal transmission, DSSS provides wider bandwidth while FHSS offers meddling avoidance. A hybrid scheme, called Hermes node, is anticipated to deal with jamming attacks. The node of Hermes performs 1,000,000 hops per second (FHSS) to evade the jammers. DSSS can make the attacker sense the data signals as white noise, which averts the detection of the communication radio band. Synchronization between nodes is important for the Hermes node to work properly, which is achieved by the sink [28, 35].

12.6.7 DEEJAM (Defeating Energy-Efficient Jamming)

This method was proposed by Wood et al. [26], and is a fresh approach to defeat jammers. It is basically used to conceal messages from the attacker, dodge its exploration and trim down the impact of corrupted messages. It allows network nodes to function effectively even in the existence of a jammer. DEEJAM can defend four classes of jamming attacks, namely, scan, pulse, activity and interrupt jamming [36, 37].

12.6.8 EM-MAC (Energy-Efficient MAC)

This method was proposed by Tang et al. [38]. It augments the employment of wireless channel and resists the intervention and jamming in wireless links by enabling every node to optimize the selection of wireless channels based on the link SNRs it senses.

12.6.9 JAM-Buster

Jam-Buster is a jam-resistant protocol proposed to stomp out the isolation between packets by using three factors: have equal sizes, randomize the wakeup times and implement multiblock payloads. These three techniques are combined to cope with an intelligent jammer and force it to spend more energy to be effective. The authors have evaluated energy consumption only on the jammer's side, whereas the lifetime of legitimate nodes was not considered. Because the system acts like a proactive defense against a jammer, it can also overcome other MAC layer constraints such as overhearing, idle listening and end-to-end delay communication [39].

12.6.10 SAD-SJ

SAD-SJ, a self-adaptive and decentralized MAC-layer, is an approach to discriminating jamming in TDMA-based WSNs. It is based on an arbitrary slot reallocation in which each node achieves an arbitrary permutation of slots. The permutation process can be done after generating a random number. The protocol was proved to be self-adaptive in that it allows the nodes to freely join and leave, yet keep the security of other nodes [40].

12.7 CONCLUSIONS

The security issues of UAV sensors have been discussed in this chapter. The small size, low memory capacity, low energy capacity and small range of communication of sensor nodes make the establishment of security protocol in WSNs a big challenge. Different types of jamming attacks with different parameters have been listed. Various countermeasures taken against the jamming attacks so far are also discussed. An intelligent jammer could use multi-layer jamming release to achieve a DoS attack. No single security protocol can resist all jamming attacks.

REFERENCES

1. R. B. Agnihotri, A. V. Singh, and S. Verma, "Challenges in wireless sensor networks with different performance metrics in routing protocols," *2015 4th International Conference on Reliability, Infocom Technologies and Optimization (ICRITO) (Trends and Future Directions)*, pp. 1–5, 2015.

2. X. Gong, H. Long, F. Dong, and Q. Yao, "Cooperative security communications design with imperfect channel state information in wireless sensor networks," *IET Wireless Sensor Systems*, vol. 6, no. 2, pp. 35–41, 2016.
3. J. Wu, K. Ota, M. Dong, and C. Li, "A hierarchical security framework for defending against sophisticated attacks on wireless sensor networks in smart cities," *IEEE Access*, vol. 4, pp. 416–424, 2016.
4. G. Mali and S. Misra, "TRAST: trust-based distributed topology management for wireless multimedia sensor networks," *IEEE Transactions on Computers*, vol. 65, no. 6, pp. 1978–1991, Jun. 2016.
5. V. J. Hodge, S. O'Keefe, M. Weeks and A. Moulds, "Wireless Sensor Networks for Condition Monitoring in the Railway Industry: A Survey," *in IEEE Transactions on Intelligent Transportation Systems*, vol. 16, no. 3, pp. 1088–1106, June 2015,
6. J. Grover and R. Rani, "Probabilistic density based adaptive clustering scheme to improve network survivability in WSN," *Fifth International Conference on Computing, Communications and Networking Technologies (ICCCNT)*, pp. 1–7, 2014.
7. X. Yi, A. Bouguettaya, D. Georgakopoulos, A. Song, and J. Willemson, "Privacy protection for wireless medical sensor data," *IEEE Transactions on Dependable and Secure Computing*, vol. 13, no. 3, pp. 369–380, 2016.
8. H. Marzi and A. Marzi, "A security model for wireless sensor networks," *2014 IEEE International Conference on Computational Intelligence and Virtual Environments for Measurement Systems and Applications (CIVEMSA)*, pp. 64–69, 2014.
9. A. Mpitziopoulos, D. Gavalas, C. Konstantopoulos and G. Pantziou, "A survey on jamming attacks and countermeasures in WSNs," in *IEEE Communications Surveys & Tutorials*, vol. 11, no. 4, pp. 42–56, Fourth Quarter 2009.
10. S. Sowmya and P. D. S. K. Malarchelvi, "A survey of jamming attack prevention techniques in wireless networks," *International Conference on Information Communication and Embedded Systems (ICICES2014)*, Chennai, 2014, pp. 1–4.
11. J. Massey and P. Mathys, "The collision channel without feedback," *IEEE Transactions on Information Theory*, vol. 31, no. 2, pp. 192–204, 1985.
12. C. S. Raghavendra, K. M. Sivalingam, and T. Znati, *Wireless Sensor Networks*. Berlin: Springer, 2006.
13. K. Lu, Y. Qian, D. Rodriguez, W. Rivera and M. Rodriguez, "Wireless Sensor Networks for Environmental Monitoring Applications: A Design Framework," *IEEE GLOBECOM 2007 - IEEE Global Telecommunications Conference*, Washington, DC, 2007, pp. 1108–1112.
14. A. Mpitziopoulos, D. Gavalas, C. Konstantopoulos, and G. Pantziou, "A survey on jamming attacks and countermeasures in WSNs," *IEEE Communications Surveys Tutorials*, vol. 11, no. 4, pp. 42–56, 2009.
15. P. Pardesi and J. Grover, "Improved multiple sink placement strategy in wireless sensor networks," *2015 International Conference on Futuristic Trends on Computational Analysis and Knowledge Management (ABLAZE)*, pp. 418–424, 2015.
16. J. Grover and Anjali, "Wireless sensor network in railway signalling system," *2015 Fifth International Conference on Communication Systems and Network Technologies*, pp. 308–313, 2015.
17. T. C. Aseri and N. Singla, "Enhanced security protocol in wireless sensor networks," *International Journal of Computers Communications & Control*, vol. 6, no. 2, pp. 214–221, 2011.
18. E. Karapistoli, P. Sarigiannidis, and A. A. Economides, "Visual-assisted wormhole attack detection for wireless sensor networks," *International Conference on Security and Privacy in Communication Networks*, pp. 222–238, 2015.

19. J. Grover and Mohit Sharma, and Shikha, "Reliable SPIN in wireless sensor network," *Infocom Technologies and Optimization Proceedings of 3rd International Conference on Reliability*, pp. 1–6, 2014.
20. D. G. Padmavathi and M. D. Shanmugapriya, "A survey of attacks, security mechanisms and challenges in wireless sensor networks," *International Journal of Computer Science and Information Security*, vol. 4, no. 1–2, 2009.
21. C. H. Tseng, S. Wang, and W. Tsaur, "Hierarchical and dynamic elliptic curve cryptosystem based self-certified public key scheme for medical data protection," *IEEE Transactions on Reliability*, vol. 64, no. 3, pp. 1078–1085, 2015.
22. S. Ghormare and V. Sahare, "Implementation of data confidentiality for providing high security in Wireless Sensor Network," *2015 International Conference on Innovations in Information, Embedded and Communication Systems (ICIIECS)*, pp. 1–5, 2015.
23. R. W. Anwar, M. Bakhtiari, A. Zainal, A. H. Abdullah, and K. N. Qureshi, "Security issues and attacks in wireless sensor network," *World Applied Sciences Journal*, vol. 30, no. 10, pp. 1224–1227, 2014.
24. S. Jaitly, H. Malhotra and B. Bhushan, "Security vulnerabilities and countermeasures against jamming attacks in Wireless Sensor Networks: A survey," *2017 International Conference on Computer, Communications and Electronics (Comptelix)*, Jaipur, 2017, pp. 559–564.
25. K. Pelechrinis, M. Iliofotou, and S. V. Krishnamurthy, "Denial of service attacks in wireless networks: the case of jammers," *IEEE Communications Surveys Tutorials*, vol. 13, no. 2, pp. 245–257, 2011.
26. A. D. Wood, J. A. Stankovic, and S. H. Son, "JAM: a jammed-area mapping service for sensor networks." *Technical Report, University of Virginia, Charlottesville, Department of Computer Science*, 2006.
27. R. Muraleedharan and L. A. Osadciw, "Jamming attack detection and countermeasures in wireless sensor network using ant system," *Wireless Sensing and Processing*, vol. 6248, article no. 62480G, 2006.
28. K. Grover, A. Lim, and Q. Yang, "Jamming and anti-jamming techniques in wireless networks: a survey," *International Journal of Ad Hoc and Ubiquitous Computing*, vol. 17, no. 4, pp. 197–215, 2014.
29. S. K. Jain and K. Garg, "A hybrid model of defense techniques against base station jamming attack in wireless sensor networks," *2009 First International Conference on Computational Intelligence, Communication Systems and Networks*, pp. 102–107, 2009.
30. W. Xu, W. Trappe, Y. Zhang, and T. Wood, "The feasibility of launching and detecting jamming attacks in wireless networks," *Proceedings of the 6th ACM International Symposium on Mobile Ad Hoc Networking and Computing*. New York: ACM, pp. 46–57, 2005.
31. S. Khattab, D. Mosse, and R. Melhem, "Jamming mitigation in multi-radio wireless networks: reactive or proactive?," *Proceedings of the 4th International Conference on Security and Privacy in Communication Networks*, pp. 27:1–27:10, 2008.
32. S. Khattab, D. Mosse, and R. Melhem, "Modeling of the channel-hopping anti-jamming defense in multi-radio wireless networks," *Proceedings of the 5th Annual International Conference on Mobile and Ubiquitous Systems: Computing, Networking, and Services, ICST*, Brussels, Belgium, pp. 25:1–25:10, 2008.
33. H. Wang, L. Zhang, T. Li, and J. Tugnait, "Spectrally efficient jamming mitigation based on code-controlled frequency hopping," *IEEE Transactions on Wireless Communications*, vol. 10, no. 3, pp. 728–732, 2011.

34. S.-U. Yoon, R. Murawski, E. Ekici, S. Park, and Z. H. Mir, "Adaptive channel hopping for interference robust wireless sensor networks," *2010 IEEE International Conference on Communications*, pp. 1–5, 2010.
35. A. Mpitziopoulos, D. Gavalas, G. Pantziou, and C. Konstantopoulos, "Defending wireless sensor networks from jamming attacks," *2007 IEEE 18th International Symposium on Personal, Indoor and Mobile Radio Communications*, pp. 1–5, 2007.
36. A. D. Wood, J. A. Stankovic, and G. Zhou, "DEEJAM: defeating energy-efficient jamming in IEEE 802.15.4-based wireless networks," *2007 4th Annual IEEE Communications Society Conference on Sensor, Mesh and Ad Hoc Communications and Networks*, pp. 60–69, 2007.
37. T. Hamza, G. Kaddoum, A. Meddeb, and G. Matar, "A survey on intelligent MAC layer jamming attacks and countermeasures in WSNs," *2016 IEEE 84th Vehicular Technology Conference (VTC-Fall)*, pp. 1–5, 2016.
38. L. Tang, Y. Sun, O. Gurewitz, and D. B. Johnson, "EM-MAC: a dynamic multichannel energy-efficient MAC protocol for wireless sensor networks," *Proceedings of the Twelfth ACM International Symposium on Mobile Ad Hoc Networking and Computing*. New York: ACM, pp. 23:1–23:11, 2011.
39. F. Ashraf, Y. Hu, and R. H. Kravets, "Bankrupting the jammer in WSN," *2012 IEEE 9th International Conference on Mobile Ad-Hoc and Sensor Systems (MASS 2012)*, pp. 317–325, 2012.
40. M. Tiloca, D. D. Guglielmo, G. Dini, and G. Anastasi, "SAD-SJ: a self-adaptive decentralized solution against selective jamming attack in wireless sensor networks," *2013 IEEE 18th Conference on Emerging Technologies Factory Automation (ETFA)*, pp. 1–8, 2013.

13 Resilient UAV Networks: Solutions and Trends

Zhiyong Xia, Fei Hu, Nathan Jeong, Iftikhar Rasheed
Electrical and Computer Engineering,
The University of Alabama, Tuscaloosa, AL, USA

CONTENTS

13.1 INTRODUCTION

Unmanned aerial vehicle (UAV) is widely used in civilian and military applications, such as military reconnaissance, rescue, data collection, payload delivery and agriculture. A UAV usually works in an open operational environment; thus it is susceptible to various disruptions. The uncertainties or unexpected events from the external environment can affect the normal operations of UAV and then degrade its performance. A system that can maintain an acceptable level of performance under the impact of disruptions is highly desirable. This ability is called resilience.

To achieve a resilient UAV system, it needs to respond to different types of disruptions and take corresponding appropriate measures to protect the system from these disruptions. The high-level modular diagram of the resilient UAV system is shown in Figure 13.1. At the beginning, the mission objectives are defined at the system level. Various constraints need to be identified at this phase. It is necessary to make sure that the mission objectives meet the design requirement of a resilient system. The resilience requirements include circumvention, recovery, and reconfiguration. There are multiple disruptions to the system, which will be discussed in detail in the following

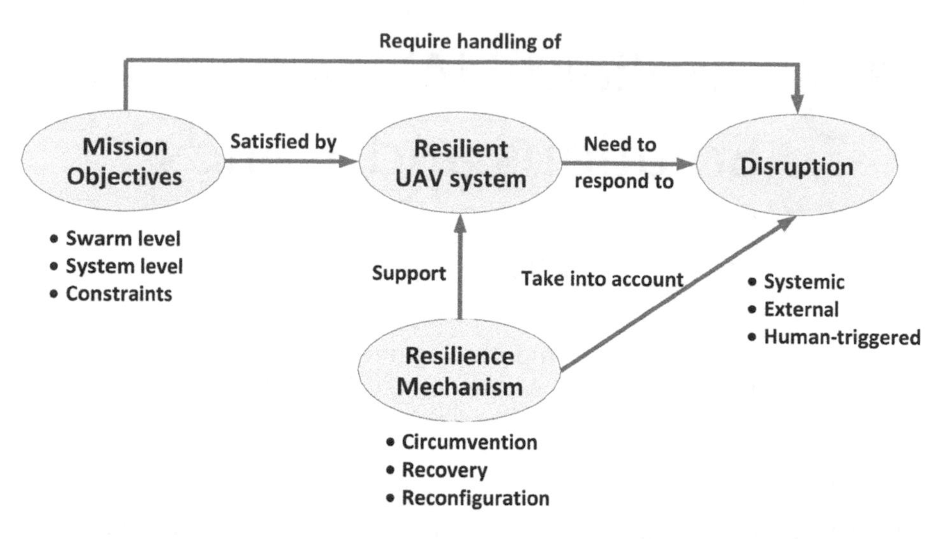

FIGURE 13.1 Big picture of resilient UAV system.

sections. The system adopts different strategies or tactics adaptively to deal with the disruptions based on the current state of the mission and remaining resources.

13.1.1 Typology of Disruptions

Disruptions can be classified into two categories, i.e., predictable or random disruptions. Predictable disruptions have the information of time and location of occurrence and triggering event. Random disruptions can only be defined by utilizing probability distributions, and it can happen unexpectedly.

On the other hand, based on the causes of disruptions, disruptions can be further classified into three different types, i.e., external, systemic and human-triggered disruptions. External disruption is associated with environmental obstacles and incidents. This type of disruptions usually occurs randomly, and the effects on the UAV system are unpredictable in terms of severity and duration. The systemic disruption occurs when the functionality, capability or capacity of internal components lead to performance degradation. However, it is easy for an intelligent system to detect this type of disruption. Human-triggered disruption refers to the disruptions from the human operators inside or outside of the system boundary.

To determine the severity of impact from the disruption, the context and duration of the disruptions must be taken into account. Disruption context characterizes the current operational usage and status of the system. The disruption duration is to decide whether the disruption is temporary or an indication of a trend. Temporary disruptions can be ignored without leading to a significant damage in the application. However, the persistent disruption should be dealt with and monitored carefully, and the corresponding resilience countermeasures need to be triggered. The response can be carried out automatically by UAVs. For the less time-critical responses, humans may be involved in the loop to perform the appropriate responses.

13.1.2 Resilience in UAV System

A UAV system is called resilient if it is able to accomplish the original mission with an acceptable level of performance in the face of disruptions [1, 2]. Some resilience heuristics have been identified by some researchers, and they can be integrated into the system during the system design stage to overcome various disruptions. The resilience heuristics guide the system to use proper strategies or mechanisms to deal with disruptions during operations. These strategies or mechanisms are explained as follows.

Redundancy and connectivity are used as two main strategies for resilience in the research community. The redundancy includes physical redundancy and functional redundancy. Physical redundancy refers to the situation in which another identical UAV is used to replace the incapacitated UAV. For instance, when a UAV happens to land unexpectedly because of disruptions, a new UAV is deployed and integrated into the system. Functional redundancy means the same functionality can be fulfilled by other means instead of only depending on a single UAV or method. Connectivity relies on the communications among nodes to share information through different pathways to keep the network from losing connection to any node or causing disruptions.

Along with redundancy and connectivity, function re-allocation and human-in-the-loop are two means to protect the system. Function re-allocation redistributes the overall functionalities (remaining tasks) among the remaining UAVs when some UAVs leave the system because of disruptions. The human-in-the-loop method allows the people in the control loop to handle the emergency when the system cannot handle the current disruptions. However, it should be noted that this method should be viewed as the last resort for an autonomous system. Pre-planned protocols should be designed and used to execute a pre-defined plan to deal with known or unknown disruptions.

13.1.3 Resilience Research Problem and Challenges

Most resilient methods focus on how to anticipate the disrupting events. They predict when, where and how disruptions can occur and then plan resilient countermeasures. However, it should be noted that in the real-life applications, UAVs usually operate in open, dynamic and uncertain environments. It is impossible for the designers to anticipate all possible disruptions and plan suitable measures to reduce the impact from the disruptions. Therefore, the system may only need to make reasonable decisions to deal with unexpected disruptions during real-time operations.

For the UAV system with resilient capability, there is one question to be solved. Even though there are multiple alternatives or mechanisms in the system that we can employ to deal with disruptions, we need to select the most appropriate solution based on the current system conditions. Therefore, there should be an evaluation method in the system to evaluate the effects of different available alternatives or mechanisms on the overall system's performance, and the one which leads to the best performance is selected to deal with the current disruptions.

TABLE 13.1
Resilient UAV Systems

Method	Characteristics	Limitations/Challenges
Model-based approach [3]	• Models the priority (mission, resources and safety) based on the current status of system • Different tactics for different priorities • Transition criteria among UAV tactics	Mainly at the simulation stage, need to be tested in the real case
Bayesian game theory–based intrusion detection for resilient UAV system [4]	• Detects malicious nodes in the network • Low overhead with high detection rate	Simulation-based work; with complicated computations
Swarm intelligent-based resilient UAV system [5]	• Adapts to various changes • Self-aware and mission-focused • Autonomous collaboration with each other	Security, privacy and trust issue when many drones are in the swarm
UAV-enabled sensor system of Adler [6]	• Achieves resilience, high-energy efficiency and low package loss ratio • Enable localization, gathering and network reconfiguration	Needs to optimize the trajectory design to cover larger-scale sensor system with minimum cost
Reliable and untethered UAV-based LTE network [8]	• Pushes evolved packet core (EPC) to the extreme edge of the core network • Support for hotspot and stand-alone multi-UAV deployment	Impact of wireless inter-UAV backhaul design and scalability for lager UAV-based LTE network

In this chapter, several examples of resilient UAV systems are discussed. Table 13.1 summarizes the methods to be covered. In the following sections, the details of each method will be provided.

13.2 MODEL-BASED APPROACH TO ENABLE RESILIENT UAV SYSTEM

When the disruptions occur, the UAV system adopts appropriate strategies or tactics to deal with them based on the current operation status and the task requirements. This is similar to the commander in the battlefield or a chess master who deals with every move made by the enemy or opponent, i.e., the disruptions. In [3] the tactics adopted by the UAVs are presented. These tactics are reviewed briefly to illustrate their basic concepts. The tactics can be classified into several categories, as shown in Table 13.2.

For Tactic T1, accomplishing the mission is the first priority, thus it is an offensive strategy. T2 and T3 try to balance the mission execution efficiency with resource

TABLE 13.2
UAV Tactics to Deal with Disruptions

Symbols	Tactics	Description
T1	Executes mission (offense)	The priority is to accomplish the mission within the required performance. Safety and resource consumption should also be considered.
T2	Resource conservation and execution of mission	The priority is to conserve resources. Executing the mission and safety have second and third priorities, respectively.
T3	Safe/secure execution of mission	The priority is safety and security. Execution of the mission and resource consumption have the second and third priorities, respectively.
T4	Safe/secure resource conservation (defense)	Safety and security and resource conservation have equal high priority, whereas executing the intended mission has low priority.

conservation and safety requirements, therefore, they use a trade-off strategy. T4 is a defensive tactic, because its first priority is safety and security instead of completing the task or mission.

In the UAV system, we need a criterion to switch among these tactics based on the operation status and the remaining resources. For instance, if the UAV system still has enough resources and in its full capacity status, but the amount of task or mission completed is small, then the system will choose the T1 tactic. Table 13.3 lists the criteria for transitioning among these several tactics.

In addition to tactics and transition criteria among these tactics, the UAV system also needs a utility function to make decisions. Utility functions are mathematical equations that can be used to evaluate or rank resilience alternatives based on the usefulness of these alternatives. Basically, the utility function takes three attributes into account, i.e., safety, resources and mission objective. The priority of these three

TABLE 13.3
Transition Criteria among UAV Tactics

Tactics	Transition Criteria
T1 (execute mission)	If (R > 0.75) and (D < 0.25) and (A < 1)
	If (0.25 < R < 0.75) and (A ← 0.25) and (L ← 0.25)
T2 (resource conservation-execute mission)	If (R < 0.25) and (A ← 0.25) and (D ← 0.5)
T3 (safe/secure-execute mission)	If (L < 0.75 and D > 0.25) and (R > 0.25) and (A > 0.25)
T4 (safe/secure-resource conservation (defense))	If (D > 0.75) and (R < 0.25) and (A > 0.25)

Note: R = resource; D = damage; A = accomplished mission. All of the values are scaled between 0 and 1.

TABLE 13.4
Priorities and Weighting Values of Utility Function under Different Tactics

Tactics	Priorities	Weighting Values
T1 (execute mission)	$W_I >> W_2$ and W_3	$W_1 = 0.66$, $W_2 = W_3 = 0.165$
T2 (resource conservation-execute mission)	$W_2 > W_1 > W_3$	$W_1 = 0.272$, $W_2 = 0.545$, $W_3 = 0.181$
T3 (safe/secure-execute mission)	$W_3 > W_1 > W_2$	$W_1 = 0.272$, $W_2 = 0.181$, $W_3 = 0.545$
T4 [safe/secure-resource conservation (defense)]	$W_2 + W_3 >> W_3$	$W_1 = 0.166$, $W_2 = 0.417$, $W_3 = 0.417$

attributes will change during mission execution. One common definition of utility function is shown in Eqn. (13.1).

$$\begin{aligned} U_{\text{commander}} &= f\left(\text{Mission, Resource, Safety}\right) \\ &= w_1\text{Mission}_{\text{Score}} + w_2\text{Resources}_{\text{Score}} + w_3\text{Safety}_{\text{Score}} \end{aligned} \quad (13.1)$$

where $\text{Mission}_{\text{Score}}$ is the score associated with the status of accomplishing the task or mission, $\text{Resources}_{\text{Score}}$ is the score associated with the status of utilizing resources (e.g., deploying UAVs to replace the incapacitated UAV) and $\text{Safety}_{\text{Score}}$ is the score associated with the safety of UAVs (e.g., collision), and W_1, W_2 and W_3 are the weights for these three types of scores. For these weights, we have $\Sigma_{i=1}^{n} w_i = 1$, $n = 3$. The values of W_1, W_2 and W_3 are based on the UAV tactics adopted in the system.

Table 13.4 shows the priority and weights under different UAV tactics that were defined earlier. Under the T1 tactic, because executing mission is the first priority, the value of W_1 is much higher than that of W_2 and W_3 (here it is set twice as high as the sum of the other two weights).

It also should be noted that utility function alone is not enough to make a decision. Multiple models and algorithms need to be integrated together to decide the effect of each alternative on the overall mission. The entire decision process is shown in Figure 13.2. Basically, the process starts by assessing the current situation, selecting the proper system tactics, then generating multiple resilience alternatives,

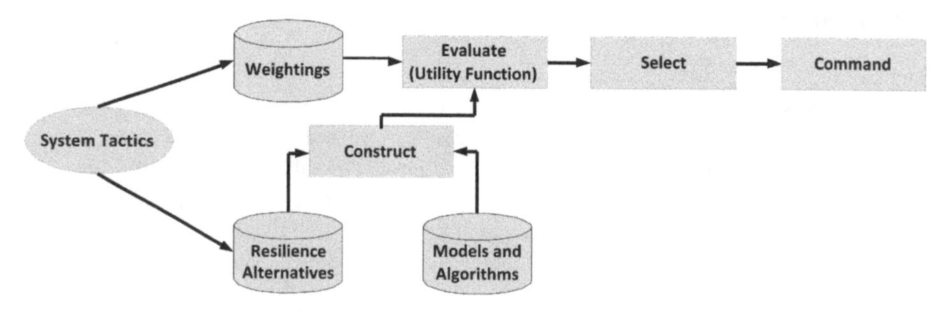

FIGURE 13.2 Decision-making process for a UAV system.

evaluating each alternative and selecting the best alternative. Once the corresponding system tactic is selected, the weighting values of the utility function are picked up from the lookup table, i.e., Table 13.4. For more details and experiment results please refer to [3].

13.3 BAYESIAN GAME THEORY–BASED INTRUSION DETECTION FOR RESILIENT A UAV SYSTEM

For the UAV to successfully accomplish its tasks, a reliable and safe communication network is required. The communication between UAVs and the ground or the aerial base station is realized through wireless communication. However, the wireless communication is usually unstable because of its weak wireless link. On the other hand, the UAV systems are usually deployed in a harsh environment or sensitive area, making the system suffer various attacks, e.g., black hole attacks or denial-of-service (DoS) attacks. These attacks can cause a series of malicious incidents, such as guiding flight routes, obtaining UAV network data, intercepting UAVs and affecting military strategies. Therefore, developing an effective intrusion detection system (IDS) is necessary for securing the UAV and its mission. Based on the IDSs presented in the literature, there are mainly three categories: classified based, behavioral prediction based and statistics and cryptography based IDS. The IDSs designed for UAV networks are very similar to the ones for ad hoc networks (VANETs).

In [4], a representative intrusion detection method based on Bayesian game theory is proposed to secure the UAV network. Bayesian game here means that the game participants do not have complete information about the profit model of its opponent. In this case incomplete information means that the IDS agent does not know the properties of the attacker and the attacker does not know whether its neighboring node uses an IDS or not.

A distributed UAV network with the LEACH routing protocol is introduced here as an example, and its network topology is shown in Figure 13.3. The UAV network is divided into several clusters. This can help to mitigate the broadcast storm and achieve a better transmission rate. For each cluster, there is a cluster head that

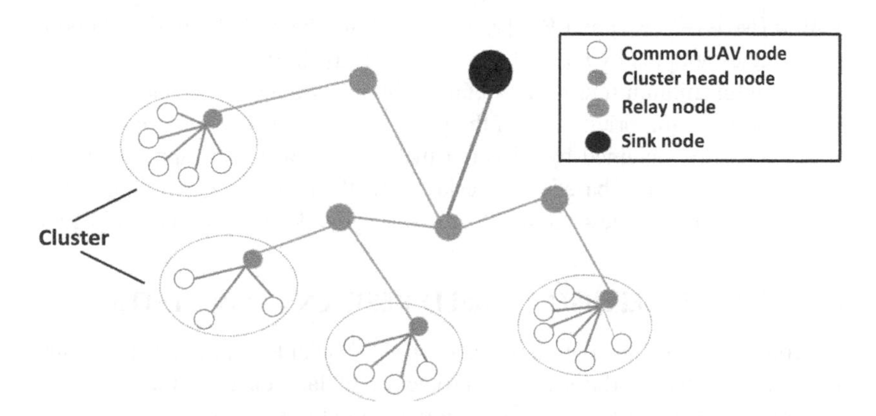

FIGURE 13.3 Clustered UAV network architecture.

receives data regularly from the common UAV nodes and transmits the aggregated data to the sink UAV node via a multi-hop routing method. The sink UAV is equipped with an advanced anti-attack system, which makes it resist to different attacks.

In this network, when one UAV node detects a suspicious event, it will broadcast an alarm message to the neighboring nodes. The UAV nodes, which receive the alarm message, will continue to broadcast this message to other UAV nodes, and this message will finally reach the destination (which could be a sink node or base station). This process will further trigger countermeasures to avoid further damage or effects on the network. The goal of the IDS system is to detect various malicious attacks and secure the UAV network.

In the network shown in Figure 13.3, the IDS embedded in the sink node monitors the relay nodes. The cluster heads are monitored by the IDS installed in the relay nodes. The IDS activated in the cluster head monitors common UAV nodes. In the IDS based on Bayesian game theory, there are two players, i.e., the IDS agent (I_{IDS}) and the attacker (I_{attack}). The IDS agent I_{IDS} will be activated to monitor the neighboring nodes, which may perform malicious behaviors. There are several strategies $\delta_{IDS} = \{\alpha_i^1 | I = 1,2,\ldots,m\}$ and $\delta_{attacker} = \{\beta_i^2 | j = 1,2,\ldots,n\}$ for I_{IDS} and I_{attack}, respectively, where m and n represent the maximum number of strategies that I_{IDS} and I_{attack} can use.

I_{IDS} and I_{attack} can randomly select any strategy (α_i^1, β_i^2), but the profit from each strategy is not the same. P_{ij} and P_{ji}' are used to represent the profits from I_{IDS} and I_{attack}. Let w_i be the probability that I_{IDS} adopts α_i^1 and v_i be the probability that I_{attack} adopts β_i^2. To achieve the best balance between the IDS high detection rate and low overhead by using Bayesian game theory, the benefits that I_{IDS} and I_{attack} can obtain are calculated by using Eqns. (13.2) and (13.3), respectively.

$$R_{IDS}(W, V) = \sum_{i=1}^{m} \sum_{j=1}^{n} P_{ij} \times w_i \times v_j \tag{13.2}$$

$$R_{attacker}(W, V) = \sum_{i=1}^{m} \sum_{j=1}^{n} P'_{ij} \times w_i \times v_j \tag{13.3}$$

where $W = \{w_1, w_2, \ldots, w_m\}$ and $V = \{v_1, v_2, \ldots, v_m\}$ are the probability distribution vectors of strategies adopted by I_{IDS} and I_{attack}, respectively. In the designed IDS, when the maximum and minimum returns are obtained by the strategies I_{IDS} and I_{attack}, respectively, the optimal equilibrium state of the game is obtained. The maximum and minimum values can be obtained by adjusting the value of w_i and v_i appropriately. This intrusion detection method based on Bayesian game theory can achieve a high detection rate and low overhead at the same time, which is highly desirable in the applications.

13.4 SWARM INTELLIGENT-BASED RESILIENT UAV SYSTEM

In the military context, the requirements for the UAVs deployed include the adaptability to the adverse filed condition and the protection of data related to the missions (e.g., flight path). A swarm of drones (SoD) is more resilient to system attacks than a single drone for most application scenarios. Because of the redundancy and collaboration

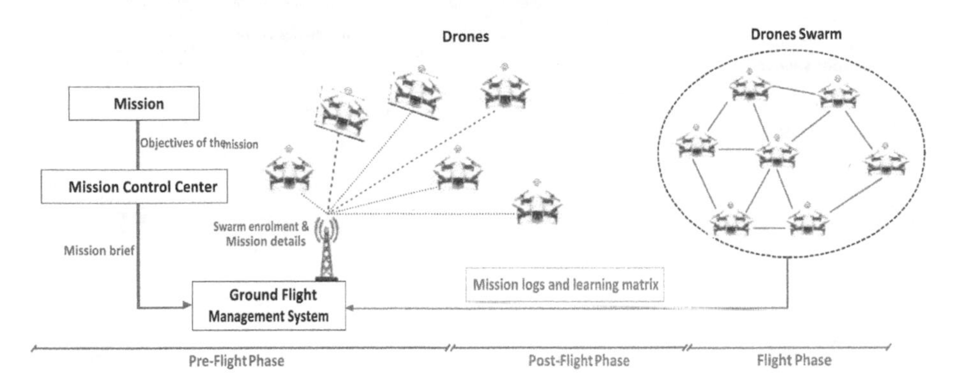

FIGURE 13.4 Construction process of an SoD.

characteristics of an SoD, it can adapt to adverse conditions, such as destruction or failure. Along with this, an SoD does not use wide range communications, reducing the chances of potential detection by adversaries. Due to the large number of drones, the workload is distributed among the swarm, and multiple tasks can be executed simultaneously. An SoD also further enables distributed sensing capabilities and an increased robustness to system failure (better than a single point system).

There are two different construction stages for the formation of an SoD, i.e., formation of an SoD during the pre-mission stage or at the post-mission stage. This process is shown in Figure 13.4. At the pre-mission stage, it starts with the formation of the mission with certain objectives and constraints. This formation is based on the mission objectives, airspace regulations, ethical principles, collaborative knowledge and security and privacy policies. Based on the requirements for the drone and organization preference, multiple drones from the inventory will be selected to participate in this mission. Once the drones are selected, the ground flight management system (GFMS) then uploads the mission brief information to each drone. After the brief is uploaded, the drones would establish secure communication among each other in the SoD. When all of the drones ae connected and GFMS gives permission to commence the mission, the SoD then initiates mission executions such as searching for specific objects.

When it comes to the end of the completion of the mission, the GFMS will set communication channels with each drone and download the mission logs and evaluation/learning matrix from each drone. The GFMS, together with the mission control center, will analyze the collected post-mission information from each drone and utilize it to improve the collaborative knowledge.

There are three basic types of SoDs, static, dynamic and hybrid.

1. Static SoD is the basic type. The members of the SoD are pre-selected at the pre-mission stage. During the execution stage of the mission, no new member joins the swarm as the system is locked at the starting point of mission commencement. The secure communication and collaboration are set up by the GFMS [Figure 13.5(a)].
2. Dynamic SoDs are open to accept new members to the swarm or allow existing members to exit the swarm at any time, i.e., at the pre-mission stage or

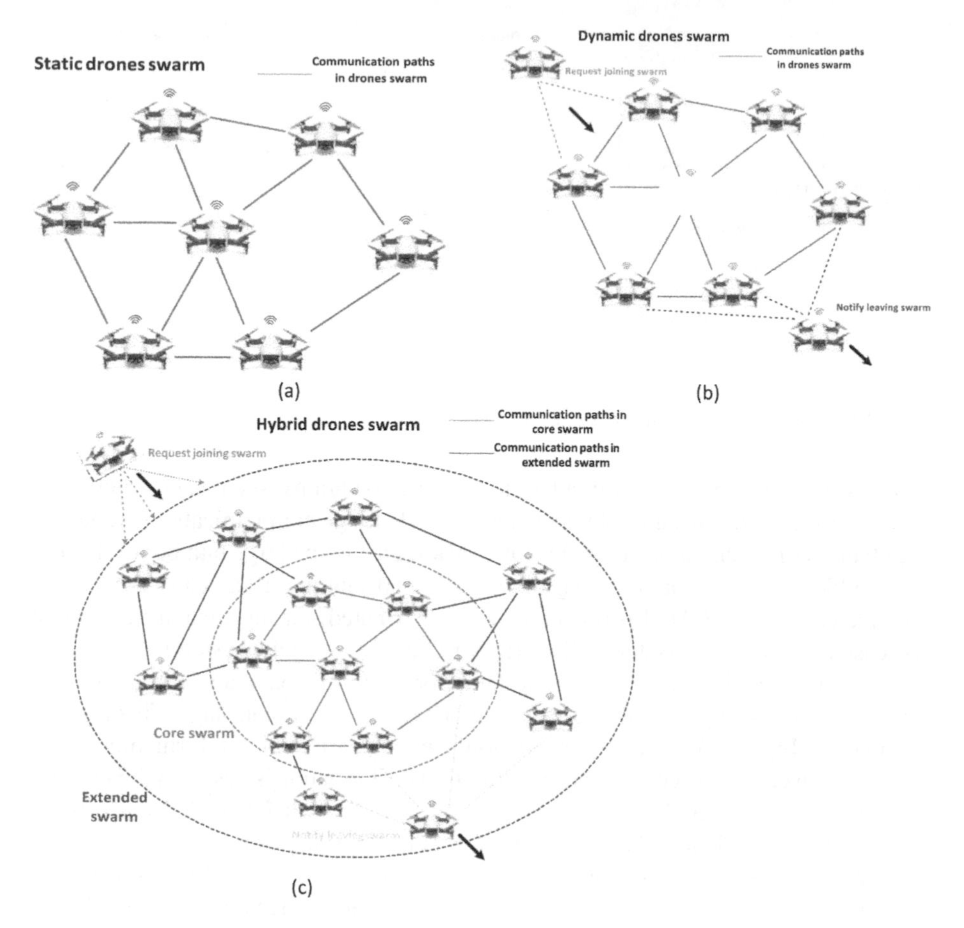

FIGURE 13.5 Types of SoD. (a) Static SoD. (b) Dynamic SoD. (c) Hybrid SoD.

execution-mission stage. The SoD can accept a new drone either from the same organization (i.e., in a closed SoD) or from the third party (i.e., from an open SoD). For this type of SoD, the secure communication, mutual trust and collaboration among drones are more challenging compared with static SoD cases. The diagram for this type of hybrid SoD is shown in Figure 13.5(b).

3. Hybrid SoDs combine the static SoD and dynamic SoD into one single collective unit. At the core of the hybrid SoD there is a static SoD, which behaves like one entity in terms of its operations. This static SoD core allows new members to enroll into the swarm. The new added members form an extended swam that behaves like a dynamic SoD. In the hybrid SoD, the core swarm has higher priority when it comes to collaborative learning, evaluation and decisions. The extended swarm joins the core swarm and provides certain services to the core swarm. Drones of the extended swarm can leave the collective at any stage. The diagram for this type of hybrid SoD is illustrated in Figure 13.5(c).

To have a well-functioning SoD, the collaboration model is necessary. Based on the current literature, there are three main types of collaboration models, centralized, decentralized and distributed. The characteristics of each type are summarized in the following:

1. *Centralized collaboration model*: In this type of model, there is a powerful master drone in the swarm that collects all the information from each individual drone and assists the swarm to do the computation and make optimal decisions.
2. *Decentralized collaboration model*: For this type there is no single master drone, but there is a small set of drones that collect the information from other drones. The small set of drones does the collaborative learning, evaluation and decision-making.
3. *Distributed collaboration model*: In this model, each drone takes more or less the equal role in the collaborative learning, evaluation and decision-making. The working load is distributed among the drones of the swarm based on their own capability, power sources and criticality to the overall mission.

Even though good progress has been achieved recently in the application of SoDs, there are still some challenges or open problems to be addressed: (1) security, privacy and trust related issues, and (2) performance and energy consumption–related issues. The first category has the following four research problems: (1) swarm authentication, attestation and secure communication; (2) fair exchange services architecture for the SoD; (3) collaborated cybersecurity deterrence mechanism and (4) detecting the mole and free-riders in the swarm. The second category has the following two issues: (1) balancing the cybersecurity with performance and energy consumption and (2) graceful degradation. For more details related to these open issues, please refer to [5].

13.5 RESILIENT UAV-ENABLED SENSOR SYSTEM OF ADLER

Sensor networks are becoming more and more popular in daily life, because they can help with monitoring tasks, such as with natural disaster, pollution, agriculture and industrial equipment. A sensor network or system that achieves resilience, high performance and energy efficiency at the same time is highly desirable. In [6], a UAV-enabled sensor system (Adler) is proposed to meet with the aforementioned requirements. Three basic fundamental applications (i.e., localization, gathering and network reconfiguration) are demonstrated using Adler.

Adler has three basic components: sensors that are mounted in the targeted area, UAVs that work with sensors and the base station for mission control and management. The diagram for Adler is shown in the Figure 13.6. In Adler, a mission is sent from the base station to UAVs together with certain input information. The input data include the target area, targeted sensors and other related application data, e.g., firmware needed to be upgraded for UAVs. The output of UAVs includes the collected sensor data and other application results related to the specific mission.

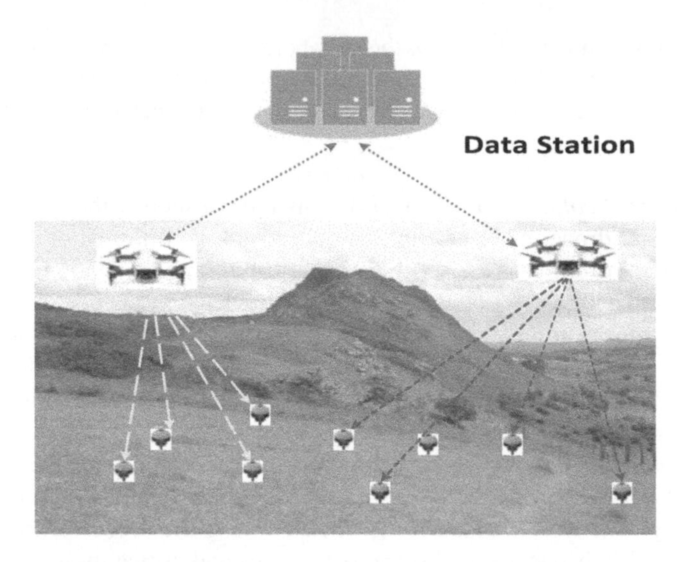

FIGURE 13.6 The Adler.

The network architecture of Adler is shown in Figure 13.7. The UAV part in Adler has three main layers: application, middleware and driver. The application layer has an application protocol and manages the flight trajectory. The middleware layer is responsible for flight and communication control. The driver layer has the UAV's software development kit (SDK) and the related communication drivers.

The application layer of a UAV has two main parts, i.e., trajectory optimization and application protocol. The trajectory optimization module calculates the trajectory for each UAV based on the targeted sensor's position and area. It specifies the flight coordinates and order of the waypoints for UAV. The application protocol is the core of execution of specific applications. Its task include (1) exchanging information with the base station, (2) creating data and application firmware, (3) broadcasting wakeup signals to the sensors and (4) arranging the communication process.

The middleware layer of the UAV consists of two parts, flight control and communication control. Flight control generates flight control commands based on the waypoint calculation from trajectory optimization. It toggles the flight state between flying and hovering. Communication control maintains the communication state or channels with sensors and other UAVs, and the communication states switches between sleeping mode and working mode.

In the driver layers, there are separate drivers for the UAVs and sensors. The UAV driver consists of UAV SDK and communication drivers, and the communication driver has a basic network layer and data link layer in an open-system interconnection model. For the sensors, the corresponding communication driver includes more modules such as an embedded operation system and power management.

The three fundamental applications of a resilient UAV-based sensor system are briefly discussed. The first application is the localization function. In the sensor network, if the sensor's position information is not available or not accurate, then the localization function has to be activated. In Adler, the UAV-based localization

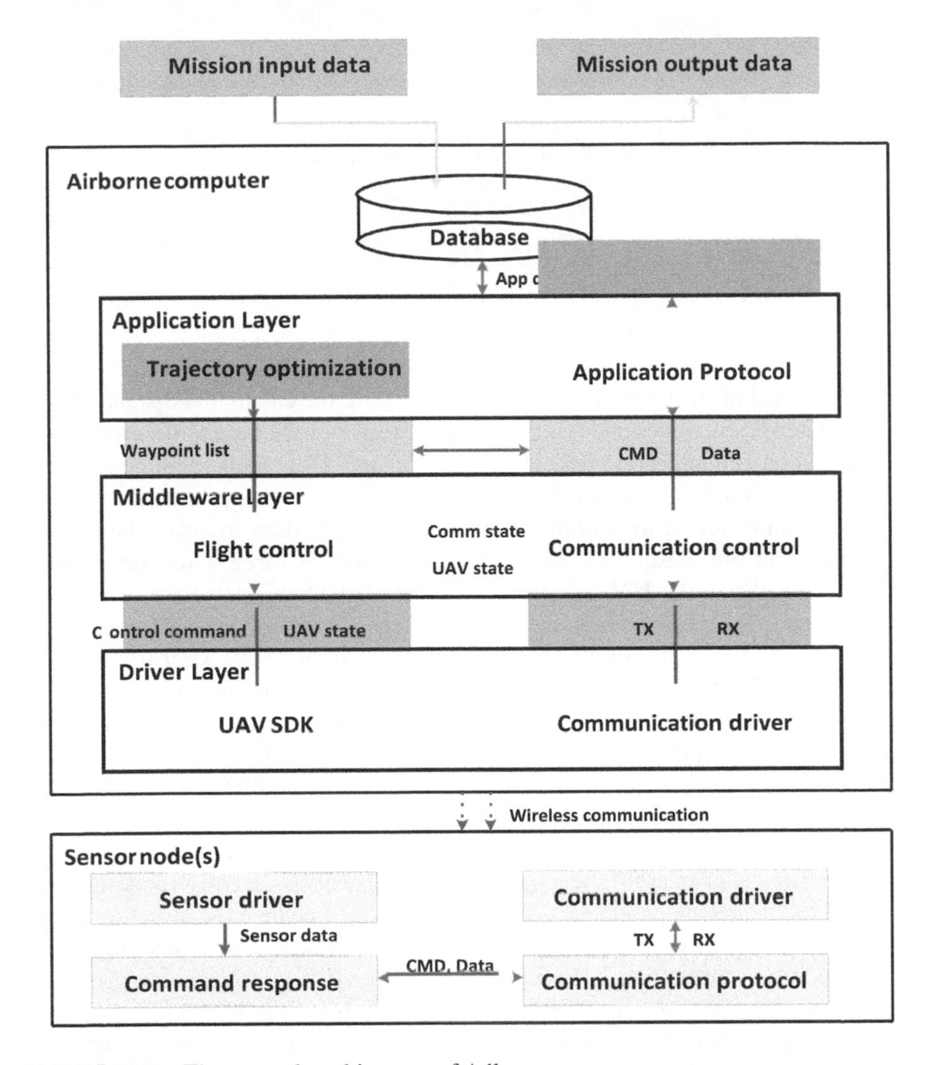

FIGURE 13.7 The network architecture of Adler.

method is proposed. Basically, it divides the target area into grids, then designs the flight trajectory of UAVs to cover all sensor points, and it finally calculates the location coordinates of each sensor via the received signal strength indicator (RSSI). The RSSI is utilized to calculate the distance between a sender and a receiver, and its formula is shown in Eqn. (13.4).

$$\text{RSSI} = A + 10n\log_{10}(d) \tag{13.4}$$

where A is the received signal strength at which the distance is 1 m away from the sender, n is the pass loss exponent variable related to the environment and d is the distance we need to measure.

When the UAV traverses the four vertices of the grid of the targeted sensor, it calculates the position coordinates of the sensor by using the following equation:

$$\begin{cases} (x-x_1)^2+(y-y_1)^2+(z-z_1)^2=r_1^2 \\ (x-x_2)^2+(y-y_2)^2+(z-z_2)^2=r_2^2 \\ (x-x_3)^2+(y-y_3)^2+(z-z_3)^2=r_3^2 \\ (x-x_4)^2+(y-y_4)^2+(z-z_4)^2=r_4^2 \end{cases} \tag{3.5}$$

where (x_1,y_1,z_1), (x_2,y_2,z_2), (x_3,y_3,z_3), and (x_4,y_4,z_4) are the coordinates of the four vertices of the grid of the targeted sensor; r_1, r_2, r_3, r_4 are the calculated distance based on the related RSSI values and (x,y,z) is the coordinate of the targeted sensor point.

The second fundamental application is gathering. This application collects the measurement data from the distributed sensors. In [6] Li et al. proposed the minimum hexagon covering algorithm for the UAV-based system to cover the sensors and then gather the sensor data. In Adler, each sensor has three states: sleep, listen and transmit. When the UAV reaches the center of the hexagon of the sensor network, it broadcasts a beacon message. After the sensor receives this beacon message, the sensor switches to transmit state and sends data to the UAV. After transmitting data between the sensor and UAV is complete, the sensor returns to the sleep or listen state. Compared with the conventional multi-hop network–based data gathering method, the proposed UAV-based method could help to solve the energy hole problem that appeared in the multi-hop network. Compared with the mobile vehicle–based method, the UAV-based method avoids the complicated ground conditions.

The third fundamental application is the network reconfiguration. Network reconfiguration is critical for the long-term running sensor system. Whenever there is a need to update the parameters of the sensors, or to change the structure of the sensor network to meet the requirements of the new task, network reconfiguration is required. The conventional multi-hop over-the-air (OTA) programming method used to upgrade the firmware of the sensor network is energy consuming, because the multi-hop method has to forward the firmware package to other sensors in a hop-by-hop fashion. Usually the size of the firmware is not small, and retransmission is often required because of the loss of packets and the issue of communication collisions. In [6], the UAV-based network reconfiguration method is proposed to mitigate this energy problem and to achieve resilient operations for the entire system. It reconfigures the sensor network using the OTA method through simple one-hop communications. This can reduce the probability of package loss compared with the multi-hop method.

A multi-UAV network has many advantages compared with its single-UAV network counterpart. The first advantage is that the multi-UAV network improves the transmission efficiency. For example, when an interruption incurs to one relay link, the relayed packets can be forwarded to other UAVs. Because of the collaboration among the UAVs, the multi-UAV network has a better information pre-processing capability and transmission efficiency [7]. Second, the multi-UAV network also enables better survivability. When some UAVs happen to be attacked, the remaining

UAVs can reconstruct the system rapidly and continue to complete the mission. Based on the above analysis, the multi-UAV network exhibits resilience to failure in a single node, so it is desirable for many application scenarios.

Compared with the conventional mobile or vehicular ad hoc network, the characteristics of UAV systems, i.e., mobility and fast response, have a large impact on their corresponding networks. Currently, the UAV system is designed based on the flying ad hoc network (FANET). In a FANET, by adopting the network-centric methodology, the UAVs are enabled to autonomously position themselves for good connectivity and to collaborate with other UAVs to achieve optimal coverage. On the other hand, there are many open challenges related to the FANET that should be addressed. In [7], Wang et al. proposed an efficient gateway selection algorithm and management mechanism for the FANET.

13.6 RELIABLE AND UNTETHERED UAV-BASED LTE NETWORK

LTE networks are widely used in our daily life, but the static nature of LTE base station deployments makes it difficult to meet certain requirements in some 5G application scenarios, for example, the hotspot service in a surging traffic situation and the emergency service for a natural disaster area. Mobility is desired for LTE networks to cater to 5G services. Because of the recent advances of UAV technology, UAV-based LTE is proposed or adopted in certain applications. The UAV-based LTE network enables excellent flexibility in the deployment and optimization of the network. Figure 13.8 depicts the diagram of the UAV-based LTE network, in which the base station can be deployed aerially on UAVs [8].

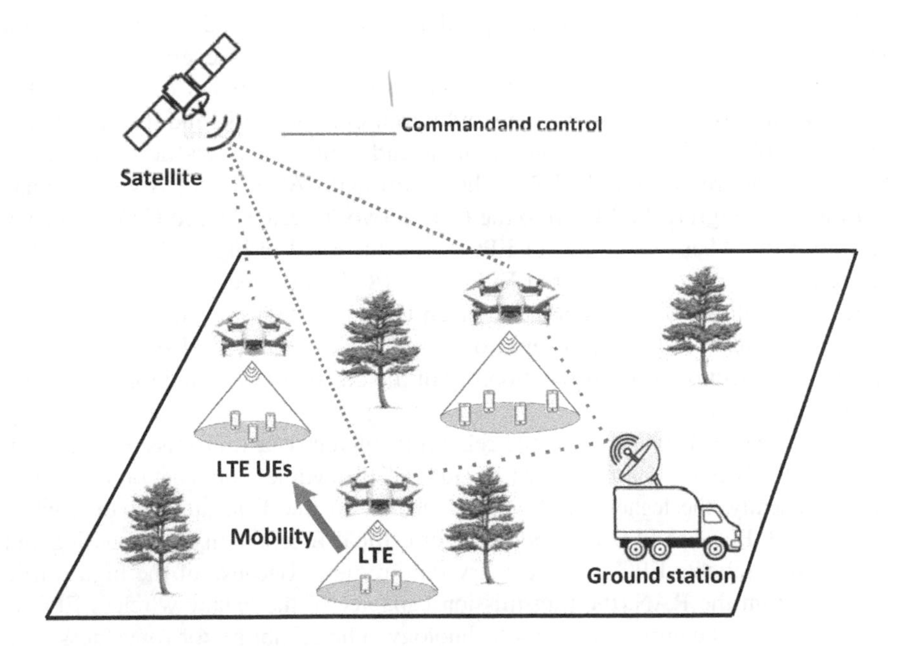

FIGURE 13.8 UAV-based LTE network.

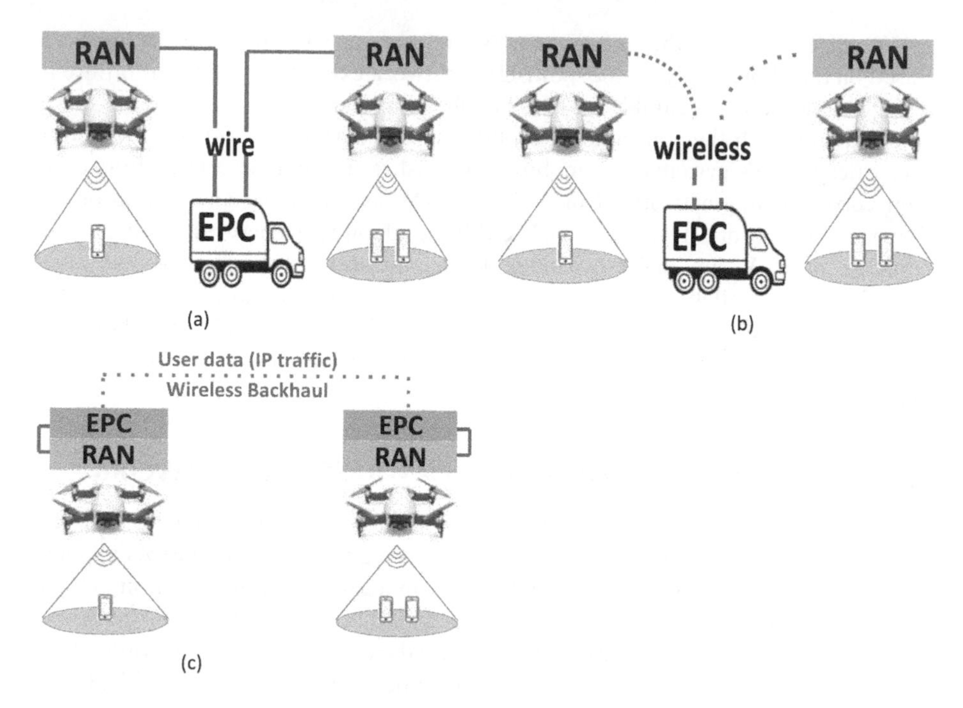

FIGURE 13.9 EPC variants of the UAV-based LTE network. (a) Legacy wired EPC. (b) Legacy wireless EPC. (c) Edge EPC.

There are two basic units in a typical LTE network: a radio access network (RAN) and an evolved packet core (EPC). RAN consists of several base stations that enable wide-area wireless connectivity services to clients (i.e., UE). EPC refers to gateways for the high-speed and wired core network, and it supports the operations of RAN. Mobility management and routing user traffic to and from the Internet are realized in the EPC. The traditional UAV-driven RAN system has two ways to integrate the EPC into the LTE network: legacy wired UAV or legacy wireless EPC. These two types of EPCs are illustrated in Figure 13.9(a) and (b), respectively. For the legacy wired EPC, the EPC is installed on the ground. It is currently the most common operator-driven UAV effort. However, it faces significant limitations in a challenging environment because the EOC-UAV is linked [8]. The wire connection restricts the mobility of the entire network and the scalability to multiple UAVs.

For the legacy wireless EPC, the reliability of the data transmissions between EPC and RAN is the main concern because of the wireless connection characteristics. Basically, the legacy wireless EPC incurs all the disadvantages of a wireless channel. Because EPC is responsible for critical functionality, e.g., routing and setting up, a reliable EPC is a necessary requirement. Because of the high traffic demands from the RAN, the transmission capacity of the legacy wireless EPC is sufficient given the current wireless technology. The reliability (or robustness) and capacity are the bottleneck of the legacy wireless EPC.

Given the aforementioned disadvantages of legacy EPC, a new EPC structure is proposed in [8] to mitigate the fundamental limitations. It is called the edge EPC, and it is illustrated in Figure 13.9(c). As its name suggests, the edge EPC pushes the EPC into the extreme edge of the core network. The basic idea is to tear down the EPC into a single and self-contained module mounted on each UAV. Completely distributed among the entire network, the edge EPC eliminates the critical EPC-RAN wireless path. This capability or characteristic overcomes the downsides of legacy EPC architecture.

To realize the untethered reliable UAV-based LET network, two associated critical modules are proposed along with the edge core to constitute the SkyCore [8], i.e., the software refactoring EPC and inter-EPC communication interface. The software refactoring EPC converts the distributed EPC interface functionalities into a single logical agent. This is realized by transforming the distributed EPC interface functionalities into multiple switching flow tables and the corresponding switching actions [8]. These switching tables store the pre-computed key attributes, e.g., security key and QoS profile, which enables the clients to access the information quickly without computations.

The inter-EPC communication interface enables efficient inter-EPC signaling and communication among UAVs [8]. For SkyCore, the mobility management is implemented at the edge of the network. Inter-EPC communication allows UAVs in the network to proactively synchronize or share their states with each other, and this advantage mitigates the effect on critical control functions from vulnerable wireless UABV-UAV links. This leads to a fast and seamless handoff of active-mode UEs and a quick track of the idle-mode UEs. On the other hand, the SkyCore in [8] aims to provide efficient and scalable services or operations for the airborne LTE network, which includes at most 10 UAVs.

13.7 CONCLUSIONS

In this chapter, the resilience and robustness related to the UAV system are discussed. In the beginning, the general definition of resilience regarding the UAV system was defined. Based on different types of disruptions to the UAV system, different tactics can be adopted to address them considering the current status and priority of the UAV system. For the sake of resilience and reliability, the multi-UAV system is preferred over the single big UAV system. Then, a Bayesian game theory–based UAV intrusion detection method was discussed. When the maximum return and the minimum return are obtained by the strategies I_{IDS} (intrusion detection player uses) and I_{attack} (attack player uses), respectively, the optimal equilibrium state of the game is obtained. The swarm of drones is a hot topic, and its basic concepts, idea and different variants were briefly introduced. Basically, the swarm of drones can adapt to external disruptions or changes, which enables good robustness for the entire system. After that, the resilient and energy-efficient UAV-enabled sensor system was presented. Compared with the conventional mobile vehicle sensor system, the UAV-based sensor system achieves better life span and performance. Three fundamental applications of the UAV-based sensor system were explained, i.e., localization, gathering and network reconfiguration. Last, to cater to several key 5G use cases, an untethered and reliable UAV-based LTE network was stated. The radical edge EPC overcomes the disadvantages of the traditional legacy EPC, and the other two associated critical techniques (i.e., software refactoring and inter-EPC communication interface) work together with the EPC to constitute the new and desirable LTE network solution, i.e., SkyCore, especially for hotspot and emergency application scenarios.

REFERENCES

1. E. Ordoukhanian and A.M. Madni, "Introducing resilience into multi-UAV system-of-systems network." In: *Disciplinary Convergence in Systems Engineering Research*. Cham, Switzerland: Springer, pp. 27–40, 2018.
2. A.M. Madni, *Transdisciplinary Systems Engineering: Exploiting Convergence in a Hyper-Connected World*. Cham, Switzerland: Springer, 2018.
3. E. Ordoukhanian and A. M. Madni, "Model-based approach to engineering resilience in multi-UAV systems." *Systems 7*, no. 1 p. 1 11, 2019.
4. J. Sun, W. Wang, Q. Da, L. Kou, G. Zhao, L. Zhang, and Q. Han. "An intrusion detection based on Bayesian game theory for UAV network." *Proceedings of the 11th EAI International Conference on Mobile Multimedia Communications*, ICST, pp. 56–67, 2018.
5. R. N. Akram, K. Markantonakis, K. Mayes, O. Habachi, D. Sauveron, A. Steyven, and S. Chaumette, "Security, privacy and safety evaluation of dynamic and static fleets of drones." *2017 IEEE/AIAA 36th Digital Avionics Systems Conference (DASC)*, IEEE, pp. 1–12, 2017.
6. D. Li, Y. Wang, Z. Gu, T. Shen, T. Wei, Y. Fu, H. Cui, M. Song, and F. C. M. Lau, "Adler: a resilient, high-performance and energy-efficient UAV-enabled sensor system." *HKU CS Tech. Report TR-2018-01*, 2018.
7. J. Wang, C. Jiang, Z. Han, Y. Ren, R. G. Maunder, and L. Hanzo, "Taking drones to the next level: cooperative distributed unmanned-aerial-vehicular networks for small and mini drones." *IEEE Vehicular Technology Magazine*, vol. 12, no. 3, pp. 73–82, 2017.
8. M. Moradi, K. Sundaresan, E. Chai, S. Rangarajan, and Z. Morley Mao, "SkyCore: moving core to the edge for untethered and reliable UAV-based LTE networks." *Proceedings of the 24th Annual International Conference on Mobile Computing and Networking*, ACM, pp. 35–49, 2018.

Part V

Hardware and Software Implementations

14 Empirical Evaluation of a Complete Hardware and Software Solution for UAV Swarm Networks

Carlos Felipe Emygdio De Melo,
Maik Basso, Marcos Rodrigues Vizzotto,
Matheus Schein Cavalheiro Correa,
T'U Lio Dapper E Silva,
Edison Pignaton De Freitas
Informatics Institute, Federal University
of Rio Grande do Sul, Brazil

CONTENTS

14.1 INTRODUCTION

The increasing use of unmanned aerial vehicles (UAVs) to perform previously unattainable missions, such as packet delivery, as well as search and rescue operations post natural disasters, is related to their ability to use multidimensional sensed data integrated into the UAV system control. Consequently, this ability enhances their navigation skills and enables autonomous (or semi-autonomous) air systems capable of completing missions with minimum human interaction [1, 2]. A mission is composed of tasks aimed to fulfill one or more common goals. On its turn, each task requires some specific resources, which might be provided by one or more systems composed of a single UAV or multiple UAVs. As a result, the latest studies have started proposing such multi-UAV systems with heterogeneous resources to operate in more complex missions [3, 4].

Multiple UAV systems can perform missions in unstructured environments, such as disaster scenarios, and achieve mission objectives with a reduced computing and time costs. Compared with a single UAV system, a multiple UAV system can benefit from greater accuracy and efficiency due to its resource heterogeneity, as well as better accessibility and robustness [5]. Also, a multiple UAV system can be considered a multi-agent system (MAS) in which an efficient consensus scheme can be implemented to accommodate any number of agents for different scenario configurations and applications [6]. In addition, the uncertainty of operating in a dynamic and uncertain environment requires an effective mission control and guidance system that must be able to cope with environmental changes, such as wind, which have a major impact on UAV battery consumption and can significantly alter mission performance and progress [7].

The mission planning process for a team of UAVs involves the generation of tactical goals, commanding structure, coordination and schedule [8]. In these cooperative systems, autonomous task allocation and planning is carried out for heterogeneous vehicle networks [9]. The proper functioning of these systems requires stable communication among the multiple devices, which brings high complexity for both simulation and real experimentation on the field [10].

In this chapter, a step-by-step tutorial on how to implement a drone-to-drone (D2D) communication system is presented. First, the hardware elements are described, which are basically composed of a set of 3DR Iris+ quadrotors, each of them having a Raspberry Pi 3 embedded computer; a Pixhawk flight controller and an XBee radio module. In addition, an application example was built on top of a robot operation system (ROS; see https://www.ros.org/), which facilitates both simulating and performing real experimentation in the field, and MAVLink (see https://mavlink.io/en/), which is a very lightweight messaging protocol developed specifically for drone communication.

14.2 ROBOT OPERATION SYSTEM

ROS is a highly regarded framework that provides support for developing robot applications. In short, ROS is a pseudo operating system as it needs a host system, such as Linux, to be executed. It contains package management tools, simulators and hardware abstractions. These features facilitate development and applications are easily extensible. Applications built on top of the ROS are distributed in the form of *packages*, which consist of sets of programs and scripts used for execution, compilation and simulation. These programs are called *nodes*, and communication among them is provided by a server named *roscore*, which enables the development of distributed applications.

The simplest type of communication among *nodes* occurs through a scheme known as *publisher-subscriber.* In this type of communication interlace, the *nodes*, which execute the *publisher* functionality, publish data in *topics*, which are reserved addresses for publishing data on the server *roscore* (either sporadically or continuously). Data are published in the form of structures known as *messages*.

Each *message* can be either a standardized type of structure or one specifically designed for a certain *package* (or application). Each *topic* is set to accept a specific type of message. On the other hand, the *nodes*, which execute the *subscriber* functionality, receive the published data through callbacks triggered by the publish event. In such a structure, there is ideally a single *node* as *publisher,* while one or more nodes perform the *subscriber* functionality by consuming data for different purposes. Having multiple publishers over the same *topic* is allowed; however, it is not recommended.

Client-server is another type of communication interlace provided by the ROS. In this communication model the client *node* sends a request to a *topic* in which this service is available and waits for a response. The server *node* receives the request, performs the requested operation and then returns a response to the client *node*. This communication model can be used in sporadic situations that do not require constant data dissemination.

Through the presented resources, the ROS framework was adopted for the development of the application proposed in the following sections.

14.2.1 ROS Package *multi-uav-xbee*

An ROS package, named *multi-uav-xbee*, was developed using the ROS framework. Designed to work on embedded hardware, this package has to be executed in combination with the *mavros* package, in which it communicates with the flight controller unit (FCU). For this, the FCU must be equipped with the latest version of *Ardupilot* firmware.

The operation of the proposed package follows the ROS communication models, which are illustrated in Figure 14.1. In this application, multiple UAVs are able to collect and modify their neighbors' positions. Each vehicle is a component of the network and runs on its embedded hardware (e.g., *roscore* and its node *multi-uav-xbee*) having only the variation in the vehicle identifier number. First, on launching, the proposed software subscribes to the vehicle's local position topics published by the *mavros* package. Immediately after obtaining this information, the message is serialized. Then, the serialized message is sent to all nodes of the network that are in range, through another communication channel. In this case the XBee S1 Pro radio was used. Messages exchanged by the communication channel between vehicles are serialized and deserialized using ROS core serialization libraries.

When a node on this network receives a message, it is stored again in ROS message structures. There are two types of messages in the proposed package. The first type

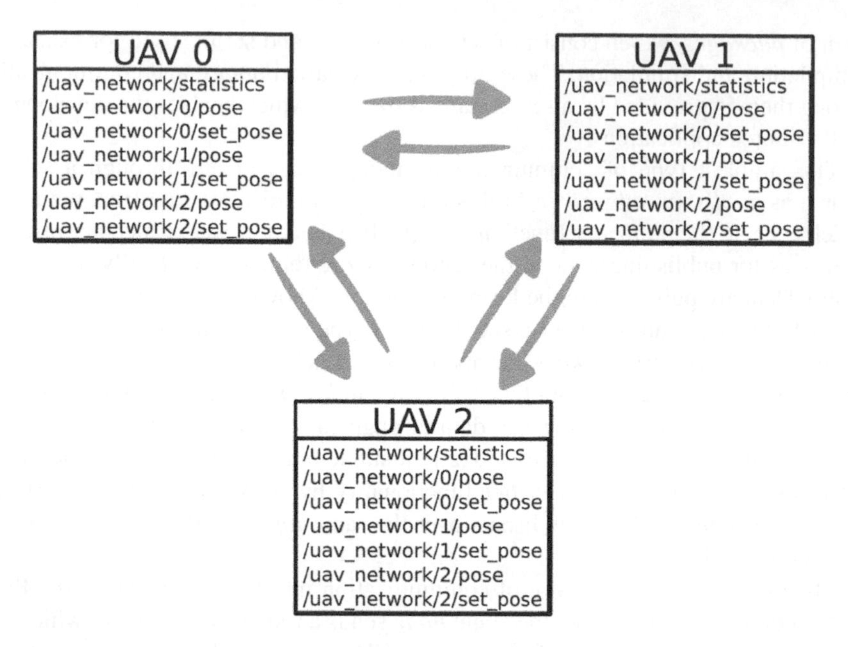

FIGURE 14.1 Functioning of the proposed ROS package.

of message refers to position beacons. This type of message contains the vehicle's pose (position and orientation) information. On receiving beacons, data obtained are automatically published to the vehicle local *roscore* server in a topic created solely to receive the positions of the sending vehicle.

Additionally, a second position command topic is created. This topic is responsible for receiving the desired position for a node on this network and then sending this command to the recipient node. This structure uses the second type of message, which is the position command message.

When an application publishes to a position command topic, the broadcast message is sent to all nodes on the network. Nodes whose identifiers are different from the identifier in the message just ignore the message. The node that has the identifier equal to the one contained in the message publishes this position command data in the *mavros* topics, which in turn will adjust the vehicle position to the position contained in the command message.

The structure of the proposed software can be described as a bridge between all *roscores*, in which, through a network channel, each vehicle knows the position of its neighbors and can change the position of its by posting to a local topic in your local ROS server.

14.3 STEP-BY-STEP IMPLEMENTATION DETAILS

14.3.1 Hardware Setup

The hardware architecture of each UAV used is comprised of an embedded hardware integrated to the UAV. The UAV used in this project is a 3DR Iris+ quadrotor. Figure 14.2 shows the hardware mounted on the 3DR Iris+ quadrotor.

FIGURE 14.2 3DR Iris+ quadrotor equipped with the embedded hardware.

The following components were considered:

- *Flight controller*: The Pixhawk Autopilot was considered, as it is an independent open-hardware project that aims to provide the standard for readily available, high-quality and low-cost autopilot hardware designs for the academic, hobby and developer communities.
- *Embedded computer*: The Raspberry Pi 3 model B was used as an embedded computer that has a 1.2-GHz 64-bit quadcore ARMv8 processor, 1 GB of RAM and a GPU VideoCore IV 3D graphics core. An SD card class 10 with 32 GB is used to store the operating system and the software that implements the developed algorithms.
- *Radio module*: To create a communication channel between UAVs, the XBee S1 Pro radio was coupled to the vehicle structure and connected using one of the USB connections available on the embedded hardware.

For the integration of all hardware and energy source to the Raspberry Pi, a board with a DC-DC converter circuit was used (Figure 14.3).

The connection between the flight controller and the Raspberry Pi 3 hardware can be seen in the Figure 14.4.

The integrated hardware was tested and configured prior to software installation and testing.

14.3.1.1 Enabling CSMA-CA in the XBee Module

The carrier sense multiple access-collision avoidance (CSMA-CA) algorithm was engineered for collision avoidance (random delays are inserted to prevent data loss caused by data collisions). Unlike the CSMA-CD, which reacts to network transmissions after collisions have been detected, the CSMA-CA acts to prevent data collisions before they occur. As soon as a device receives a packet to be transmitted, it checks if the channel is clear (no other device is transmitting). If the channel is clear, the packet is sent over the air. If the channel is not clear, the device waits for

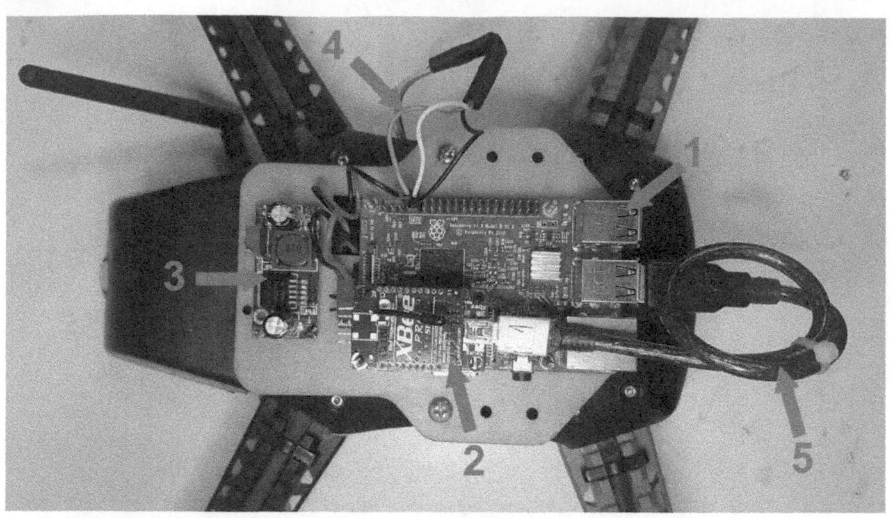

1 - Raspberry Pi 3 (RPi3) 2 - XBee Pro S1 3 - DC-DC Converter
4 - Serial connection PX4 to RPi3 5 - USB connection RPi3 to XBee

FIGURE 14.3 3DR Iris+ quadrotor equipped with the embedded hardware.

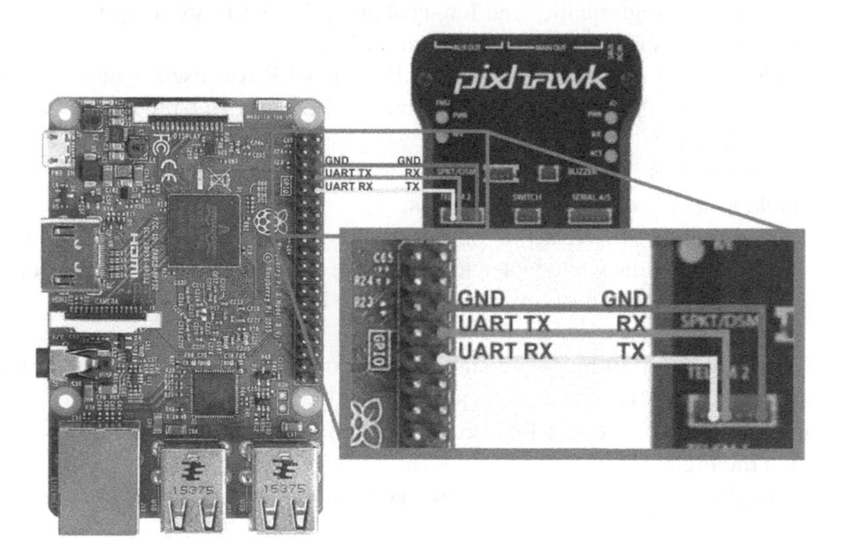

TELEM2 in Pixhawk		GPIO in Raspberry	
Pin number	**Pin name**	**Pin number**	**Pin name**
2	TX	10	UART RX
3	RX	8	UART TX
6	GND	6	GND

FIGURE 14.4 Connection between flight controller and Raspberry Pi 3.

a randomly selected period of time, then checks again to see if the channel is clear. After a time, the process ends and the data are lost.

In Network and Security section of the protocol, *random delay slots* should be set between 1 and 3, as it is the minimum value of the backoff exponent in the CSMA-CA algorithm. As a default (set to zero), there is no delay between a request to transmit and the first iteration of CSMA-CA, therefore, the CSMA-CA is not enabled.

14.3.2 Software Setup

14.3.2.1 Setting Up the Raspberry Pi 3

To use the proposed ROS package, the following components are required to be installed in the Raspberry Pi 3: Linux, ROS and *mavros*. Therefore, a version of Ubuntu MATE from Ubiquity Robotics (see https://downloads.ubiquityrobotics.com/), which already includes ROS, is a preferable choice. Download the "2018-01-13-ubiquity-xenial-lxde" Ubuntu image by using the following command:

```
wget https://ubiquity-pi-image.sfo2.cdn.digitaloceanspaces
.com/2018-01-13-ubiquity-xenial-lxde-raspberry-pi.img.xz
```

For a reliable and fast SD card to support image transfer, use an SD formatter application, such as BalenaEtcher (https://www.balena.io/etcher), to burn the Ubuntu image mentioned previously.

14.3.2.2 Enabling Serial Communication

The serial communication port needs to be enabled in Raspberry Pi 3, as it is the communication interface with the flight controller Pixhawk. To do this, access the Raspberry Pi 3 configuration interface by using the following command:

```
sudo raspi-config
```

In the configuration interface, access "Interfacing Options" and then "Serial". The system requests: "Would you like a login shell to be accessible over serial?"; set "No". It enables the serial port by setting "Yes" to the request "Would you like the serial port hardware to be enabled?".

14.3.2.3 Installing Additional ROS Packages

The following ROS packages are required for the application. For each package, its description and respective installing commands are presented below:

- *rosserial*: It is a protocol for wrapping standard ROS serialized messages and multiplexing multiple topics and services over a character device such as a serial port or network socket. In the developed application, it is used for serializing and deserializing topics to and from string, as well as transmitting to the XBee modules.

  ```
  sudo apt-get install ros-kinetic-rosserial
  ```

- *mavros, mavlink and others*: These libraries are essential for using MAVLink protocol for UAV communication.

  ```
  sudo apt-get install ros-kinetic-mav*
  ```

- *rosconsole**: it is a ROS package that supports console output and logging. It provides a macro-based interface, which allows both printf and stream-style output.

  ```
  sudo apt-get install ros-kinetic-rosconsole*
  ```

14.3.2.4 Installing *multi-uav-xbee* ROS Package

The next steps should be to install and build our suggested application. To do so, the source code folder of the ROS workspace needs to be accessed.

```
cd ~/catkin_ws/src
```

Then, clone the git repository in the source code folder.

```
git clone https://github.com/maikbasso/multi-uav-xbee.git
```

Go back to the root workspace folder.

```
cd ~/catkin_ws
```

Finally, compile the project.

```
catkin_make
```

If all goes well, the installation is complete. Otherwise, execute the command below to remove the cache files and then try to compile the code again.

```
sudo rm -rf build devel
```

14.3.2.5 Running the Application

To run our suggested application, we have developed a launch file with the following parameters:

- <droneId>: Each node has a different ID number. Default value is 0.
- <serialPort>: The serial port needs to be indicated. Default value is" /dev/ttyUSB0".
- <baud>: The serial baud rate needs to be set. Default value is 9600.

After that, the following commands should be used to run our suggested application:

```
source devel/setup.bash
roslaunch multi_uav_xbee multi_uav_xbee_network.launch
_droneId:=<droneId> _serialPort:=<serialPort> _baud:=
```

14.3.2.6 Saving Network Data in a ROS Bag File

To store all network data to be used in a post-mission analysis, a folder to store the bag files needs to be created. For this, the following command is used:

```
mkdir -p ~/catkin_ws/src/multi-uav-xbee/bag
```

Then, the next step is to execute the application, including a new parameter.

The parameter <recordBag> needs to be set as the value 1.

```
roslaunch multi_uav_xbee multi_uav_xbee_network.launch
_droneId:=<droneId> _serialPort:=<serialPort> _baud:=
_recordBag:=1
```

All of the logs will be stored in the folder named bag that was created in a previous step. The name of the files will be the current node identifier plus the current time stamp to prevent overwriting the file.

14.3.2.7 Published Network Topics

After you run the network on multiple nodes, some topics are created to access shared information. The topics are

- /uav network/statistics: Demonstrates the local network statistics.
- /uav network/<droneId>/pose:Local pose of node <droneId>.
- /uav network/<droneId>/set pose: Wish pose of node <droneId>.

From this information, an application, for example, the application presented in [11], can connect to these topics and perform mission control by positioning the various vehicles. Other proposals that may benefit from data sharing are [12, 13]. Sharing information in precision farming applications can make it easier to divide tasks between robots and open up a new range of possible operations.

14.3.2.8 Graphical Display of Node Placements

First, libraries to display the graphs need to be installed. To do this, execute the following commands:

```
pip install drawnow
pip install matplotlib
pip install numpy
```

With the application running on another terminal, execute the following commands in a new terminal, in any of the nodes, to view a graphical representation of the current position of nodes connected to the network.

```
cd ~/catkin_ws/src/multi-uav-xbee/python
python graphGenerator.py
```

FIGURE 14.5 Aerial view of the experiments area.

14.4 EXPERIMENTS AND RESULTS

Experiments were performed in the field on December 12, 2019, at CAAERO Club, in the city of Alvorada, in the state of Rio Grande do Sul, Brazil. During the experiments, three vehicles equipped with the described embedded hardware performed independent autonomous missions in a region of approximately 40 × 120 m and 10 m of altitude. Figure 14.5 represents the aerial view of the area in which the experiments were performed.

At the beginning of the experiment, the UAVs were placed in line, with 10 m of offset in position (Figure 14.6).

The mission executed by the UAVs flying on the field is demonstrated in Figure 14.7. During the mission, all the vehicles were flying in straight lines with 100 m of the distance from the home point and 10 m of altitude from the ground.

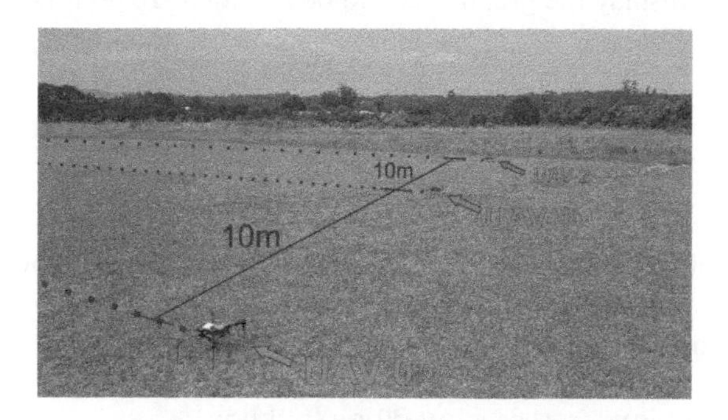

FIGURE 14.6 Home points of all UAVs in the experiment scenario.

FIGURE 14.7 Mission performed by the UAVs during the field experiments.

The trajectory traveled by the UAVs during the experiment can be seen in the Figure 14.8. This trajectory was defined by the nodes moving linearly. The defined trajectory also prevents the nodes from colliding with each other.

In this experiment, the UAVs were set to transmit their position by broadcasting over a wireless channel. Figure 14.9 presents the number of messages that have been transmitted along the execution of the experiment. The position packet is transmitted at a rate of 1 packet per second. The duration of the experiment was approximately 1000 seconds, therefore, in the end, about 1000 packets were transmitted per UAV. Each drone has a different flying start time as represented by the lag in the lines on the graph. UAV0 was the first one to start flying, next the UAV1 was started and then UAV2 was initialized. The initialization sequence was used because the same person was responsible for starting each drone, avoiding operator error during the initialization, which would have contaminated the result from the simulation round.

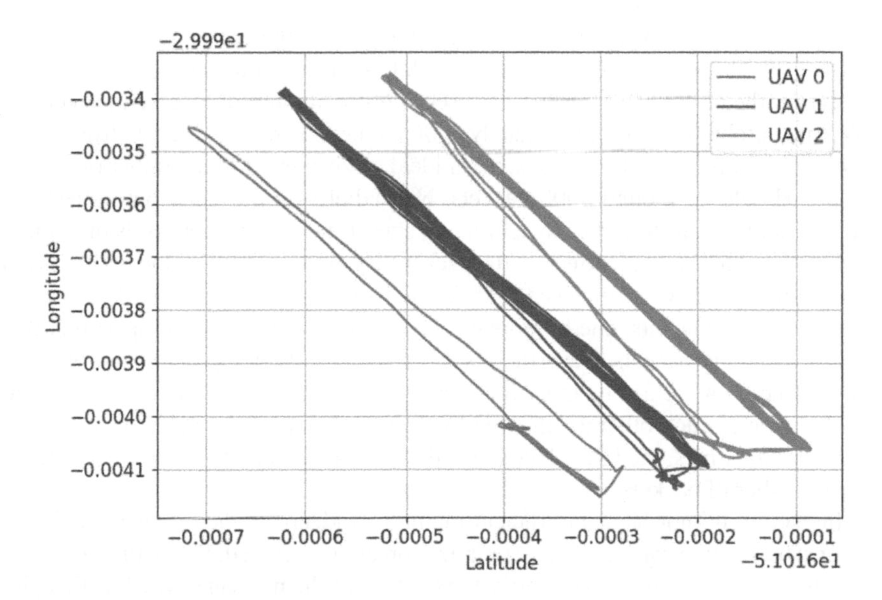

FIGURE 14.8 Trajectory traveled by the UAVs during the experiment.

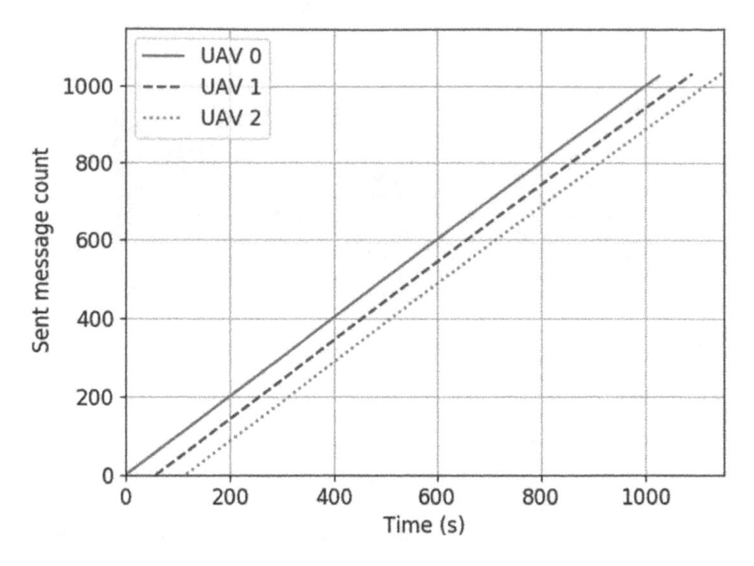

FIGURE 14.9 Number of packets sent.

The same sequence was used to end the data record in each drone. The continuous lines in the graph showed that, throughout the experiment, all three UAVs transmitted without interruption. Interruptions in the position packet broadcast would have a negative effect on the results of the experiment, causing the applications that use the communication package to have outdated data. Outdated data in distributed applications can somewhat affect the performance of a mission control application, such as causing collisions among the vehicles.

The number of packets received for the duration of the experiment is presented in Figure 14.10. As presented previously, each UAV transmitted approximately 1000 packets for the entire experiment, but observing the amount of received packets per UAV, this number is also approximately 1000 packets. However, if each drone transmitted approximately 1000 packets, in an ideal environment, the number of packets received should be around 2000 packets. Nevertheless, in a real environment, the distance between drones, the asynchronous start time and packet collision, among other factors, have a direct effect on the packet loss, which is noticeable in this graph. For instance, in this experiment each UAV had its mission started at a different time, then at certain moments, one or more drones could have been so far apart that they stopped receiving packets. Nevertheless, because they fly going back and forth in a straight line, at a certain moment they will be in communication range again and start to listen to each other again. Due to the mobility model and the UAV's asynchronous start times, the UAVs face intermittent disconnection, which is reflected in a high number of packets lost.

This setup did not have an efficient method to identify when a packet collision happens; however, occasionally a deserialization error occurred. When this error happens, the receiver is unable to identify the content of the message, which is then discarded and the packet is considered lost. Figure 14.11 presents the number of packets

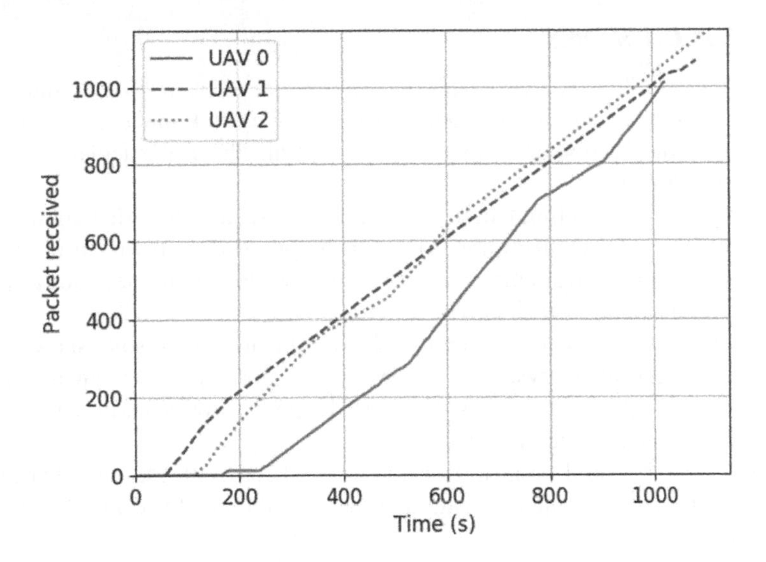

FIGURE 14.10 Number of packets received.

lost to its error per UAV. UAV0 has a higher deserialization error rate than the other two UAVs because UAV0 presents a higher receive rate, as shown in Figure 14.10.

The results have shown the feasibility for using the proposed hardware architecture in an application involving communication among multiple UAVs in a real scenario. More complex applications can be developed using this complete hardware and software architecture.

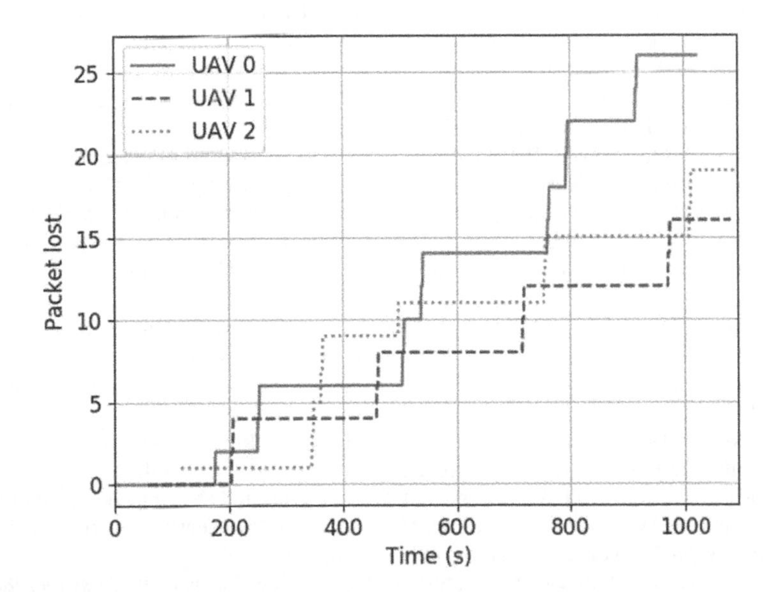

FIGURE 14.11 Number of packets lost because of a deserialization error.

14.5 CONCLUSIONS

This chapter describes an alternative to hardware and software for communicating multiple UAVs in a scenario in which all vehicles need to know each other's positions and have the freedom to control their movement. This is particularly useful for UAV-swarm systems.

The proposed architecture can be used on commercial drones that contain some variation of *Ardupilot* as firmware. Also, all the proposed software uses standard packages and libraries available in the ROS core, so there is no need for any other installation and complexities in the solution assembly.

The system was evaluated through field experiments. The results show that the proposed architecture hardware has acceptable performance for synchronization of the two specified message types. Keeping the pose of vehicles in constant sync between all elements of the network.

In the future, it would be ideal to use a higher bandwidth radio so that all topics could be fully synchronized, bringing other possibilities for using the proposed application.

REFERENCES

1. A. Al-Kaff, D. Martín, F. García, A. de la Escalera, and J. M. Armingol, "Survey of computer vision algorithms and applications for unmanned aerial vehicles," *Expert Systems with Applications*, vol. 92, no. Supplement C, pp. 447–463, 2018.
2. C. Kanellakis and G. Nikolakopoulos, "Survey on computer vision for UAVs: current developments and trends," *Journal of Intelligent & Robotic Systems*, vol. 87, no. 1, pp. 141–168, 2017.
3. A. C. Trujillo, J. Puig-Navarro, S. B. Mehdi, and A. K. McQuarry, "Using natural language to enable mission managers to control multiple heterogeneous UAVs," In: P. Savage-Knepshield and J. Chen, Eds. *Advances in Human Factors in Robots and Unmanned Systems: Proceedings of the AHFE 2016 International Conference on Human Factors in Robots and Unmanned Systems*, Cham, Switzerland: Springer International Publishing, pp. 267–280, 2017.
4. R. S. De Moraes and E. P. De Freitas, "Multi-UAV based crowd monitoring system," *IEEE Transactions on Aerospace and Electronic Systems*, pp. 1–1, 2019.
5. W. Zhao, R. Li, and H. Zhang, "Finite-time distributed formation tracking control of multi-UAVs with a time-varying reference trajectory," *IMA Journal of Mathematical Control and Information*, p. dnx028, 2017. [Online]. http://dx.doi.org/10.1093/imamci/dnx028
6. L. Han, X. Dong, K. Yi, Q. Tan, Q. Li, and Z. Ren, "Circular formation tracking control for time-delayed second-order multi-agent systems with multiple leaders," *2016 IEEE Chinese Guidance, Navigation and Control Conference (CGNCC)*, pp. 1648–1653, 2016.
7. L. Evers, T. Dollevoet, A. I. Barros, and H. Monsuur, "Robust UAV mission planning," *Annals of Operations Research*, vol. 222, no. 1, pp. 293–315, 2014.
8. C. Ramirez-Atencia, G. Bello-Orgaz, M. D. R-Moreno, and D. Camacho, "Solving complex multi-UAV mission planning problems using multi-objective genetic algorithms," *Soft Computing*, vol. 21, no. 17, pp. 4883–4900, 2017.
9. S. S. Ponda, L. B. Johnson, A. Geramifard, and J. P. How, *Cooperative Mission Planning for Multi-UAV Teams*. Dordrecht: Springer Netherlands, pp. 1447–1490, 2015.

10. M. Basso, I. Zacarias, C. E. Tussi Leite, H. Wang, and E. Pignaton de Freitas, "A practical deployment of a communication infrastructure to support the employment of multiple surveillance drones systems," *Drones*, vol. 2, no. 3, 2018.
11. T. Dapper e Silva, C. F. Emygdio de Melo, P. Cumino, D. Rosa´rio, E. Cerqueira, and E. Pignaton de Freitas, "STFANET: SDN-based topology management for flying ad hoc network," *IEEE Access*, vol. 7, pp. 173, 2019.
12. M. Basso and E. P. de Freitas, "A UAV guidance system using crop row detection and line follower algorithms," *Journal of Intelligent and Robotic Systems*, vol. 97, pp. 605–621, 2020.
13. M. Basso, D. Stocchero, R. Ventura Bayan Henriques, A. L. Vian, C. Bredemeier, A. A. Konzen, and E. Pignaton de Freitas, "Proposal for an embedded system architecture using a GNDVI algorithm to support UAV-based agrochemical spraying," *Sensors*, vol. 19, no. 24, 2019.

15 Efficient and Robust Unmanned Aircraft C-Band Communication System

Dominik Rieth
Institute for Communication Technologies and Embedded Systems, RWTH Aachen University, Germany

CONTENTS

This chapter explains the details of an efficient and robust UAV communication system, starting with the derivation of low-complexity algorithms and their efficient implementation in a complete system, and ending with results of the system in realistic flight scenarios.

To derive the low-complexity algorithms, shaped offset QPSK (SOQPSK) demodulation simplifications that help to increase the usability of the continuous phase modulation based waveform in resource constraint environments are explained. Based on a multifunctional frame design, supporting burst and continuous mode, the derivation of a decision-directed synchronization method for SOQPSK modulation to correct timing, phase and frequency offsets is discussed. Together with the proposed combination of start of frame detection and phase ambiguity resolution with nested Barker unique words, a fully coherent demodulator algorithm is presented, which is highly suitable for low-complexity implementations.

This is followed by a descriptive overview of the developed data link system and a detailed implementation description of its main components on the FPGA part of its software-defined radio platform, the SOQPSK demodulator. The implementation results show superior hardware efficiencies compared with existing implementations in literature.

Toward the end of the chapter, several performance evaluations, such as throughput, Doppler frequency robustness, realistic flight scenario BER performance and a link budget of the final airborne communication systems are reported. With a number of different waveform and transceiver configurations, the included ability for a flexible parametrization is highlighted, which is an important prerequisite for the proposition of a scenario-adaptive waveform.

In total, the chapter shows a big step toward a communication system that supports the integration of a large number of future unmanned aircraft into non-segregated airspace by providing a sophisticated solution for the key component of unmanned aerial vehicle (UAV) operations. Extensions to this chapters' content can be found in [1].

15.1 LOW-COMPLEXITY SOQPSK SYNCHRONIZATION AND DEMODULATION ALGORITHM

Shaped offset QPSK (SOQPSK) is a highly bandwidth and power efficient modulation scheme, especially with the parameter set from [2]. It is a binary waveform from the family of continuous phase modulation (CPM). Usually, maximum likelihood (ML) sequence detection is used to optimally detect the transmitted data because they outperform symbol-by-symbol detectors by 1–2 dB [3], but the receiver's complexity grows exponentially with the signal's memory [4].

To tackle this inherent drawback of SOQPSK and to make the modulation even more attractive for applications like the Internet of Things (IoT), machine to machine (M2M) and unmanned aerial vehicle (UAV) communication or telemetry links, a novel demodulator architecture with superior hardware efficiency is presented in this chapter. At the same time, the proposed implementation helps to keep the energy consumption very low, which is a desirable property for mobile or battery powered devices.

There are multiple simplifications in the receiver's CPM scheme that can reduce its demodulation complexity. A method called pulse truncation (PT) approximates partial response CPMs by reducing the phase pulse to a length of one symbol time [5], allowing much simpler trellis representations. The pulse amplitude modulation (PAM) approximation of SOQPSK is introduced and used in [6–8] and comparisons of PT and PAM simplifications are presented in [9, 10]. However, in one approach, the SOQPSK trellis was even reduced to two states [3, 11], with a bigger bit error rate (BER) performance degradation.

Powerful synchronization mechanisms are key enablers for robust and reliable data links. For fully coherent reception, it is necessary to gain knowledge about carrier frequency and phase offsets, which occur due to Doppler shifts, imperfect local oscillators and varying undetermined distances between transmitter and receiver. Recovery of incorrect sample times, a result of the freely running clocks, has to be taken into account as well. A preamble, which is repeated regularly, can support the parameter estimation process. This sequence of known information is also appropriate to detect a start of frame (SoF), which is highly important for several error correction techniques. Synchronization is a task that usually adds a remarkable amount of complexity to the receiver. Therefore, it is also worth simplifying symbol timing, phase, frequency and frame synchronization as much as reasonably possible. For example, [12, 13] describe a low-complexity timing error correction loop and [14] combines timing and phase correction for CPM signals. Finding the SoF with a known data sequence and an unknown frequency offset is the topic of [15]. A burst mode SOQPSK synchronization of all unknown parameters is derived in [16]. Different algorithms are used for CPM frequency estimation of large offsets (e.g., [17, 18]). However, assuming moderate frequency errors allows the implementation of a much more hardware friendly method, which is presented in Section 15.1.1.

As UAV communication has not been fully regulated or standardized, the definition of an access scheme is still open. It is also conceivable that uplink and downlink will use different access schemes. To meet these uncertainties, this chapter presents a multifunctional approach that is capable of burst and continuous mode communication. With this hybrid synchronization type, time division multiple access (TDMA) and frequency division multiple access (FDMA) applications can be realized.

15.1.1 Symbol Timing, Carrier Phase and Frequency Synchronization

Synchronization is a prerequisite for coherent reception, but at the same time, it leads to a considerable amount of the receiver's total complexity. Basically, there are three parameters to adjust, carrier *frequency offset*, *phase offset* and *symbol timing offset*, and one event to detect, which is the start of a new frame. A brief explanation for the cause of these offsets is given below:

- Due to the Doppler effect and inaccuracies in the transmitter and receiver clocks, a frequency offset might impair the communication. Frequency recovery counteracts these phenomena.
- A phase offset is the result of an unknown path length from the transmission to the receiving antenna and its modulo lies in the interval $[0; 2\pi)$. Another source of offset is the unknown and possibly different starting point of the local oscillators of transmitter and receiver. The impact a of phase error is a rotated constellation diagram. Estimation or detection of this offset with subsequent countermeasures is done by the phase recovery.
- Remembering the fact that transmitter and receiver and thus their clocks are decoupled, one can easily imagine asynchronous transmission and analog to digital conversion in the receiver. Therefore, the received signal might

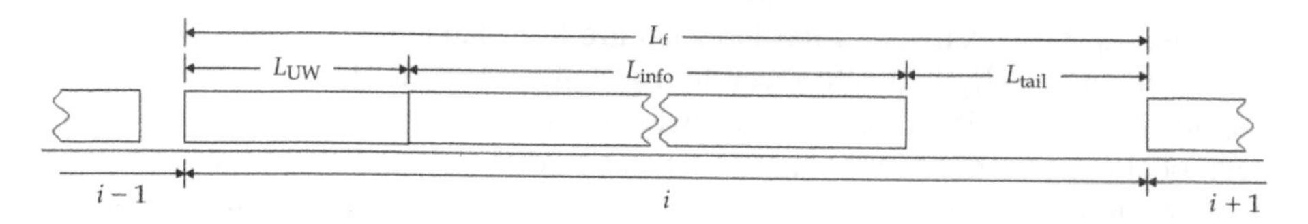

FIGURE 15.1 Multifunctional frame structure with unique word, information part and tail. Three subsequent frames $i-1$, i and $i+1$ are indicated.

not be sampled at its optimal positions, which would result in a shifted eye-diagram. Correcting this effect is the task of symbol timing recovery.

- When the receiver has no knowledge about the start of data transmission, the SoF is an unknown that has to be detected to propagate this information to layers above the physical one or to distinguish between the unique word (UW) and actual data. There are also forward error correction (FEC) mechanisms that need exact SoF notification to be able to work properly.

At this point, it is useful to introduce the underlying multifunctional frame structure of the data link, as seen in Figure 15.1, with three indicated frames $i-1$, i and $i+1$. The first part of each frame with length L_f is a UW, also called preamble, with L_{UW} bits. It is followed by L_{info} information bits and a gap to the next frame of length L_{tail}. During the tail sequence, it is assumed that only white Gaussian noise (WGN) is received. Multifunctionality is accomplished by means of a variation of L_{tail} as follows: $L_{tail} > 0$ is a burst mode configuration and $L_{tail} = 0$ enables continuous mode transmission.

In the presence of additive WGN (AWGN), the complex baseband representation of the receive signal is

$$r(t) = \sqrt{\frac{E_s}{T_s}}\, e^{(j2\pi f_d t + \theta)} s(t-\tau) + n(t)$$

with $n(t) \sim \mathcal{N}\left(0, \frac{N_0}{2}\right)$ and the parameters frequency offset f_d, phase offset θ and symbol timing offset τ, which have to be estimated by the synchronizer to properly detect the desired signal.

Two different methods of operation are distinguished, *closed loop* and *feedforward*. The first one recursively estimates and diminishes the error, trying to approach the error to zero. In contrast, the latter one performs error estimations for a feed-forward actor without any feedback structure. In the proposed detector, frequency, phase and symbol timing errors are mitigated by closed-loop approaches.

Depending on the type of prior data knowledge for parameter error estimation, there are basically three classes of synchronizers, data aided (DA), non-data aided (NDA) and decision directed (DD) (cf. [16]):

- Data-aided algorithms use prior knowledge of the transmitted data, which can be a known preamble or a UW.

- Without any prior knowledge of the data, non–data-aided synchronization is used.
- Decision directed algorithms work with estimated data instead of prior known data. Hence, the tasks of synchronizing and data estimation are interconnected in this class.

A DD timing and phase synchronization, based on [19], with an extension to remove frequency offsets, is proposed for the detector. The goal of the frequency add-on is adding the lowest possible amount of complexity to time and phase correction. Omitting constant factors, an ML estimation of the unknown parameters is obtainable starting with the log-likelihood function (LLF)

$$\Lambda(r \mid \tilde{a}) = \sum_{n=1}^{L_0-1} \mathrm{Re}\left\{ e^{-j\tilde{\theta}} e^{-j2\pi n \tilde{f}_d / T} z_n(\tilde{a}_n, \tau) e^{-j\tilde{\theta}_{n-1}} \right\}$$

Included is a hypothetically known bit sequence $\tilde{a} = \{\tilde{a}_n\}$ with a length of L_0. To estimate frequency, phase and timing with this function, the strategy is to consider all parameters individually and assume that the others are known, respectively. Computing the partial derivative of the LLF without constant factors results in

$$\frac{\partial}{\partial \tilde{\tau}} \Lambda(r \mid \tilde{\tau}) = \sum_{n=0}^{L_0-1} \mathrm{Re}\left\{ e^{-j\theta} e^{-j2\pi n f_d / T} y_n(a_n, \tilde{\tau}) e^{-j\theta_{n-1}} \right\}$$

for symbol timing,

$$\frac{\partial}{\partial \tilde{\theta}} \Lambda(r \mid \tilde{\theta}) = \sum_{n=0}^{L_0-1} \mathrm{Im}\left\{ e^{-j\tilde{\theta}} e^{-j2\pi n f_d / T} z_n(a_n, \tilde{\tau}) e^{-j\theta_{n-1}} \right\}$$

for the phase parameter, and finally

$$\frac{\partial}{\partial \tilde{f}_d} \Lambda(r \mid \tilde{f}_d) = \sum_{n=0}^{L_0-1} \mathrm{Im}\left\{ n e^{-j\theta} e^{-j2\pi n \tilde{f}_d / T} z_n(a_n, \tau) e^{-j\theta_{n-1}} \right\}$$

for the frequency, where $y_n = \frac{\partial}{\partial \tilde{\tau}} z_n$. The nulls of the three derivatives indicate the maxima of the LLFs. In a closed-loop realization, the maximization task is performed iteratively, requiring knowledge of the transmitted symbols $\{a_n\}$. Supposing a DD algorithm, the estimation of this sequence $\{\tilde{a}_n\}$ is sufficient and is obtained from the inherent Viterbi detector. Introducing the integer delay D, the error functions without constants are defined as

$$e_\tau[n-D] \equiv \mathrm{Re}\left\{ e^{-j(\hat{\theta}[n-D] + 2\pi(n-D)\tilde{f}_d[n-D]/T)} \cdot y_{n-D}\left(\hat{a}_{n-D}, \tilde{\tau}[n-D]\right) e^{-j\theta_{n-D-1}} \right\}$$

for sample timing,

$$e_\theta[n-D] \equiv \mathrm{Im}\left\{e^{-j(\hat{\theta}[n-D]+2\pi(n-D)\hat{f}_d[n-D]/T)} \cdot z_{n-D}\left(\hat{a}_{n-D}, \hat{\tau}[n-D]\right)e^{-j\hat{\theta}_{n-D-1}}\right\}$$

for carrier phase offsets and

$$e_{f_d}[n-D] \equiv \mathrm{Im}\left\{(n-D)e^{-j(\hat{\theta}[n-D]+2\pi(n-D)\tilde{f}_d[n-D]/T)} \cdot z_{n-D}\left(\hat{a}_{n-D}, \hat{\tau}[n-D]\right)e^{-j\hat{\theta}_{n-D-1}}\right\}$$

for carrier frequency shifts. Due to the Viterbi characteristic of the detector, the reliability of the symbol estimation $\hat{a}_{n-D}$ increases with the delay D. However, a value as low as $D = 1$ has shown satisfactory results in [19]. As expected, e_θ and e_{fd} have a similar structure, which allows the implementation of a tracking filter for both of them with a single second-order loop.

To illustrate how the error signals are used to mitigate symbol timing, phase and frequency offsets, the schematic in Figure 15.2 is explained. A digital baseband signal $r[n]$ is interpolated and rotated to correct symbol timing, phase and frequency errors, respectively. The following *matched filters* (MFs) feed the *Viterbi detector* and the *phase & freq. error calculator.* A derivative MF output y_n is also computed and forwarded to the *timing error calculator.* Both error calculators receive symbol estimations $\hat{a}$ from the Viterbi detector, which is a crucial demand of the underlying DD algorithms. The error calculators are followed by *loop filters*, a multiplication with a loop filter constant in the simplest case for timing, a second-order filter for phase and frequency and, finally, the *estimators.* They try to approach the three

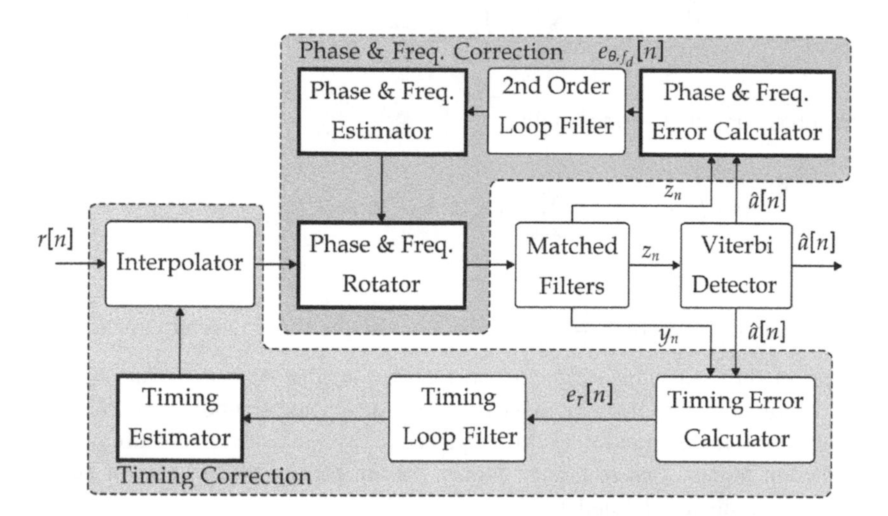

FIGURE 15.2 SOQPSK detector with recursive timing, phase and frequencycorrection. The output of the Viterbi detector is a stream of estimated bits still affected by phase ambiguity.

synchronization parameters and feed them back to the *interpolator* and *phase & freq. rotator*, which close the loops.

15.1.2 Joint SoF Detection and PAR

After the phase correction loop and Viterbi detection, the estimated symbols still suffer from a $\frac{\pi}{2}$ phase ambiguity due to the rotational symmetry of the underlying constellation diagram. The detection of a UW is a well-known procedure used to resolve this remaining uncertainty [20]. A known preamble is also helpful for frame synchronization [21] and to keep hardware complexity as low as possible, reusing the UW for SoF detection is proposed. The same argument holds for choosing a hard output instead of a soft output variant. In the first case, two copies of the UW are saved, the original one and a $\frac{\pi}{2}$ phase shifted version. When the sum of a negated XOR operation of the estimated bits and the saved ones exceeds a fixed threshold, a SoF signal is generated together with a signal to correct phase ambiguity. As the entire procedure works on registered samples from the past, no latency is added in the respective hardware unit. To finally resolve the detected ambiguity, the phase of the estimated output stream is shifted as follows. Depending on which of the four threshold values has been reached, all values are just forwarded (phase shift of 0), negated (phase shift of π) or alternating values are negated (phase shifts of $\pm\frac{\pi}{2}$). A more detailed explanation of this phase ambiguity resolution (PAR) principle can be found in [22].

A desired design criteria for UWs are low sidelobes of their nonperiodic autocorrelation function [23]. Inspired by [24], the long preambles used in this chapter are generated by nesting original binary Barker codes [25]. Starting with a code of length L, each one is replaced by an original copy and each zero by a negated copy of the sequence. In total, this results in nested codes with lengths L_2, still having low sidelobe characteristics.

The total length of the UW can be selected depending on the required detection performance. Being highly modest in terms of the amount of utilized logic, the presented combination of SoF detection and PAR is very attractive for demodulators with constrained resources and energy. However, in case of burst mode with no data between the bursts, a decoupling sequence (DS) enlarges the total preamble length for good synchronization results. The outcome of this effect is visible in Figure 15.3, where the SoF probability is measured for low E_b/N_0 values. Barker codes of lengths seven, 11 and 13 have been used to create nested UWs of lengths 49, 121 and 169. Two different DS lengths have been chosen to demonstrate superior performance when the demonstrator's DD feedback loops have some data in front of the UW to settle, which explains the bad results of all three burst mode versions with a DS length of zero. Nevertheless, this effect is not present in either continuous mode or burst mode with random data in between different bursts to keep all loops settled. However, it is interesting that even in the continuous mode, the versions with DS outperform the one without, which has only random data between two subsequent UWs.

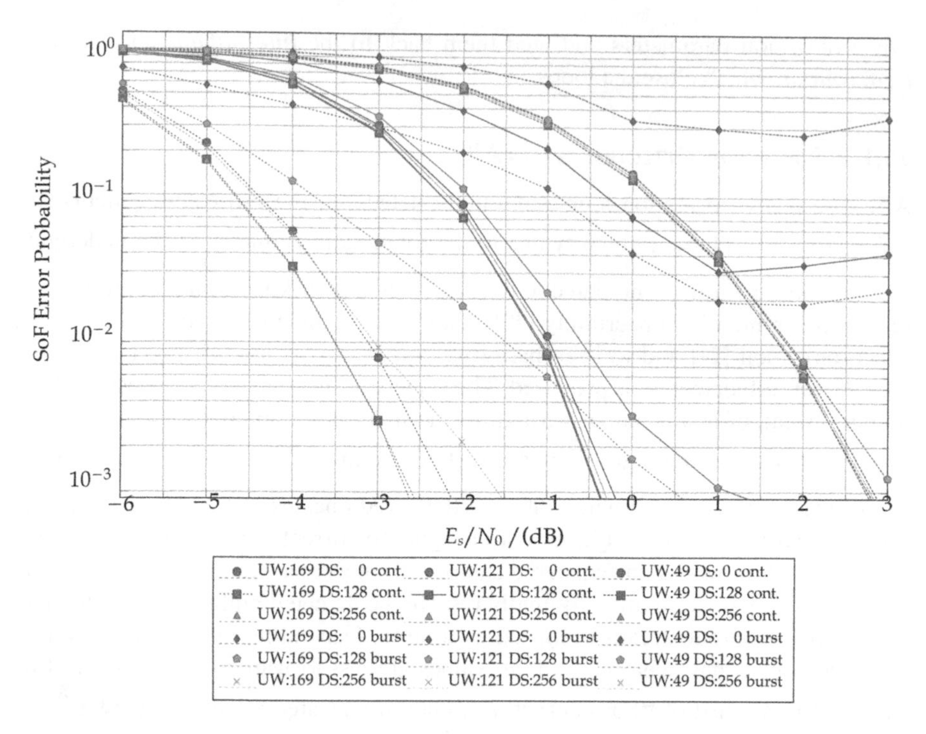

FIGURE 15.3 SoF error probability for burst and continuous mode, for two different decoupling sequences (128 and 256) and three different nested Barker code lengths (49, 121 and 169). This is the result from more than 1e5 simulated frames per point with random SoF value in the presence of random frequency $\left(e_{fd} \in [0, 10^{-4} f_s\right)$, phase ($e_\theta \in [0, 2\pi]$) and timing errors ($e_\tau \in [0,1]$).

15.2 AERONAUTICAL DATA LINK SYSTEM IMPLEMENTATION

Within the following section, a complete aeronautical data link demonstrator is presented, which is optimized for UAV applications in C-band. The system concept mainly follows a modular software-defined radio (SDR) approach, ensuring a high level of configuration flexibility. Even though this flexibility comes with an amount of overhead that reduces the overall efficiency, the lack of standards regarding UAV control and non-payload communication (CNPC) waveforms and operating frequencies, lower costs for small quantities and the ability to rapidly prototype still justify this approach.

The digital baseband signal processing is realized within the FPGA or the programmable logic (PL) part of a Xilinx Zynq system on chip (SoC). The demodulator of the waveform is the key component of an efficient data link system. Hence, a novel low-complexity architecture is derived and its FPGA implementations are described. Another integral part of the data link is a very flexible transceiver SoC, which is capable of transforming the baseband signals into the radio-frequency (RF) band for transmission and vice versa for reception.

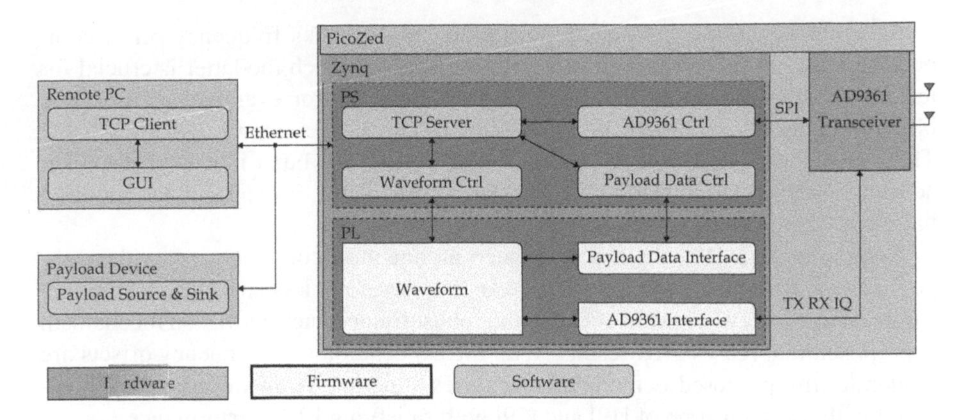

FIGURE 15.4 Aeronautical data link system demonstrator simplified functional block diagram. The modem functionality is implemented in the PS and PL part of a Zynq [27], which is part of the PicoZed board [26]. This board also contains an AD9361 transceiver chip [28] that is connected to the Zynq SoC for control and baseband signals and to radio-frequency ports for transmit and receive antennas. With a remote PC, the waveform and the transceiver chip can be configured at runtime. A generic payload device is also shown to highlight the data link's transparent IP functionality via Ethernet.

15.2.1 Data Link System Overview

This section serves as an introduction overview of the aeronautical data link system demonstrator's main parts, which are depicted in Figure 15.4. The demonstrator is realized on a PicoZed-SDR board [26] that combines a Zynq SoC [27] with an AD9361 transceiver [28]. Payload data are generated and received from a generic payload device that communicates with the payload data control via the Ethernet. Within the waveform unit are modulator, demodulator and FEC mechanisms. Digital baseband transmit and receive signals are exchanged between waveform and transceiver with the aid of a firmware interface. For simplicity reasons, power amplififers and filters are not shown in the RF domain after the AD9361 device. Instead, two antenna pictograms indicate the possibility of using FDMA for transmission and reception.

From a flexibility point of view, most waveform and transceiver parameters can be configured at runtime with a graphical user interface (GUI) running on an Ethernet-connected remote PC. It communicates via Transmission Control Protocol (TCP) with the corresponding control software blocks. The transceiver registers are read and set over a serial peripheral interface (SPI) connection. In contrast, status registers and baseband debug samples from the AD9361 and the PL part are collected from the TCP server and transmitted to the remote PC GUI.

15.2.2 Low-Complexity and Low-Energy Architecture

A continuous mode SOQPSK detector field programmable gate array (FPGA) implementation with phase and timing recovery loop is presented in [19]. Its architecture is heavily pipelined leading to high clock frequencies and for this reason the architecture is not energy friendly. Furthermore, [19] lacks frequency offset compensation and frame synchronization mechanisms, in which the latter is crucial for low-density parity check (LDPC) coded transmissions, for example, because this block code–based error correction procedure requires exact knowledge of the SoFs. Those two shortages have been addressed by a SOQPSK burst mode synchronizer add-on in [29], yet, compared with only the demodulator, it increases the required hardware resources significantly.

Compared with [19], the now presented architecture combines a data throughput increase with much lower clock speeds and, therefore, less energy consumption. Additionally, a very hardware efficient and robust frame synchronization mechanism is proposed that reuses PAR results. With the expectation that frequency offsets are moderate, the proposed demodulator achieves superior hardware efficiency compared with a combination of [19] and [29] with satisfying BER performance.

To increase hardware and energy efficiency, a new architecture for fully coherent SOQPSK demodulation is proposed that is suitable for continuous and burst mode transmission. It contains DD synchronization loops for frequency, phase and timing offsets and a low complex method combining robust SoF detection with PAR based on nested Barker codes. A coarse grained pipeline structure aims for minimal clock speeds and energy consumption while keeping the overall throughput high. Large complexity reductions are achieved by a multiplier-free MF design. Computer simulations and FPGA implementation results show that the complexity-accuracy trade-offs have been reasonably chosen in terms of close-to-optimal BER performance and that hardware efficiency gains of more than 90% compared with implementations from literature are achievable.

15.2.2.1 General Architecture Decisions

To minimize the required energy, a low oversampling rate (OSR) of 4 is chosen and the design will be clocked exactly at this rate f_{OSR}. Due to the feedback loops, consisting of four processing units each, this is also a lower bound to match a symbol estimation delay of one. On the one hand, minimum oversampling prohibits resource sharing; however, the results in Section 15.2.2 revealed that this decision still allows very hardware efficient implementations. An architecture overview is drawn in Figure 15.5, with the complex baseband receive signal input R_OS and the hard

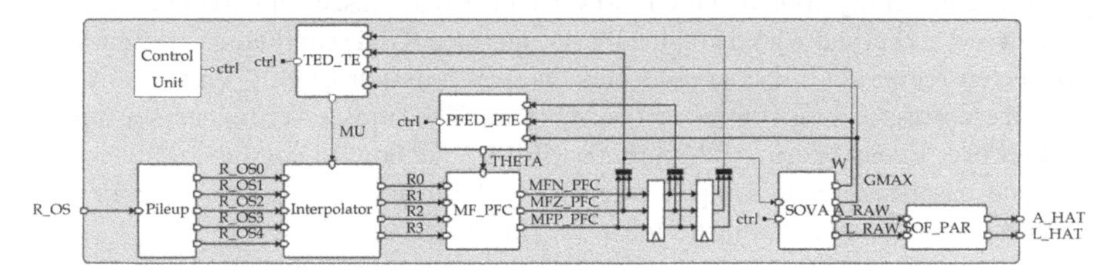

FIGURE 15.5 Demodulator architecture overview with oversampled complex baseband input R_OS, and estimated hard and soft outputs A_HAT and L_HAT. MF_PFC, matched filter with successive phase and frequency correction; SOVA, soft output Viterbi algorithm; SoF_PAR, SoF and phase ambiguity resolution and {T,PF}ED_{T,PF}E, timing, phase and frequency error detection and estimation.

and soft output estimations A_HAT and L_HAT. The *Pileup* unit simply consists of registers for a serial to parallel conversion of N_{OSR}+ 1 samples and the *Control Unit* is the detector's state machine. All other units are designed as input registered processing units (IRPUs), which latch all inputs in advance to otherwise solely combinatorial logic. Basically, each stage of the coarse-grained pipeline does all computations required for a complete symbol in one clock cycle, a method enabled by the parallelization of the *Pileup* unit. The granularity of the separation into the different units is determined by the computational complexity in each of them. Linear interpolation removes the estimated time error MU, and phase correction is done together with matched filtering in the subsequent unit. Two registers save the three MF outputs to have early, late and on-time versions. The *early late* method from [20] is used as an approximation to compute the MF derivative for synchronization. Timing, phase and frequency error detection, together with loop filtering and estimation, are combined in two additional units. An implementation of an either soft or hard output Viterbi algorithm with a final SoF detection and PAR concludes the demodulator architecture.

15.2.2.2 Multiplier-Free MF Implementation

A careful realization of MF banks is a crucial part when the reduction of implementation costs is desired, as they are among the main contributors to the receiver's overall complexity. In [19], the multipliers of the MFs are in the critical path, which highlights their impact on the overall design and the maximum achievable throughput. The method described here can be generalized to other CPM schemes, but for the sake of simplicity, it is solely explained tailored to SOQPSK.

Exploiting the complexity reduction PT [9], the time discrete MF output of each possible transmit symbol $\tilde{a} \in \{-1,0,1\}$ is given by

$$z_n = \sum_{k=0}^{N_{\text{OSR}}-1} r[n+k]e^{j\pi\tilde{a}q_{\text{PT}}[k]} \tag{15.1}$$

where the $q_{\text{PT}}[k]$ are the NOSR samples of $q_{\text{PT}}(t)$ (cf. Figure 15.6 for an example with NOSR = 4). As is the placement for the complex multiplication by $e^{j\pi\tilde{a}q_{\text{PT}}[k]}$, the approach is to approximate the coefficients with sums of powers of 2. Then, all MF outputs are realizable with sum operations and binary shifts, whereby the latter costs nothing in hardware.

Starting with the exploitation of the coefficients' symmetry in Eqn. (15.1), they can be defined as

$$\begin{aligned}
\exp\{\pm j\pi q_{\text{PT}}[0]\} &=: q_1 \mp jq_2 \\
\exp\{\pm j\pi q_{\text{PT}}[1]\} &=: q_3 \mp jq_4 \\
\exp\{\pm j\pi q_{\text{PT}}[2]\} &=: q_4 \mp jq_3 \\
\exp\{\pm j\pi q_{\text{PT}}[3]\} &=: q_2 \mp jq_1
\end{aligned}$$

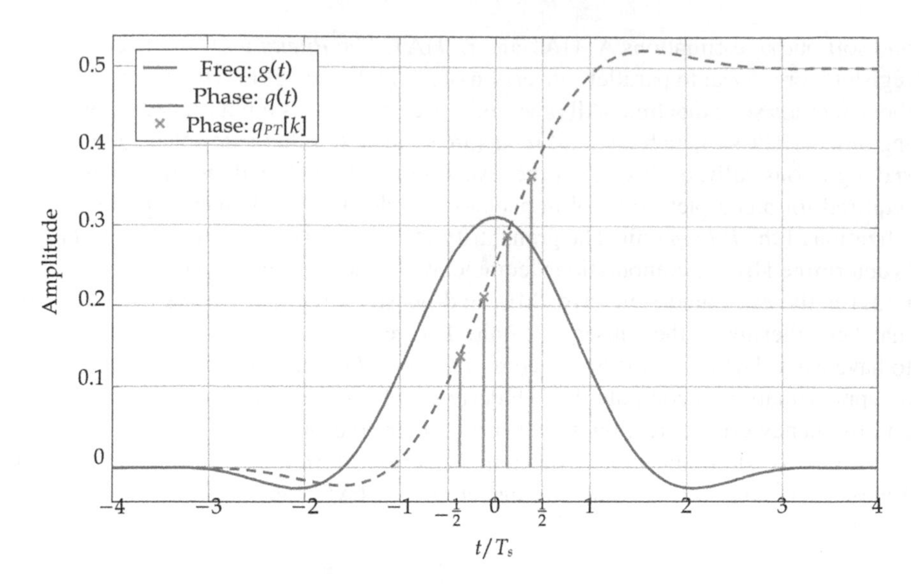

FIGURE 15.6 Time-shifted frequency and phase pulse of SOQPSK-TG together with the sampled version of q_{PT} for NOSR = 4.

for $\tilde{a} = \pm 1$. So, four different real numbers q_1 to q_4 are sufficient for an exact coefficient representation. The MF output of the remaining symbol $\tilde{a} = 0$ simplifies to a summation of all N_{OSR} signal samples because all exponential terms are equal to 1.

With these relations, different levels of complexity reduction L_{0-4} have been studied. L_0 is the reference case, where PT and no further approximation is used. In L_1 the coefficients q_{1-4} are reduced to a sum of two power of 2 values and in L_2, only one power of 2 value per coefficient approximates the exact values of q_{1-4}. For completeness, L_3 is the case where $q_{1-4} = 1$. Table 15.1 lists the real q_{1-4} values for all complexity levels. Just for the visualization in the table, the values of L_0 have been rounded to four decimal places. Simulations have been performed with double precision for L_0 and the results are shown in Figure 15.7. It is known that the curve of PT is close

TABLE 15.1
Different Levels of MF Coefficient Approximation for $N_{\text{OSR}} = 4$

Level	q_1	q_2	q_3	q_4
L_0	0.9085	0.4178	0.7880	0.6156
L_1	$1-\frac{1}{16}$	$\frac{1}{4}+\frac{1}{16}$	$\frac{1}{2}+\frac{1}{4}$	$\frac{1}{2}+\frac{1}{8}$
L_2	1	$\frac{1}{2}$	1	$\frac{1}{2}$
L_3	1	1	1	1

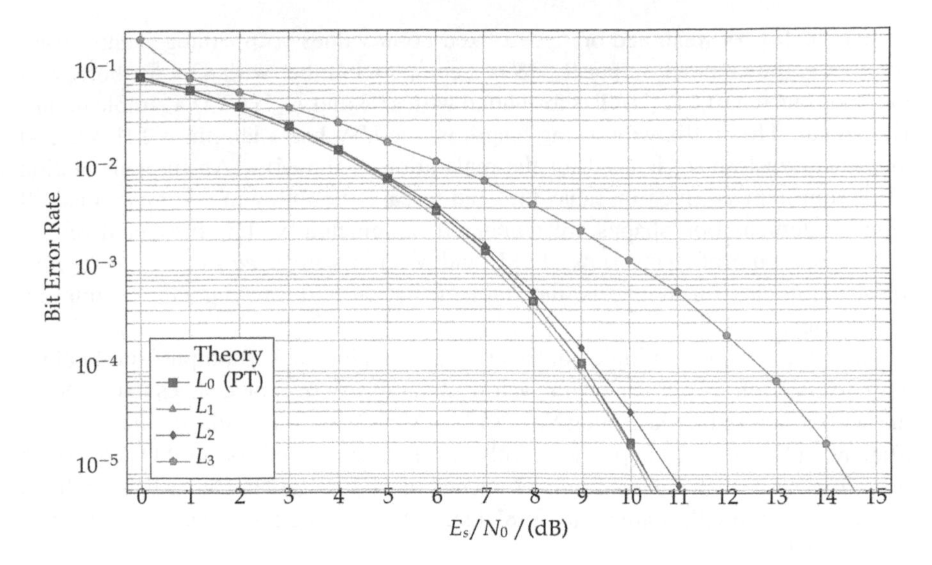

FIGURE 15.7 BER for different levels of MF approximation. For comparison, the theory curve shows the calculated BER performance for SOQPSK.

to the theoretical optimum (e.g., [9]). Additionally, Table 15.2 contains the E_b/N_0 performance losses at fixed BER values for all approximation levels, showing almost the same losses of L_1 and L_2. Reduction level L_1 almost perfectly matches the curve of L_0, showing that the level of approximation is reasonable. The curve of the highest degree of simplification, L_3, shows the largest offset and at $E_b/N_0 = 0$ dB, there is a discontinuity in the BER plot resulting from the fact that the MF outputs are not precise enough to keep the timing loop in a locked state.

15.2.2.3 Demodulator Implementation Results

To showcase efficiency enhancements achievable with the proposed demodulator architecture, different proof-of-concept FPGA designs have been implemented on two target platforms (Xilinx Virtex 5 XC5LX110T and Z-7020 SoC XC7Z020).

For the results in this section, the complete demodulator has been described in VHDL and implemented with the design suites ISE 14.7 for the Virtex 5 and Vivado 14.4 for the Z-7020 device. Some important parameters are set as follows. With the

TABLE 15.2
Losses for MF Approximation Levels at Fixed BER

BER	ΔL_0/dB	ΔL_1/dB	ΔL_2/dB	ΔL_3/dB
10^{-4}	0.15	0.15	0.41	3.84
10^{-5}	0.11	0.10	0.61	4.13

Note: Δ is the absolute value of the difference to the theoretical value.

format {u,s}i.f for *u*nsigned or *s*igned fixed point values with *i* integer and *f* fractional bits, the channel output R_OS is a complex number with s3.9 for both, real and imaginary parts. L_HAT is formatted with s3.3 and A_HAT is a simple binary u1.0 signal. The trellis windowing length is, the UW has a length of 121 bits and simplification level L_1 is used for the realization of the MFs. An implementation result summary is given in Table 15.3 and to ease the following comparisons, all eight implementation shapes are numbered incrementally. The presented results are the outcome of many runs of ISE and Vivado with *area* and *speed optimization* strategies for synthesis and implementation and varying target sampling frequencies.

The table includes multiple implementation runs for the soft output Viterbi algorithm (SOVA) on Z-7020: one with the lowest number of occupied slices (No. 5: 857), one with a maximum data rate (No. 7: 12.70 Mbps) and one with the highest slice efficiency (No. 6: 14.20 kbps/slice). Only two lines are listed for the SOV Aversion on Virtex 5, because the maximum efficiency of 11.40 kbps/slice goes hand in hand with either the smallest area (No. 3: 879 slices) or the highest throughput of 10.40 Mbps (No.4).

As a reference, the most efficient SOVA architecture from [19] is listed with No. 1, which supports time and phase correction. A hypothetical combination of this one and burst mode architecture [29] adds frame and frequency synchronization (No. 2). No. 3 and No. 4 show a slice hardware (HW) efficiency gain of more than 93% and register HW efficiency improvements of more than four compared with design No. 1, and at the same time synchronization functionalities are improved. When No. 2 is considered, slice efficiencies of this work's implementations are almost an order of magnitude higher, while supporting both burst and continuous modes. This reveals, that it is worth having a closer look into the really required frequency compensation range. If a moderate correction is sufficient, the proposed architecture can help to gain roughly 10 times slice HW efficiency. Generally, the efficiency increase from Virtex 5 to Z-7020 is based on the advanced capabilities of a newer PL. From an architecture point of view, the slices in Virtex 5 [30] and in the PL of the Zynq family [31] are very similar, which both have four lookup tables (LUTs). Only the number of storage elements (flip-flops) is different, four in the case of Virtex 5 and eight in the Zynq slices. In the end, No. 8 reveals that additional 30% efficiency can be gained compared with No. 6, in which hard output values are sufficient.

A real multiplication in the hardware description can be mapped to one digital signal processor (DSP) slice with an included hardware multiplier. All designs from this work occupy only half of the resources of No. 1 and less than a third of the resources of No. 2. This result is of high importance as DSP slices are relatively large on silicon and their number is limited.

To get a more fine-grained view, Table 15.4 lists the implementation results of No. 6 for all units in the demodulator. It reveals that 8 real multiplications are used in the parallel interpolator and 12 for phase and frequency shift of 3 complex values in MF_PFC. As expected, the largest amount of slices and registers is occupied by SOVA because of the implementation of the exact so-called *register exchange method.* Table 15.5 lists the resource usage of the demodulator.

TABLE 15.3
Implementation Results of Different SOQPSK Demodulator Architectures Together with Hardware Utilization and Efficiency Numbers

Architecture	Synchronization	No.	FPGA	DSP Slices	Max. Data Rate/(Mbps)	Slices (Utilization)	Slice HW Efficiency/ (kbps/slice)	Slice Registers	Register HW Efficiency/ (kbps/reg)	Δ/dB
[19]	Time, phase	1	Virtex 5	40[a]	8.8	1488 (8.6%)	5.9	3764	2.3	0.30[b]
[19, 29]	Time, phase, frequency, frame	2	Virtex 5	61[a]	8.8	7271 (42.1%)	1.2			0.95[c]
		3	Virtex 5	20	10.1	879 (5.1%)	11.4	1077	9.3	0.35
	Time	4	Virtex 5	20	10.4	913 (5.3%)	1.4	1076	9.7	0.35
This	Phase	5	Z-7020	20	9.7	857 (6.4%)	11.4	1207	8.1	0.35
work	Frequency[c]	6	Z-7020	20	12.3	868 (6.5%)	14.2	1207	10.2	0.35
(SOVA)	Frame	7	Z-7020	20	12.7	990 (7.4%)	12.9	1214	10.5	0.35
This work (HOVA)	Time, phase Frequency[c] Frame	8	Z-7020	20	11.8	640 (4.8%)	18.4	1093	10.8	0.35

[a] Assuming one DSP Slice per real multiplication in the demodulator.
[b] Estimated values from figures of [19, 29].
[c] Moderate.
Note: Δ is the absolute value of the implementation loss at a BER of 10^{-4}.

TABLE 15.4
Zynq 7020 Utilization Results of Fully Coherent Demodulator Architecture and Its Processing Units Achieved with Xilinx Vivado 2014.4

Unit	DPS Slices	Slice Registers	Slices
Demodulator total	20	1207	868
Pileup	0 (0%)	125 (10%)	54 (6%)
Interpolator	8 (40%)	134 (11%)	39 (4%)
MF_PFC	12 (60%)	108 (9%)	136 (16%)
SOVA	0 (0%)	253 (21%)	346 (40%)
PFED_PFE	0 (0%)	124 (10%)	117 (13%)
TED_TE	0 (0%)	162 (13%)	119 (14%)
SOF_PAR	0 (0%)	191 (16%)	203 (23%)

Note: The percentages in brackets indicate the relations to the total demodulator numbers.

TABLE 15.5
FPGA Resource Usage Results for SOPQSK demodulator

Resource	Used (Total)	Used (%)
DSP slices	20	9.09
Slice registers	1207	1.13
Slices	868	6.53

Note: The third column is the percentage of the total resources in the Z-7020.

In this context, the notation $\tilde{x}$ of a variable x stands for a parameter that is varied while using the configuration set. The waveform parameters influence the implementation in the PL part of the Zynq SoC, whereas the transceiver parameters are related to the configuration of the AD9361SoC.

15.3 AERONAUTICAL DATA LINK PERFORMANCE EVALUATION

The achievable performance, robustness and efficiency of the implemented system is shown in this section. The realistic scenarios and conditions for these analyses are derived from the results of the air-to-ground (AtG) channel investigations in [32].

15.3.1 Gross Throughput

To reveal the maximum gross throughput of the communication system, configuration No. 1 from Table 15.6 was used, f_c was set to multiple values from 1–6 GHz

TABLE 15.6
Parameter Sets of Several Data Link Example Configurations

Configuration	No. 1	No. 2	No. 3
Waveform			
L_{UW}	121	121	121
L_{info}	1944	1944	19440
L_{tail}	100	100	10
$\|k_{\tau}\|$	2^{-9}	2^{-9}	2^{-9}
$\|k_{\varphi}\|$	2^{-7}	k_{φ}	2^{-7}
μ_{SoF}	24	24	24
FEC	None	none	LDPC(1944, $\frac{1}{2}$)
Transceiver			
f_s	$\tilde{f}_s$	30 MHz	40 MHz
f_c	$\tilde{f}_c$	5.00 GHz	5.03 GHz
B_{RX}	15 MHz	8 MHz	12 MHz
B_{TX}	15 MHz	8 MHz	12 MHz
AGC mode	Manual($\tilde{G}_{agc}$)	Manual (31 dB)	Fast-attack

(including 5.03 GHz and 5.91 GHz, the potential lower and upper bound of the UAV C-band section) and f_s was increased until the BER became greater than zero. To perform this measurement, the board was used in a loop, where one transceiver output is directly connected to one input by a cable, as this avoids the influence of unwanted disturbances, such as clock imprecisions and interference. The final results for the maximum throughput are between $f_s = 54.40$ MHz at $f_c = 1$ GHz and $f_s = 54.00$ MHz at $f_c = 6$ GHz (Table 15.7). With these sample clock frequencies, more than 10^9 net information bits were measured without a single bit error. This translates into the gross throughputs of $f_s/N_{OSR} = 13.60$ and 13.50 Mbps, respectively. From the different results it can be concluded that the transceiver plays only a subordinated role in reaching these maximum throughputs. For higher carrier frequencies, the throughput is slightly deteriorated. In general, the outcome is higher than the synthesis results

TABLE 15.7
Maximum Gross Throughput for Several Example Configurations with $N_{OSR} = 4$

f_c/GHz	G_{agc}/dB	$f_{s,max}$/MHz	Gross Bit Rate/Mbps
1.00	14	54.40	13.60
2.00	15	54.40	13.60
5.03	31	54.20	13.55
5.91	31	54.20	13.55
6.00	48	54.00	13.50

for different platforms in Table 15.3 and it is quite close to the maximum achievable sample frequency of the AD9361 transceiver, which is 61.44 MHz.

15.3.2 Frequency Recovery Capability

Due to the capabilities of the data link demonstrator, it is possible to measure the limits of the frequency recovery algorithm that was derived in Section 15.1.1 and has been used for the implementation in Section 15.2. In avoidance of unnecessary hardware complexity overhead, the selected idea for the implementation is a closed-loop synchronization mechanism capable of estimating and correcting frequency deviations in a satisfactory range. Some realizations in literature can handle larger offsets, but at the cost of high hardware resource demands.

Configuration No. 2 from Table 15.6 was set to get a feeling of the steady-state synchronization capability of the real system. Like in Section 15.3.1, the demonstrator was used with a closed TX-RX-loop. The AD9361 transceiver allows the selection of the frequencies for transmit and receive local oscillators independently, which helps to perform the frequency offset measures. Its C-band results are shown in Figure 15.8 for three different values of the linear loop filter factor k_φ. The BER is drawn over the relative value of frequency shift f_d and sample rate f_s.

Aiming for an arbitrary BER of less than 10^{-6} the receiver can handle f_d/f_s relations of roughly 5.83 10^{-4}, 9.78 10^{-4} and 1.66 10^{-3} for $k_\varphi = 2^{-7}$, $k_\varphi= 2^{-6}$ and $k_\varphi= 2^{-5}$, respectively. The interpretation of these values for the given AtG communication scenario is the most interesting. Some examples for different sampling frequencies are shown in Table 15.8. Obviously, the maximum aircraft velocity increases with rising sample rates. However, even for the lowest f_s value that is listed, relative velocities

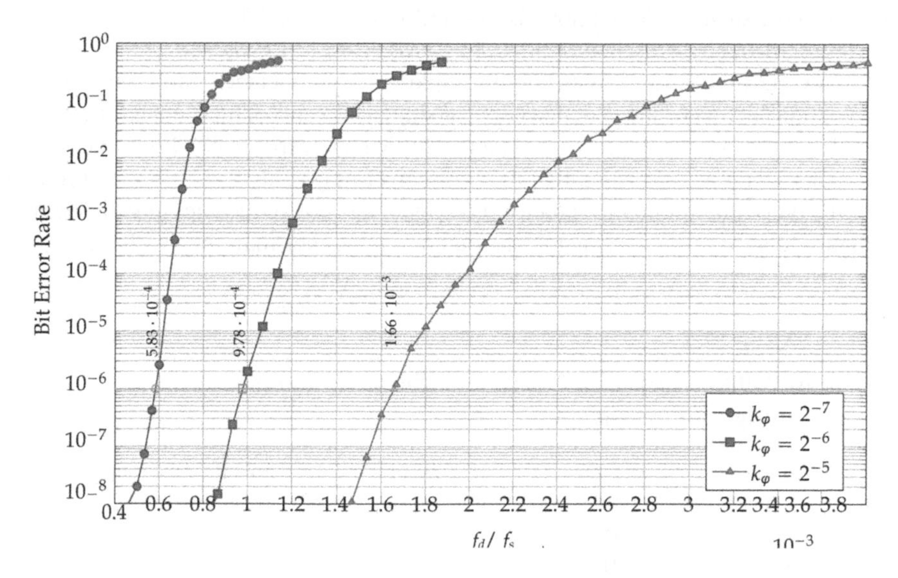

FIGURE 15.8 Frequency correction capability showing BER over frequency shift fdin relation to sampling frequency f_s for multiple values of the frequency and phase correction linear loop filter coefficient k_φ. Configuration No. 2 from Table 15.6 was used for this measurement.

TABLE 15.8
Maximum Frequency Offsets and Corresponding Aircraft Velocities for Several Example Configurations Assuming the C-Band Carrier Frequency $f_c = 5.09$ GHz and an Arbitrarily Chosen Acceptable BER of 10^{-6}

f_c/MHz	$k\varphi$	$f_{s,max}$/MHz (at BER = 10^{-6})	$v_{max/(m/s)}$ (at $f_c = 5.09$ GHz)
	2^{-7}	2.33	137
4	2^{-6}	3.91	231
	2^{-5}	6.64	391
	2^{-7}	17.49	1031
30	2^{-6}	29.34	1729
	2^{-5}	49.80	2935
	2^{-7}	29.73	1752
51	2^{-6}	49.88	2939
	2^{-5}	84.66	4989

of up to 137 m/s are supported. If faster airplanes are equipped with the data link, f_s and the filter coefficient must be adapted accordingly. In summary, these examples show that the presented frequency synchronization mechanism is able to handle the shifts of airborne-typical relative velocities between transmitter and receiver. Again, emphasis is put on the fact that no fast Fourier transform (FFT) has been required for the implementation, which is a key enabler to achieving the competitive efficiency results in Table 15.3.

15.3.3 Scenario Performance

To complete the functional validation on the basis of the scenario characteristics presented in [32], the data link prototype was subject to a performance investigation with combinations of K-factor and delay spread values. Transceiver configuration No. 3 from Table 15.6 was used with a superimposed AWGN of $E_b/N_0 = 10$ dB and also fading (with the option fine delay of 30 MHz) that originates from the K and σ_τ columns in Table 15.9. A two-path configuration was chosen as it represents the vast majority of all relevant channel profiles. The first path was not modified and the second has a delay of σ_τ and attenuation of K.

Table 15.9 shows the achieved BER results for three different K-value and delay spread combinations for many UAV-related scenarios. A coarse clustering has led to the given upper limits.

Supposing an arbitrary threshold of 1 10^{-6}, all table entries with a poor error performance are highlighted. As a general observation, the data link seems more sensitive to lower K-factors than to higher delay spreads. A few performance outliers can be spotted only for parking and taxiing channel conditions. In total, this table reveals that this work's communication system is able to establish a robust link for almost all of the investigated UAV operation scenarios.

TABLE 15.9
Data Link Demonstrator Aircraft to Ground Scenario Performance

Scenario	K_{mean}/dB	K_{min}/dB	στ,mean/ns	στ,max/ns	BER @ (*K*mean,στ,mean)	BER @ (*K*mean,στ,max)	BER @ (*K*min,στ,mean)
November 18							
North runway	19.63	4.55	23	1053	$<1\cdot10^{-8}$	$<5\cdot10^{-7}$	$<5\cdot10^{-5}$
North taxiway	20.10	6.02	19	1086	$<1\cdot10^{-8}$	$<5\cdot10^{-7}$	$<1\cdot10^{-5}$
South runway	23.67	9.21	16	235	$<1\cdot10^{-8}$	$<5\cdot10^{-8}$	$<5\cdot10^{-7}$
South taxiway	17.28	2.42	31	1213	$<1\cdot10^{-8}$	$<1\cdot10^{-6}$	$<5\cdot10^{-4}$
November 23							
Parking, taxiing,	11.77	7.42	133	1789	$<1\cdot10^{-6}$	$<1\cdot10^{-4}$	$<5\cdot10^{-3}$
take-off	18.05	6.53	42	2184	$<5\cdot10^{-8}$	$<5\cdot10^{-7}$	$<1\cdot10^{-6}$
Flight pattern	28.04	12.62	7	64	$<5\cdot10^{-8}$	$<5\cdot10^{-8}$	$<1\cdot10^{-7}$
	23.75	11.66	4	52	$<5\cdot10^{-8}$	$<5\cdot10^{-8}$	$<1\cdot10^{-7}$
Low-approach	28.24	11.66	7	52	$<5\cdot10^{-8}$	$<5\cdot10^{-8}$	$<1\cdot10^{-7}$
Landing **literature**	23.23	10.55	8	58	$<5\cdot10^{-8}$	$<5\cdot10^{-8}$	$<1\cdot10^{-7}$
[33]	29.60	21.80	11	109	$<5\cdot10^{-8}$	$<1\cdot10^{-8}$	$<5\cdot10^{-8}$
Near-urban							
RX1	29.40	23.10	10	177	$<5\cdot10^{-8}$	$<5\cdot10^{-8}$	$<5\cdot10^{-8}$
[34]							
Mountainous							
Hilly (Latrobe) [34]	28.80	22.20	18	371	$<5\cdot10^{-8}$	$<1\cdot10^{-7}$	$<5\cdot10^{-8}$
Hilly (Palmdale) [34]	29.80	23.60	19	1044	$<5\cdot10^{-8}$	$<5\cdot10^{-8}$	$<5\cdot10^{-8}$
Over sea [35]	31.30	11.10	10	365	$<5\cdot10^{-8}$	$<5\cdot10^{-8}$	$<5\cdot10^{-8}$
Over fresh water [35]	27.30	12.40	10	73	$<5\cdot10^{-8}$	$<5\cdot10^{-8}$	$<5\cdot10^{-8}$

Note: BERs larger than $1\cdot10^{-6}$ are in bold.

15.3.4 Link Budget

A common measure to characterize the operational capability of a wireless communication systems is its link budget. With transmit/receive power $P_{TX/RX}$, transmit/receive antenna gain $G_{TX/RX}$, transmit/receive chain loss $L_{TX/RX}$, coding gain G_C and receiver sensitivity S_{RX}, the acceptable free space path loss is

$$LFS = PTX + GTX - LTX - LRX + GRX + GC - SRX$$

With the single-stage prototype C-band front end, measurements have led to

$$\begin{aligned} \text{LFS} &= 13\text{ dBm} + 32.30\text{ dBi} - 0.50\text{ dB} - 0.50\text{ dB} + 3\text{ dBi} + 6.67\text{ dB} + 85\text{ dBm} \\ &= 138.97\text{ dB} \end{aligned}$$

where the chain losses are estimated with $L_{TX/RX} = 0.50$ dB and $G_C = 6.67$ dB (a measured value of the system under test). For the aircraft and the ground station (GS), a hemispherical antenna with 3 dBi and a GS C-band antenna gain of 32.20 dBi are assumed. To measure S_{RX}, a chain of two data link transceivers was used, one acting as the transmitter and the other as the receiver with disabled transmission to avoid crosstalk. In between, variable attenuators helped to determine the minimum sensitivity value with which uncoded demodulation was still possible.

Using the path loss outcome with $f = 5091$ MHz, the maximum link range of the first prototype can be determined to roughly 41.60 km. For a future version of the front end, a two-stage amplifier design is envisaged with a total output power of PTX = 27 dBm. Based on that, the possible UAV to GS distance increases to approximately 209 km.

15.4 CONCLUSIONS

Based on a multifunctional frame design, supporting burst and continuous mode, the derivation of a DD synchronization method for SOQPSK modulation to correct timing, phase and frequency offsets is shown. Together with the proposed combination of SoF detection and PAR with nested Barker UWs, a fully coherent demodulator algorithm is presented, which is highly suitable for low-complexity implementations. Despite its simplicity, the novel SoF detector shows a robust performance even for low signal-to-noise ratios (SNRs) and random timing, phase and frequency errors.

A novel architecture implementation is presented for a fully coherent SOQPSK demodulator supporting burst and continuous mode. Aiming at highest hardware efficiency, a coarse-grained pipeline architecture has been designed with so-called IRPUs, combining the outcome of several complexity reduction evaluations, such as the combination of SoF detection and PAR. MFs are reported to be a bottleneck in CPM receivers, which is why the proposed and admittedly simple measure to remove all multiplications in the MFs results in such a noticeable efficiency gain. Computer simulations of the fixed-point demodulator with all simplifications show that the BER performance is fairly close to its optimum, and PL implementation results show superior hardware efficiencies (at least 93%) compared with existing implementations. Future single-carrier data link systems based on SOQPSK, or CPM in general, can benefit from the provided set of simplification methods resulting in a feature fusion of spectrum, hardware and energy efficiency.

Within the present chapter, several performance evaluations of the final airborne communication systems are reported. With a number of different waveform and transceiver configurations, the included ability for a flexible parametrization is highlighted, which is an important prerequisite for the proposition of a scenario-adaptive waveform.

Starting with the achievable gross throughput of more than 13.50 Mbps, the data link shows its applicability to payload data and telemetry use cases, e.g., the Doppler shift robustness of the novel and highly hardware efficient frequency correction mechanism has revealed more than sufficient performance to even handle the consequences of very fast aircraft.

To round up the scenario performance studies, one example configuration of the link prototype was tested under the influence of AWGN and fading with parameters from the channel measurement campaigns. With only very few outliers, the obtained BERs verify the communication system's robustness, even while being exposed to the harsh effects of the desired AtG scenarios.

Finally, a link budget analysis shows a maximum range of 41.60 km, using the first prototype of a matching C-band front end. With an updated version, which is expected to provide an output power of 27 dBm, a more than 200-km link distance will become feasible. The power-efficient design of such a front end, which does not make demands regarding linearity, is an important engineering topic for the future to beneficially contribute to the efficiency of the overall communication system.

REFERENCES

1. D. Rieth, "Efficient and robust data link for wireless aeronautical communications," Dissertation, RWTH Aachen University, Aachen, 2018.
2. Telemetry Group, "Telemetry Standards IRIG Standard 106-04 Part I," 2004.
3. E. Perrins and B. Kumaraswamy, "Decision feedback detectors for SOQPSK," *IEEE Transactions on Communications*, vol. 57, no. 8, pp. 2359–2368, 2009.
4. C. Sundberg, "Continuous phase modulation," *IEEE Communications Magazine*, no. 4, pp. 25–38, 1986.
5. A. Svensson, C. Sundberg, and T. Aulin, "A class of reduced-complexity Viterbi detectors for partial response continuous phase modulation," *IEEE Transactions on Communications*, vol. 32, no. 10, pp. 1079–1087, 1984.
6. T. Nelson, E. Perrins, and M. Rice, "Near optimal common detection techniques for shaped offset QPSK and Feher's QPSK," *IEEE Transactions on Communications*, vol. 56, no. 5, pp. 724–735, 2008.
7. E. Perrins and M. Rice, "Simple detectors for shaped-offset QPSK using the PAM decomposition," *GLOBECOM '05. IEEE Global Telecommunications Conference, 2005*, IEEE, vol. 1, pp. 408–412, 2005.
8. E. Perrins and M. Rice, "PAM representation of ternary CPM," *IEEE Transactions on Communications*, vol. 56, no. 12, pp. 2020–2024, 2008.
9. E. Perrins, T. Nelson, and M. Rice, "Coded FQPSK and SOQPSK with iterative detection," *Military Communications Conference (MILCOM)*, pp. 2–8, 2005.
10. E. Perrins and M. Rice, "Reduced-complexity approach to iterative detection of coded SOQPSK," *IEEE Transactions on Communications*, vol. 55, no. 7, pp. 1354–1362, 2007.

11. B. Kumaraswamy and E. Perrins, "Simplified 2-state detectors for SOQPSK," *MILCOM 2007 - IEEE Military Communications Conference*, IEEE, pp. 1–7, 2007.
12. P. Chandran and E. Perrins, "Decision-directed symbol timing recovery for SOQPSK," *IEEE Transactions on Aerospace and Electronic Systems*, pp. 781–789, 2009.
13. P. Chandran and E. Perrins, "Decision directed timing recovery for SOQPSK," *Military Communications Conference (MILCOM)*, 2007.
14. M. Morelli, U. Mengali, and G. Vitetta, "Joint phase and timing recovery with CPM signals," *IEEE Transactions on Communications*, vol. 45, no. 7, pp. 867–876, 1997.
15. Z. Y. Choi and Y. H. Lee, "Frame synchronization in the presence of frequency offset," *IEEE Transactions on Communications*, vol. 50, no. 7, pp. 1062–1065, 2002.
16. E. Hosseini and E. Perrins, "Timing, carrier, and frame synchronization of burst-mode CPM," *IEEE Transactions on Communications*, vol. 61, no. 12, pp. 5125–5138, 2013.
17. A. N. D'Andrea, A. Ginesi, and U. Mengali, "Digital carrier frequency estimation for multilevel CPM signals," *Proceedings IEEE International Conference on Communications ICC'95*, vol. 2, no. 2, pp. 1041–1045, 1995.
18. A. N. D'Andrea, A. Ginesi, and U. Mengali, "Frequency detectors for CPM signals," *IEEE Transactions on Communications*, vol. 43, no. 2, pp. 1828–1837, 1995.
19. E. Hosseini and E. Perrins, "FPGA implementation of a coherent SOQPSK-TG demodulator," *2011 - MILCOM 2011 Military Communications Conference*, pp. 471–476, 2011.
20. M. Rice, *Digital Communications*, Upper Saddle, NJ: Prentice Hall, 2009.
21. A. Nowbakht and J. Bergmans, "Design of optimum sync and detection patterns for frame synchronisation," *Electronics Letters*, vol. 37, no. 24, pp. 1437–1439, 2004.
22. G. P. E. R. Zanabria, "A hardware implementation of a coherent SOQPSK-TG demodulator for FEC applications," Ph.D. dissertation, University of Kansas, 2011.
23. M. Grayson and M. Darnell, "Optimum synchronization preamble design," *Electronics Letters*, no. 1, pp. 36–38, 1991.
24. N. Levanon, "Cross-correlation of long binary signals with longer mismatched filters," *IEEE Proceedings - Radar, Sonar and Navigation*, vol. 152, no. 6, p. 377, 2005.
25. R. Barker, "Group synchronizing of binary digital systems." In: W. Jackson, Ed. *Communication Theory*. New York: Academic Press, pp. 273–287, 1953.
26. AVNET, "*PicoZed SDRZ7035/AD9361 SOM User Guide Version 1.5*," pp. 1–63, 2016.
27. XILINX, "*Zynq-7000 All Programmable SoC Overview*," pp. 1–23, 2016.
28. Analog Devices, "*AD9361 Reference Manual UG-570*," pp. 1–128, 2015.
29. E. Hosseini and E. Perrins, "FPGA Implementation of burst-mode synchronization for SOQPSK-TG," *Proceedings of the 2014 International Telemetering Conference*, pp. 1–9, 2014.
30. XILINX, "*Virtex-5 FPGA User Guide*," pp. 1–385, 2012.
31. XILINX, "*7Series FPGAs Configurable Logic Block – User Guide*," pp. 1–74, 2016.
32. D. Rieth, C. Heller, and G. Ascheid, "Aircraft to ground-station c-band channel—small airport scenario," *IEEE Transactions on Vehicular Technology*, vol. 68, no. 5, pp. 4306–4315, 2019.
33. D. W. Matolak and R. Sun, "Air-ground channel characterization for unmanned aircraft systems: the near-urban environment," *IEEE Transactions on Vehicular Technology*, vol. 66, no. 8, pp. 6607–6618, 2017.
34. R. Sun and D. Matolak, "Air-ground channel characterization for unmanned aircraft systems—Part II: Hilly & mountainous settings," *IEEE Transactions on Vehicular Technology*, vol. 66, no. 3, pp. 1913–1925, 2016.
35. D. W. Matolak and R. Sun, "Air-ground channel characterization for unmanned aircraft systems—Part I: Methods, measurements, and models for over-water settings," *IEEE Transactions on Vehicular Technology*, vol. 68, no. 1, pp. 26–44, 2016.

16 An Analysis of the SwarmOS Framework for UAV Swarms: Usage in Mission Plan for Rural Applications

Pablo César C. Ccori
University of São Paulo, Brazil (pcalcina@ime.usp.br)

Flávio Soares Corrêa da Silva
University of São Paulo, Brazil (fcs@ime.usp.br)

Laisa C. Costa de Biase
University of São Paulo, Brazil (laisa@lsi.usp.br)

Arthur Miyazaki
University of São Paulo, Brazil (arthurmiy@gmail.com)

Marcelo Knörich Zuffo
University of São Paulo, Brazil (mkzuffo@usp.br)

Paul Chung
Loughborough University, Loughborough, UK
(P.W.H.Chung@lboro.ac.uk)

CONTENTS

16.1 INTRODUCTION

The use of unmanned aerial vehicles (UAVs) for civil and military applications emerged as an alternative to traditional manned aircraft operations and brought some advantages such as lower costs, minimization of safety risks to personnel [1] and waypoint-mission planning [2]. Due to the growing interest in UAVs, some efforts developed technology in different areas such as surface reconnaissance, disaster assistance and borderline surveillance [3].

The operation of multiple drones to achieve a common purpose represents a challenging evolution of the individually controlled UAVs. The use of UAV swarms is especially advantageous in scenarios with time constraints and where missions could be performed more efficiently than with individual drones [2]. As stated in [4], the cooperation of multiple autonomous agents allows for better results compared with a similar scenario in which the participants do not cooperate. Thus, the concept of UAV swarms, in a wide manner, is not solely related to the control of multiple aircraft but also to the task cohesion of the group.

There are several advantages related to the use of UAV swarms. First, the redundant characteristics of the system, usually composed by a homogeneous group of UAVs, permits the execution of missions despite individual failures. The comparison of single and multi-UAV systems presented in [5] states that the latter has higher scalability and survivability with faster mission speed and lower costs. The gain in speed is mainly related to the task division, whereas the cost reduction is due to the use of smaller and simpler UAVs. In military applications, attacks using dispersed UAVs make the defense onerous and expensive while the attack itself remains cheap and efficient [6].

The advantages of UAV swarms come with a trade-off linked to the increase of control complexity. A higher complexity stimulates the search for solutions that transfer the control task from the user to the system [7]). The UAV swarm control, contrary to traditional single UAV, focuses on the behavior of the group without individual-level intervention [6], which may cause some difficulties for the user interface design. Despite technological advances, UAV flights are mostly manual or planned beforehand using waypoints in known areas [3]). The high level of dynamism offered by the environment and the interaction of UAV units to find a group solution for the tasks leads to nonlinear and unpredictable results [1]. Further challenges for UAV swarms include (1) the limited number of sensors in UAVs [7], (2) establishing and maintaining efficient communication between UAV units [5] and (3) the necessary adaptation to changes in environment and group formation with on-the-fly mission replanning [1].

Several solutions were developed over the years to overcome the challenges of UAV swarm control, including both centralized and decentralized models for decision-making. According to [4], centralized solutions optimize timing and task constraints demanding intensive computation and robust communication, whereas decentralized solutions have better robustness and adaptation to environment changes despite lower optimization and predictability. Due to the computational cost from data fusion and the number of individual decisions, distributed control with relaxed autonomy is increasingly targeted [8].

As previously stated, some level of autonomous control is required in decentralized systems. [3] gave an overview of autonomous control of UAVs, suggesting three levels of task division: (1) organizational (decision-making and highest intelligence), (2) coordination (intermediate level, interface between the other two levels) and (3) execution (direct UAV control with lower intelligence). The coordination level is basic for every UAV flight (even for manual flights during which human intelligence is used to replace the other two levels), and for that reason the implementation challenges in UAV swarms are usually related to the organizational and coordination level.

There are some algorithms used to manage UAV swarms, and many of them are inspired by nature, physics and even graph theory to keep the task cohesion. Some bio-inspired algorithms like ant systems, artificial bee colonies, particle swarm optimization and wolf search algorithms are briefly summarized by [7]. Each bio-inspired algorithm tries to imitate cooperative groups found in nature like groups of ants, bees, birds, fish and wolves. Another approach is the use of artificial physics as presented by [9], which brings an efficient mathematical analysis. Last, there are also some solutions based on consensus control, as described by [10].

The presented solutions for the control problem can be applied in different areas. In the military field, the most common usages include attack, defense and support functions [6]. Uses of UAV swarms in civilian applications include aerial power line inspections, image understanding and recognition, 3D reconstruction, crowdsensing and data transmission [11]. UAV groups are already commercially used for entertainment and light shows, although, a centralized architecture and pre-defined path are necessary to control aircraft, thus depending on a known position of the UAV [12]. Decentralized swarms also are being tested for performance and behavior under environment uncertainties. [13] presented a preliminary performance evaluation and test results of a nature-inspired swarm composed by 25 UAVs. In [14] another decentralized, multifunction swarm was field tested. For this multifunction swarm, the performance was measured assuming a collision-free operation for different functions such as search tasks and network maintenance. The work presented by [14] also shows a more natural way to interact with UAV swarms by focusing on the selection of an available function instead of the traditional micromanaging.

On the other hand, the SwarmOS [15] platform was introduced to leverage the creation of a decentralized network of heterogeneous IoT devices. The SwarmOS provides a middleware called SwarmBroker, which mediates the communication among devices. Additionally, the SwarmOS platform features a lightweight semantic service discovery mechanism, a framework for policy-based access control, an economic model based on price and trust and a blockchain-based mechanism for storing transactions and reputation. In this chapter, we explore the application of the SwarmOS platform to UAV swarms, including some potential novel applications

The rest of this chapter is structured as follows. In Section 16.2 we present the SwarmOS platform, its architecture and main components. In Section 16.3 we discuss opportunities for the application of the SwarmOS platform to UAV Swarms. Section 16.4 describes in detail three cases of UAV swarms in rural applications, and in Section 16.5 we discuss mission specifications, using a formal notation based on soft institutions. Finally, in Section 16.6 we summarize and discuss the chapter, including perspectives for future work.

16.2 THE SwarmOS PLATFORM

The term swarm was first coined by [16], who used it to name the whole sensory at the edge of the cloud, and identified the opportunity for the realization of cyber-physical and cyber-biological systems, immersive computing and augmented reality. Later work led to a more concrete definition of the Swarm, in particular, the architecture definition for the Swarm in the context of a larger project called TerraSwarm [17]. They also outlined a common framework for devices to communicate and share resources, called SwarmOS. The architecture of the SwarmOS framework was further developed [15], which complements the distributed storage system (data plane) with a module for sharing and management of resources (control plane).

The SwarmOS is an IoT platform for smart objects that creates a heterogeneous network of cooperating devices. Devices in the Swarm do not rely on the cloud for storage and processing, thus adopting an edge computing approach, with variable computing power and energy capabilities. In Figure 16.1 we illustrate the general structure of the Swarm network. In a parallel with swarms of bees, the SwarmOS platform is composed of specialized devices (bees) whose interaction contributes to solving a common problem. The Swarm network behaves like an organism and shows an organized behavior that results in an emergent collective intelligence.

16.2.1 Architecture

The SwarmOS is a service-oriented platform, where device functionalities are exposed through RESTful services. Swarm services are specialized in specific functionalities, and many services can perform similar or equivalent operations. The true potential of the Swarm resides in the composition of services, which dramatically extends the network capabilities. Additionally, the SwarmOS shares many key concepts with the microservices architecture style, such as the use of services as the main building block, loose coupling, high cohesion, decentralized governance, decentralized data management and evolutionary design [18].

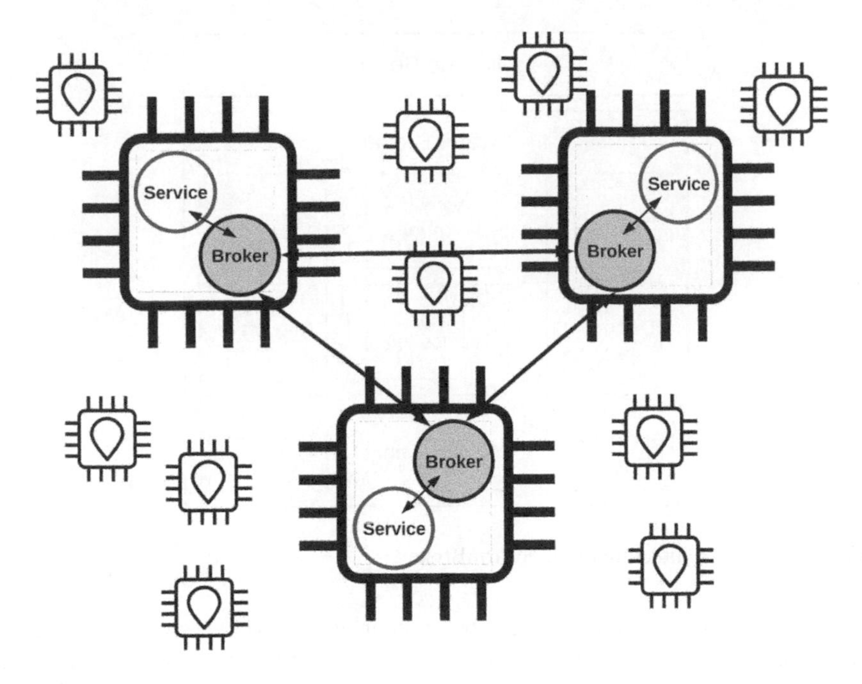

FIGURE 16.1 Structure of the Swarm network.

The connection among devices in the Swarm is established in real time, based on the availability of devices in the network. Thus, interaction occurs opportunistically, with no prior agreement. In response to external events, devices form groups to perform an action. Although those groups are transient, the success of each interaction is learned by the network and serves to build a reputation score for each device.

A central component of the Swarm network is the SwarmBroker, which mediates communication among services. All devices in the Swarm have a SwarmBroker installed, as illustrated in Figure 16.1.

16.2.1.1 The SwarmBroker

The SwarmOS platform is based on a lightweight middleware installed in every IoT device called SwarmBroker, which acts as a communication facilitator. Some functions provided by the SwarmBroker include service registry, semantic service discovery, access control policy enforcement, service contracting and blockchain-based reputation. Additionally, the SwarmBroker provides a transaction model that mediates contracts between service consumers and providers that are chosen by a combination of price and reputation. That interaction creates an economic model for resource sharing in the IoT.

We define two categories of services in the Swarm: platform service, which constitutes the core functionalities of the Swarm network, and application services, which are all other services that participate in the Swarm. Thus, the SwarmBroker can be seen as the collection of platform services. Figure 16.2 illustrates the landscape of platform services that constitute the Broker. As the Swarm network is

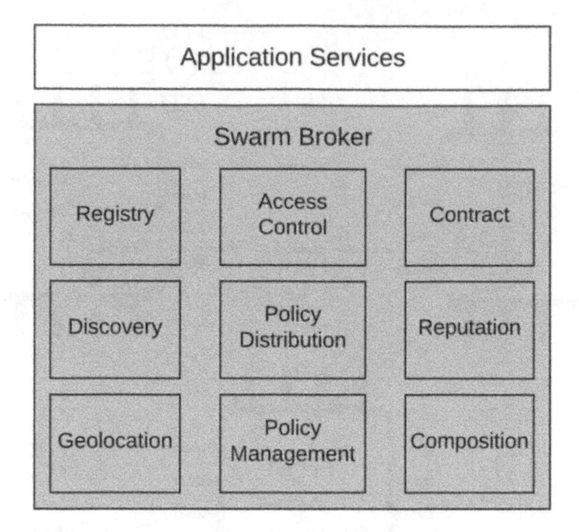

FIGURE 16.2 Architecture of the SwarmBroker.

composed of heterogeneous devices, the number of platform services varies across devices, according to their capabilities. Less powerful devices will provide fewer platform services. Currently, we have four implementations, C, Lua, Java, and Elixir programming languages. Further implementations are expected, to cover a wider range of devices.

16.2.2 The Minimum SwarmBroker

To overcome the challenge of heterogeneity in the IoT, a Minimum SwarmBroker contains the core features necessary for a device to be part of the Swarm network. That way, devices with resource-constrained resources can use an external software proxy to translate communication [19]. The basic requirements for Swarm participants are being discoverable and usable. Thus, a simplified implementation can provide support for requests through different protocols, such as HTTP and SSDP. Figure 16.3 illustrates the interaction between Minimum SwarmBroker and common Swarm Broker, focusing on the discovery process.

16.2.3 Semantic Swarm

The use of semantics in the IoT has increased over time in several areas such as data exchange, device interoperability, information integration, data access, resource discovery and logical reasoning from sensors information [20, 21].

One of the main applications of semantics in the Swarm is service discovery. Finding a suitable service to interact in the Swarm is a problem of major importance. The SwarmOS architecture uses a discovery mechanism for services based on semantic reasoning. A service requester searches for another service that matches the expected functionality. A non-semantic discovery mechanism would base the

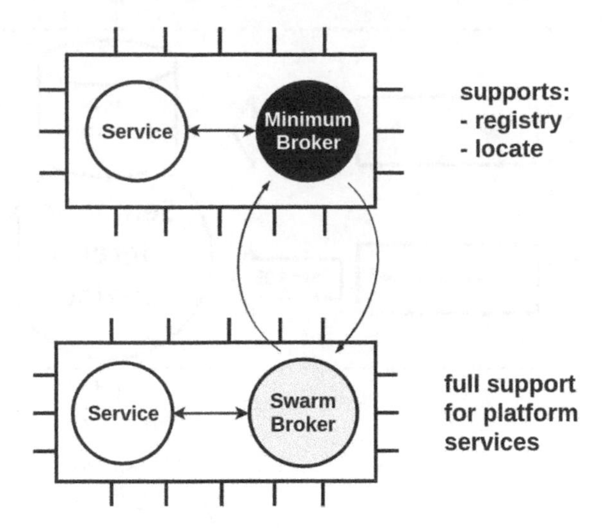

FIGURE 16.3 Minimum SwarmBroker.

search on string comparison, which poses severe limitations. In contrast, the semantic discovery mechanism of the Swarm searches by equivalent concepts in a knowledge base, thus increasing the flexibility of the search.

Figure 16.4 shows the semantic discovery process, which is divided into two stages. In the *registry stage*, services register in the semantic registry service. During the *discovery* stage, a service request is sent to the broker and that request

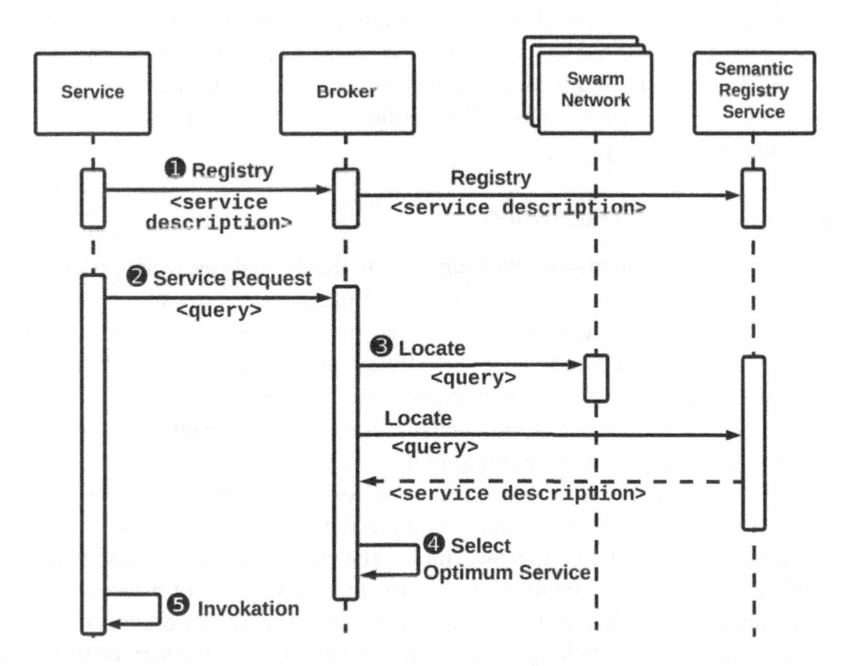

FIGURE 16.4 Process of semantic discovery in the SwarmOS platform.

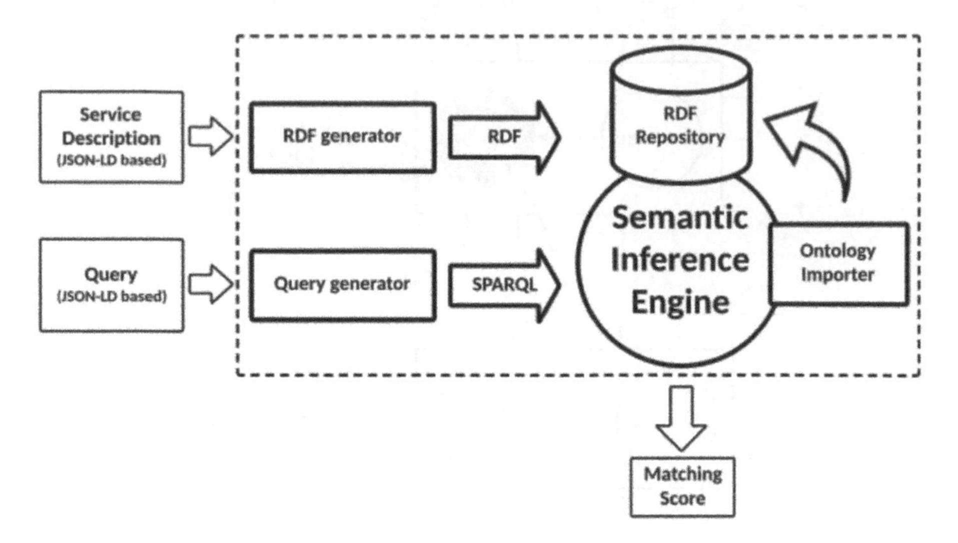

FIGURE 16.5 Architecture of the semantic registry service.

is forwarded to the semantic registry, which searches in its knowledge base. If a semantic registry is not available in the network, a syntactic search is still performed, guaranteeing an acceptable answer from the SwarmOS platform.

Figure 16.5 shows the architecture of the semantic registry, according to the two stages of the discovery process. This service receives a service description based on JSON for linked data (JSON-LD), which is translated to the resource description framework (RDF), a model used for the semantic description of Web resources [22], and later incorporated to the service knowledge base. In the discovery stage, a query based on JSON-LD format is translated into SPARQL query language and later applied to the knowledge base. The candidate services are ranked according to the relevance to the query.

16.2.4 Swarm Economic Model

The rich interaction among the participants of the Swarm network poses important challenges to guarantee the soundness and integrity of device interactions. As a result, [23] proposed an economic model for the SwarmOS that aims to regulate the transactions of services in the network. Relevant concepts are part of this model, such as trust, reputation, rewarding mechanism and billing.

The introduction of a rewarding mechanism in the SwarmOS platform is of major importance for the service ecosystem, as it leverages a concrete incentive for resource sharing. Devices can share their functionalities either in exchange for the use of other devices, or to accumulate credits in the SwarmCoin currency.

In the economic model of the SwarmOS, the Swarm Broker is responsible for linking the parties and for facilitating transactions. The price of a resource represents the number of credits necessary for the service provider to grant access to the service consumer. Credits belong to the owner and are used by the service to contract or purchase any service on behalf of its owner.

The SwarmOS economic model is based on the price of a service and the reputation of both service consumer and provider, hence, it is called the price-reputation model. A service provider defines the number of credits necessary to allow a third party to use it. On a service request, candidate providers are ranked by the lowest price according to the following formula:

$$P = \begin{cases} P_{min} + \dfrac{P_{max} - P_{min}}{T_{th}}\left(T_{th} - T_{pc}\right), & T_{pc} < T_{th} \\ P_{min}, & T_{pc} < T_{th} \end{cases}$$

During the transaction process, reputation points evaluate the success of the operation. The price-reputation transaction is the simplest transaction defined for the Swarm framework: consumers get the service by paying a number of credits settled by the provider, depending on their behavior, and they both get reputation points during the process.

Regarding security, the distributed and decentralized nature of the Swarm requires a new model for transactions. Traditional technologies, such as the public key infrastructure (PKI), are not suitable as they demand a centralized certificate authority (CA). Accordingly, the SwarmOS adopted the blockchain technology, used in Bitcoin cryptocurrency, to create a decentralized mechanism for trust in the economic model of the Swarm network [23].

The advances in an economic model for the IoT go in the same direction as a growing trend in the world economy, and this is called sharing economy. As in the physical world, the Swarm favors the digital sharing of resources over the acquisition of dedicated devices. Consequently, it produces a reduction of device consumption and better global use of resources.

16.2.5 Security and Access Control

The resource sharing vision of the Swarm can only be implemented if security is built into the system. This includes both the use of appropriate algorithms and protocols to protect exchanged messages and a flexible access control mechanism to govern which interactions are allowed. Additionally, managing the access control of a large number of devices becomes a significant challenge.

The access control mechanism of the SwarmOS platform is based on the attribute-based access control (ABAC) model. In ABAC, a subject request to perform operations on objects is granted or denied based on the attributes of the participants. The referred attributes can be related to the subject, the object, environment conditions and policies. The ABAC-based model created in the SwarmOS platform is called ABAC-them and focuses on combining simplicity and expressiveness. The main characteristics of the ABAC-them model include the following:

- *Enumerated policies*: Attribute enumeration allows the creation of policies that are easy to parse and to embed into small devices.

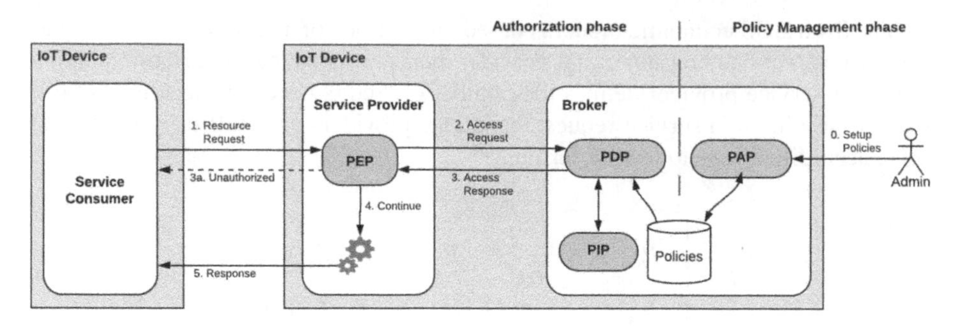

FIGURE 16.6 Architecture of the ABAC-them access control model for the SwarmOS platform.

- *Hierarchical attributes*: Allow the creation of high-level policies that are easier to write and understand. During execution time, low-level attributes present in access requests benefit from attribute hierarchies, which allow them to match with the high-level policies.
- *Typed attributes*: Provide a counterbalance to policies that can grow large when using enumeration, such as those involving numerical ranges.
- *Multiple attributes*: Very specifically, these features allow easy creation of conjunctions when using enumerated policies.

The ABAC-them model was implemented within an architecture based on the National Institute of Standards and Technology (NIST) recommendation for ABAC systems [24]. The architecture contemplates four main components. The *policy decision point* (PDP) evaluates policies that are managed through the *policy administration point* (PAP), while the *policy information point* (PIP) is responsible for gathering context and other attributes. The *policy enforcement point* (PEP) intercepts requests and verifies their permission with the PDP. Although the original NIST architecture proposes that PDP, PIP and PAP reside in an authorization server, the SwarmOS implementation puts all points inside the IoT device, enhancing its autonomy and security. Figure 16.6 illustrates the current architecture of the ABAC-them model in the SwarmOS [25].

16.3 SwarmOS FOR UAV SWARMS

In this section, we describe the opportunities and implications of using the SwarmOS framework to enrich the operation of UAV swarms. Three important aspects from the SwarmOS are useful for UAV cooperation: semantic communication, the Swarm economic model for trust and reward and the interaction with other devices beyond UAVs.

16.3.1 Semantic Communication of UAVs

Previous work showed the importance of using artificial cognitive mechanisms for controlling multiple UAV units, such as semantics. Uhrmann and Schulte [26] proposed a task-based guidance method for UAV that relies on authority delegation and sharing a common goal. The authors of that work claim an improvement of the overall mission performance, tactical advantages and moderate resource consumption of operator resources. Their work is based on a cognitive process aided by a knowledge base, which in turn stores information about the environment, supervisory control, mission, cooperation, task synthesis, task scheduling and role management.

Alirezaie et al. [27] proposed a path planner to reduce the planning time in highly constrained environments. They use an ontology to model the outdoor environment and propose a representation for path connections through a reusable ontology design pattern. They created the OntoCity ontology to model the path segments, relying on RDF data and GeoSPARQL.

16.3.1.1 Mission Plan for UAV

Two important activities related to the operation of UAVs are mission planning and mission control. Mission planning refers to the detailed design of the behavior of UAVs to complete a mission and reach a goal, including reactive behavior in case of unexpected or undesired events. Mission control refers to procedures to monitor a mission to ensure that a plan is performed as expected.

To prevent accidents, mission planning and mission control are highly regulated by appropriate organizations such as the national aviation authorities of specific countries.

To bid for certification it is essential to have a careful design and documentation of UAV specification, tools and mechanisms for mission planning and control and mission plans themselves.

In previous related work, a diagrammatic language was specified to design, verify and document mission plans, based on the concept of *soft institutions*, aiming specifically at hazard prevention [28].

We suggest that the same language be used to design mission plans for UAV swarms in such a way that specific scenarios can be selected and potential hazards in each scenario can be solved in advance.

16.3.1.2 Ontology for UAV

A necessary step to take advantage of semantics for UAV communication is the development of an ontology to model this domain.

Preece et al. [29] created a framework for sensor information processing that can be used in intelligence, surveillance and reconnaissance for mission planning. Barbieri et al. [30] explored a high-level minimum information framework for drone data. Lammerding [31] created an ontology that aims to represent the necessary information for an autonomous UAV mission.

Semantics in the SwarmOS platform is present in many aspects such as discovery, information exchange, automatic service execution and service composition. For all of these tasks, ontologies are necessary to model the platform and application domains. To achieve a successful integration between SwarmOS and UAV swarms, a mandatory step is to integrate existing UAV ontologies to the SwarmOS ontology to leverage logic inference a unified knowledge base.

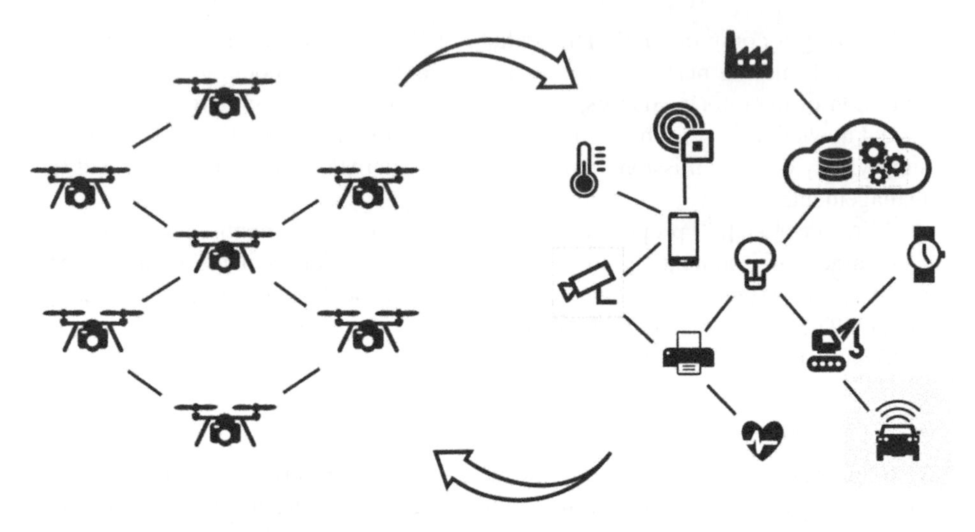

FIGURE 16.7 Cooperation between UAV swarms and IoT swarms.

16.3.2 SwarmOS Economy for Opportunistic UAV Recruiting

The SwarmOS economic model can be applied to a UAV swarm, particularly the retribution mechanism, as it acts as an incentive for UAV participation and cooperation in the network.

16.3.3 Beyond UAV Swarms: Cooperation With Other IoT Networks

Along with the cooperation of multiple UAV units to achieve a common task, a UAV swarm can benefit from the full SwarmOS network, creating an extensive cyber-physical system, with heterogeneous sensors and actuators. Through the implementation of a constrained version of the SwarmBroker middleware, UAVs can use other IoT services in the SwarmOS. Conversely, other SwarmOS participants can benefit from the functionalities provided by these UAV units.

To register in the SwarmOS network, any device only needs to communicate with a Broker and send a JSON-LD service description [19, 32]. Figure 16.7 shows the opportunity of cooperation between UAV swarms and IoT swarms from the SwarmOS platform. The seamless cooperation among these devices in the SwarmOS network represents a potential for relevant applications that benefit and complement mutually. Some of these scenarios are explored in Section 16.4.

16.4 USE CASES FOR RURAL APPLICATIONS

16.4.1 UAV Swarms for Livestock

UAVs have been successfully used to prevent human–wildlife conflicts in risk areas, namely, to drive away elephants in Tanzania [33], India [34] and

Thailand [35] to avoid conflicting situations among elephants and humans in urban areas. On the other hand, several human–wildlife conflicts grew in Brazil in the past decade involving wild species such as jaguars, pumas, wild boars and capybaras [36].

To solve these human–animal conflicts in rural areas of Brazil, we propose the use of the SwarmOS platform to communicate UAV units in coordination with other IoT sensors. In this configuration, some sensors are needed to detect the presence of the target wild animals, such as microphones in the forest. After the wild animal is detected, a nearby UAV unit is recruited to approach the target animal and emit a specific sound that drives away the animal. UAVs must advertise their capabilities for emitting those sounds in the network, and this information will be used to recruit them.

16.4.2 UAV Swarms for Farming

Extensive efforts have been developed toward the use of drones for farming applications. Examples include the evaluation of UAV-acquired images for precision farming of vineyard and tomato crops [37], crop and weed classification [38] and automatic UAV data interpretation for precision farming [39].

As farming is one of the most important economic activities in Brazil, it poses some challenging demands. Farming lands extend over wide areas and need several UAV units to cover the entire area, which favors UAV swarm solutions. Current efforts in using UAVs include the image-based detection of crop diseases.

Additionally, we observe a strong potential for the use of UAV swarms of medium and large sizes for lightweight irrigation of varied substances, which can cover an entire area in an organized way and shorter time.

16.5 A MISSION SPECIFICATION

Due to the critical aspect of UAV swarm control, we include a mechanism to switch from decentralized to centralized and manual control of vehicles in the model to respond to emergencies in the mission.

Using the diagrammatic language, which was previously designed to specify mission plans as soft institutions [28], Figure 16.8 presents a high-level diagram depicting how the switch will behave. The syntax and semantics of that language are not presented in detail here for the sake of conciseness. Instead, we walk through the diagram in Figure 16.2 and explain, in general terms, how each element in the diagram can be interpreted.

This diagram depicts a snippet of an interaction protocol between a UAV (*UAV001*) and a control tower (*ATC001*). Any item in a list of *emergency pre-conditions* can trigger the *control handoff request*, which then starts a sequence of actions. In this simplified example, the sequence contains only one action, which is a request for the control tower to enter the *control handoff accept* state. In this state, the control tower

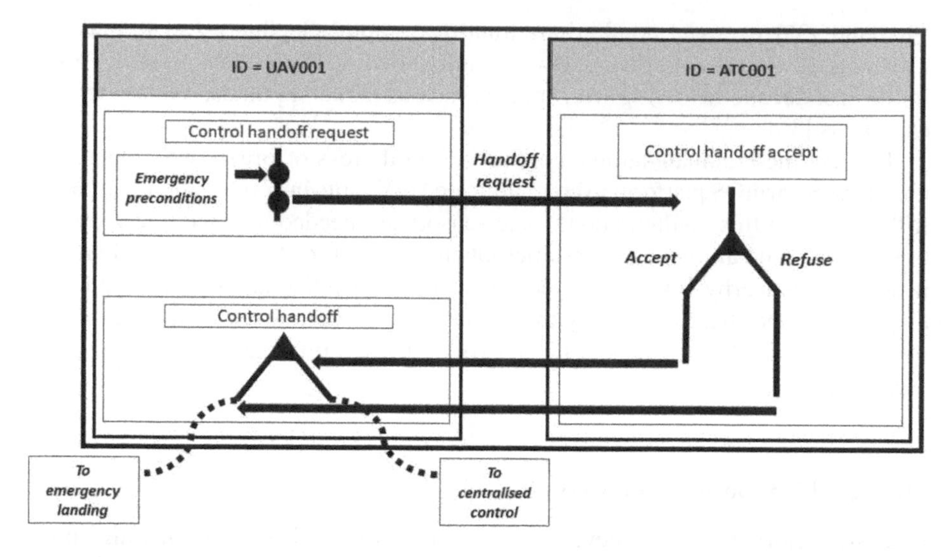

FIGURE 16.8 Graphical specification of the details of a mission plan.

can accept the request and assess it. Depending on the result of the assessment, two outcomes can occur:

1. If the control tower accepts the handoff, it sends a message back to the UAV, requesting it to move to the *control handoff* state. In this case, the UAV will take actions to perform an internal move to the *centralized control* state.
2. If the control tower refuses the handoff, it sends the corresponding message back to the UAV, requesting it to move to the *control handoff* state. In this case, however, because the content of the request message is a refusal, the UAV must move to the *emergency landing* state.

The point of formal specification of mission plans using a language such as the one depicted in Figure 16.8 is that such a specification can be translated into an executable form, which can then be verified through model checking to ensure that every critical state has been considered. This verification can be useful in minimizing the possibility that potential hazards that can be either costly or frequent remain unnoticed during the design and implementation of mission plans.

16.6 CONCLUSIONS

In this chapter we described the opportunities of integration between the SwarmOS platform and UAV swarms. Our goal was to provide a high-level description of the implications of the convergence of these areas. The SwarmOS platform is being used actively in general purpose IoT applications and provides the necessary features for use in UAV swarms.

As future work, we see an important opportunity to automate the communication and coordination of the participants in UAV swarms. This effort will open the door for the creation of intelligent cyber-physical systems with the potential of solving complex tasks that are not available today.

Another relevant way to explore is to use UAV swarms for applications of high impact on the local economy, such as those described in this chapter for Brazilian farming and livestock.

REFERENCES

https://doi.org/10.1108/IJIUS-01-2013-0001

1. R. McCune, R. Purta, M. Dobski, A. Jaworski, G. Madey, A. Madey, Y. Wei, and M. B. Blake, "Investigations of DDDAS for command and control of UAV swarms with agent-based modeling," *2013 Winter Simulations Conference (WSC)*, pp. 1467–1478, 2013. https://doi.org/10.1109/WSC.2013.6721531
2. S. Aldo, B. Jaimes, and M. Jamshidi, "Consensus-based and network control of UAVs," *2010 5th International Conference on System of Systems Engineering*, pp. 1–6, 2010. https://doi.org/10.1109/SYSOSE.2010.5544106
3. H. Chen, X. Wang, and Y. Li, "A survey of autonomous control for UAV," *2009 International Conference on Artificial Intelligence and Computational Intelligence*, pp. 267–271, 2009. https://doi.org/10.1109/AICI.2009.147
4. H. Cheng, J. Page, and J. Olsen, "Cooperative control of UAV swarm via information measures," *International Journal of Intelligent Unmanned Systems*, 1, 256–275, 2013. https://doi.org/10.1108/IJIUS-01-2013-0001
5. L. Gupta, R. Jain, and G. Vaszkun, "Survey of important issues in UAV communication networks," *IEEE Communications Surveys and Tutorials*, 18, 1123–1152, 2016. https://doi.org/10.1109/COMST.2015.2495297
6. I. Lachow, "The upside and downside of swarming drones," *Bulletin of the Atomic Scientists*, 73, 96–101, 2017. https://doi.org/10.1080/00963402.2017.1290879
7. P. de Sousa Paula, M. F. de Castro, G. A. L. Paillard, and W. W. F. Sarmento, "A swarm solution for a cooperative and self-organized team of UAVs to search targets," *2016 8th Euro American Conference on Telematics and Information Systems (EATIS)*, pp. 1–8, 2016. https://doi.org/10.1109/EATIS.2016.7520118
8. H. Qiu, C. Wei, R. Dou, and Z. Zhou, "Fully autonomous flying: from collective motion in bird flocks to unmanned aerial vehicle autonomous swarms," *Science China Information Sciences*, 58, 1–3, 2015. https://doi.org/10.1007/s11432-015-5456-x
9. Q. Luo and H. Duan, "An improved artificial physics approach to multiple UAVs/UGVs heterogeneous coordination," *Science China Technological Sciences*, 56, 2473–2479, 2013. https://doi.org/10.1007/s11431-013-5314-2
10. J. A. Fax and R. M. Murray, "Information flow and cooperative control of vehicle formations," *IEEE Transactions on Automatic Control*, 49, 1465–1476, 2004. https://doi.org/10.1109/TAC.2004.834433
11. M. Bacco, S. Chessa, M. Di Benedetto, D. Fabbri, M. Girolami, A. Gotta, D. Moroni, M. A. Pascali, and V. Pellegrini, "UAVs and UAV swarms for civilian applications: communications and image processing in the SCIADRO Project." In: P. Pillai, K. Sithamparanathan, G. Giambene, M. Á. Vázquez, and P. D. Mitchell, Eds. *Wireless and Satellite Systems, Lecture Notes of the Institute for Computer Sciences, Social Informatics and Telecommunications Engineering*. Cham, Switzerland: Springer International Publishing, pp. 115–124, 2018.
12. K. Z. Y. Ang, X. Dong, W. Liu, G. Qin, S. Lai, K. Wang, D. Wei, S. Zhang, S. K. Phang, X. Chen, M. Lao, Z. Yang, D. Jia, F. Lin, L. Xie, and B. M. Chen, "High-precision multi-UAV teaming for the first outdoor night show in Singapore," *Unmanned Systems*, 6, 39–65, 2018.

13. H. Hildmann, E. Kovacs, F. Saffre, and A. F. Isakovic, "Nature-inspired drone swarming for real-time aerial data-collection under dynamic operational constraints," *Drones*, 3, 71, 2019. https://doi.org/10.3390/drones3030071
14. S. Engebraten, K. Glette, and O. Yakimenko, "Field-testing of high-level decentralized controllers for a multi-function drone swarm," *2018 IEEE 14th International Conference on Control and Automation (ICCA)*, IEEE, pp. 379–386, 2018. https://doi.org/10.1109/ICCA.2018.8444354
15. L. C. P. Costa, J. Rabaey, A. Wolisz, M. Rosan, and M. K. Zuffo, "Swarm OS control plane: an architecture proposal for heterogeneous and organic networks," *IEEE Transactions on Consumer Electronics*, 61, 454–462, 2015. https://doi.org/10.1109/TCE.2015.7389799
16. J. M. Rabaey, "The swarm at the edge of the cloud - A new perspective on wireless," *2011 Symposium on VLSI Circuits (VLSIC)*, pp. 6–8, 2011.
17. E. A. Lee, J. Rabaey, B. Hartmann, J. Kubiatowicz, K. Pister, A. Sangiovanni-Vincentelli, S. A. Seshia, J. Wawrzynek, D. Wessel, T. S. Rosing, D. Blaauw, P. Dutta, K. Fu, C. Guestrin, B. Taskar, R. Jafari, D. Jones, V. Kumar, R. Mangharam, G. J. Pappas, R. M. Murray, and A. Rowe, "The swarm at the edge of the cloud," *IEEE Design & Test*, 31, 8–20, 2014. https://doi.org/10.1109/MDAT.2014.2314600
18. J. Lewis and M. Fowler, "Microservices," 2014. https://martinfowler.com/articles/microservices.html. Accessed October 25, 2019.
19. L. C. De Biase, P. C. Calcina-Ccori, G. Fedrecheski, D. Navarro, R. Y. Lino, and M. K. Zuffo, "Swarm minimum broker: an approach to deal with the Internet of Things heterogeneity," *2018 Global Internet of Things Summit (GIoTS)*, pp. 1–6, 2018. https://doi.org/10.1109/GIOTS.2018.8534433
20. P. Barnaghi, W. Wang, C. Henson, and K. Taylor, "Semantics for the Internet of Things: early progress and back to the future," *International Journal of Semantic Web and Information Systems*, 8, 1–21, 2012.
21. I. Szilagyi and P. Wira, "Ontologies and semantic web for the Internet of Things - a survey," *IECON 2016 - 42nd Annual Conference of the IEEE Industrial Electronics Society*, pp. 6949–6954, 2016. https://doi.org/10.1109/IECON.2016.7793744
22. O. Lassila and R. R. Swick, 1998. "Resource description framework (RDF) model and syntax specification," W3C Consortium. https://www.w3.org/1998/10/WD-rdf-syntax-19981008/
23. L. C. C. D. Biase, P. C. Calcina-Ccori, G. Fedrecheski, G. M. Duarte, P. S. S. Rangel, and M. K. Zuffo, "Swarm economy: a model for transactions in a distributed and organic IoT platform," *IEEE Internet of Things Journal*, 6, 4561–4572, 2019. https://doi.org/10.1109/JIOT.2018.2886069
24. V. Hu, D. Ferraiolo, R. Kuhn, A. Schnitzer, K. Sandlin, R. Miller, and K. Scarfone, "Guide to attribute based access control (ABAC) definition and considerations," Report No. SP 800-162. Gaithersburg, MD: National Institute of Standards and Technology, 2014.
25. G. Fedrecheski, L. C. C. De Biase, P. C. Calcina-Ccori, and M. K. Zuffo, "Attribute-based access control for the swarm with distributed policy management," *IEEE Transactions on Consumer Electronics*, 65, 90–98, 2019. https://doi.org/10.1109/TCE.2018.2883382
26. J. Uhrmann and A. Schulte, "Concept, design and evaluation of cognitive task-based UAV guidance," *International Journal of Advances in Intelligent Systems*, 5, 2012.
27. M. Alirezaie, A. Kiselev, F. Klügl, M. Längkvist, and A. Loutfi, "Exploiting context and semantics for UAV Path-finding in an urban setting," *CEUR Workshop Proceedings*, pp. 11–20, 2017.

28. F. S. Correa da Silva, P. W. Chung, M. K. Zuffo, P. Papapanagiotou, D. Robertson, and W. Vasconcelos, "Hazard prevention in mission plans for aerial vehicles based on soft institutions," *Civil Aircraft Design and Research*, 126, 105–116, 2017.
29. A. Preece, M. Gómez Martínez, G. Mel, W. Vasconcelos, D. Sleeman, S. Colley, G. Pearson, T. Pham, and T. Porta, "Matching sensors to missions using a knowledge-based approach," *Proceedings Defense and Security Symposium*, 2008. https://doi.org/10.1117/12.782648
30. L. Barbieri, A. Thomer, and J. Wyngaard, "Drone data & the semantic web," 2018. https://esipfed.github.io/stc/symposium/2018/talks/ModESIPWinter18(1).pdf
31. D. M. Lammerding, "Dronetology, the UAV Ontology," 2019. http://www.dronetology.net/. Accessed October, 24, 2019.
32. P. C. Calcina-Ccori, L. C. C. De Biase, G. Fedrecheski, F. S. Corrêa da Silva, and M. K. Zuffo, "Enabling semantic discovery in the swarm," *IEEE Transactions on Consumer Electronics*, 65, 57–63, 2019. https://doi.org/10.1109/TCE.2018.2888511
33. C. Goldbaum, "Watch how drones keep elephants away from danger in Tanzania," 2015. https://qz.com/africa/456772/watch-how-drones-keep-elephants-away-from-danger-in-tanzania/. Accessed October 24, 2019.
34. L. Shekhar, "First forest drone in Tamil Nadu launched," 2017. https://www.deccanchronicle.com/nation/current-affairs/230517/first-forest-drone-in-tamil-nadu-launched.html. Accessed October 24, 2019.
35. The Nation, "Speaker-equipped drone helps drive away wild elephants," 2018. https://www.nationthailand.com/breakingnews/30345274. Accessed October, 24, 2019.
36. S. Marchini and P. G. Crawshaw, "Human–wildlife conflicts in Brazil: a fast-growing issue," *Human Dimensions of Wildlife*, 20, 323–328, 2015. https://doi.org/10.1080/10871209.2015.1004145
37. S. Candiago, F. Remondino, M. De Giglio, M. Dubbini, and M. Gattelli, "Evaluating multispectral images and vegetation indices for precision farming applications from UAV images," *Remote Sensing*, 7, 4026–4047, 2015. https://doi.org/10.3390/rs70404026
38. P. Lottes, R. Khanna, J. Pfeifer, R. Siegwart, and C. Stachniss, "UAV-based crop and weed classification for smart farming," *2017 IEEE International Conference on Robotics and Automation (ICRA)*, pp. 3024–3031, 2017. https://doi.org/10.1109/ICRA.2017.7989347
39. J. Pfeifer, R. Khanna, D. Constantin, M. Popovic, E. Galceran, N. Kirchgessner, A. Walter, R. Siegwart, and F. Liebisch, "Towards automatic UAV data interpretation for precision farming," *CIGR-AgEng Conference*, 2016.

Index

V

W